OXFORD SERIES ON NEUTRON SCATTERING IN CONDENSED MATTER

General Editors

S.W. Lovesey, E.W.J. Mitchell

OXFORD SERIES ON NEUTRON SCATTERING IN CONDENSED MATTER

1. W.G. Williams: *Polarized neutrons*
2. E. Balcar and S.W. Lovesey: *Theory of magnetic neutron and photon scattering*
3. V.F. Sears: *Neutron optics*
4. M.F. Collins: *Magnetic critical scattering*
5. V.K. Ignatovich: *The physics of ultracold neutrons*
6. Yu. A. Alexandrov: *Fundamental properties of the neutron*
7. P.A. Egelstaff: *An introduction to the liquid state*
8. J.S. Higgins and H.C. Benoît: *Polymers and neutron scattering*
9. H. Glyde: *Excitations in liquid and solid helium*
10. V. Balucani and M. Zoppi: *Dynamics of the liquid state*
11. T.J. Hicks: *Magnetism in disorder*
12. H. Rauch and S. Werner: *Neutron interferometry*
13. R. Hempelmann: *Quasielastic neutron scattering and solid state diffusion*
14. D.A. Kean and V.M. Nield: *Diffuse neutron scattering from crystalline materials*

QUASIELASTIC NEUTRON SCATTERING AND SOLID STATE DIFFUSION

Rolf Hempelmann
Institute of Physical Chemistry
University of Saarbrücken

CLARENDON PRESS • OXFORD
2000

This book has been printed digitally and produced in a standard specification in order to ensure its continuing availability

OXFORD
UNIVERSITY PRESS

Great Clarendon Street, Oxford OX2 6DP

Oxford University Press is a department of the University of Oxford.
It furthers the University's objective of excellence in research, scholarship,
and education by publishing worldwide in

Oxford New York

Auckland Cape Town Dar es Salaam Hong Kong Karachi
Kuala Lumpur Madrid Melbourne Mexico City Nairobi
New Delhi Shanghai Taipei Toronto
With offices in
Argentina Austria Brazil Chile Czech Republic France Greece
Guatemala Hungary Italy Japan South Korea Poland Portugal
Singapore Switzerland Thailand Turkey Ukraine Vietnam

Published in the United States
by Oxford University Press Inc., New York

Reprinted 2007

ISBN 978-0-19-851743-6

Preface

Quasielastic neutron scattering (QENS) has made many important contributions to the atomistic elucidation of diffusion processes in solids. The aim of this book is to inform researchers in solid state physics, solid state chemistry and inorganic materials science of the potential of QENS. Although the book has been written for experimentalists, some theoretical background both in neutron scattering and in solid state diffusion is essential for a proper use of QENS. This is presented in Chapters 3–6. Chapter 2 deals with neutron sources, and Chapter 7 with some experimental details of neutron scattering. The second part of the book is then devoted to special materials systems: hydrogen in metals, metals and alloys, intercalation compounds, solid state proton conductors, and other solid state ionic conductors.

Every nuclear method in condensed matter research exhibits its best performance only with few nuclei: for instance Mössbauer spectroscopy relies on ^{57}Fe, but since Fe is a widespread element (in particular in biological systems) Mössbauer spectroscopy is a very general method. By far the most favourable element for QENS is hydrogen: (i) hydrogen is very mobile in metals and in ceramics, so the restricted energy/time resolution of QENS is no limitation; (ii) hydrogen has a huge scattering cross-section, so the restricted intensity of the existing neutron sources is no limitation; and (iii) last, but not least, hydrogen leads to almost purely incoherent scattering which facilitates the theoretical treatment and interpretation of neutron scattering events appreciably. For all these reasons 'Hydrogen in metals' and 'Solid state proton conductors' are the more voluminous chapters in the second part of this book.

This coincides with the scientific interests of the author, and perhaps it is for this reason that the other materials systems (metals, intercalation compounds, and solid ionic conductors) happen to be described in less extensive chapters. We apologize for these deficiencies to those colleagues whose work we have failed to quote appropriately.

For several reasons the writing of this book took more than six years. The author owes a great debt to the publishers and particularly to Dr. S. Adlung for his patience and encouragement during these many years. The author is indebted to Prof. Springer, Forschungszentrum Jülich, for the preparation of the basic versions of Chapters 2, 3, 6, 9, and 10 and for his encouragement towards completion of the manuscript. Furthermore, the author thanks his secretary Mrs. Petra Theobald for the careful typing of the manuscript and, again, for her patience, and

Joschi Wilhelm for technical help. Dr. Joachim Wagner has carefully checked the whole book for inconsistencies and has done all the figure work; many thanks to him for this outstanding contribution to this book.

Saarbrücken R.H.
July 2000

Acknowledgements

We are grateful to the following publishers and learned societies for permission to use their illustrations as figures in this book: Academic Press, EDP Sciences S.A., Elsevier Science B.V., Elsevier Sequoia S.A., Institut Max von Laue–Paul Langevin, IOP Publishing Ltd, Oldenbourg Wissenschaftsverlag, Pergamon Press Inc., Rutherford Appleton Laboratory, Taylor & Francis Ltd, Springer-Verlag, Wiley-VCH, The American Physical Society, The Institute of Physics.

The following authors have also kindly allowed us free use of their original material: F. Altorfer, A.V. Belushkin, R. Blinc, P. Böni, C.J. Carlile, C.J. Cook, A.J. Dianoux, K. Funke, M.J. Gillan, O. Hartmann, M.T. Hutchings, S. Janssen, E. Karlsson, K.W. Kehr, R.E. Lechner, A. Magerl, J. Mesot, W. Petry, D. Richter, D.K. Ross, W. Schirmacher, H.R. Schober, T. Springer, G. Vogl, G. Wahnström, R. Wäppling, H. Wenzl, H. Wipf, H. Zabel.

It is a pleasure to record our sincere thanks to them all.

Contents

Glossary of symbols

$\text{Å} = 10^{-10}\,\text{m}$	(angström)
b	bound scattering length
b^+	scattering length for $I + \frac{1}{2}$ state
b^-	scattering length for $I - \frac{1}{2}$ state
D	diffusion coefficient without specification
D_E	Einstein diffusion coefficient
D_{bulk}	bulk diffusion coefficient
$D_{\text{c}}(Q)$	Q-dependent collective diffusion coefficient
D_{chem}	chemical or Fick's diffusion coefficient
D_{s}	self-diffusion coefficient
D_{t}	tracer diffusion coefficient
D_σ	conductivity diffusion coefficient
$G(\boldsymbol{r}, t)$	space-time correlation function: $G(\boldsymbol{r}, t) = G_{\text{d}}(\boldsymbol{r}, t) + G_{\text{s}}(\boldsymbol{r}, t)$
$G_{\text{d}}(\boldsymbol{r}, t)$	'distinct' space-time correlation function
$G_{\text{s}}(\boldsymbol{r}, t)$	'self'-space-time correlation function
$g(\boldsymbol{r})$	pair correlation function: $g(r) = \rho(\boldsymbol{r})/\rho_0$
$I(\boldsymbol{Q}, t)$	intermediate scattering function
$I_{\text{s}}(\boldsymbol{Q}, t)$	'self'-intermediate scattering function
$S(\boldsymbol{Q})$	static structure factor
$S(\boldsymbol{Q}, \omega)$	coherent scattering function
$S_{\text{i}}(\boldsymbol{Q}, \omega)$	incoherent scattering function
σ	bound total cross-section (scattering plus absorption)
σ	site occupancy (in Chapter 6)
σ_{c}	bound coherent scattering cross-section
σ_{i}	bound incoherent scattering cross-section
σ_{s}	bound scattering cross-section (coherent plus incoherent)
σ_{f}	free scattering cross-section
σ_{a}	absorption cross-section
$\text{d}\sigma/\text{d}\Omega$	differential cross-section
$\text{d}^2\sigma/\text{d}\Omega_1\,\text{d}\varepsilon_1$	double differential cross-section

Neutron properties are characterized by lower case letters:

n	chemical symbol of neutron as well as number of neutrons
$\tilde{n}$	particle number density of neutrons (in m^{-3})

m	mass of neutron
ψ	wavefunction of neutron
$\varepsilon_0, \varepsilon_1$	neutron energy before, after scattering event
$\boldsymbol{k}_0, \boldsymbol{k}_1$	wavevector of neutrons before, after scattering event
$\lvert k_0\rangle, \lvert k_1\rangle$	unnormalized neutron wavefunctions

Sample properties are characterized by upper case letters:

N	number of nuclei
$\tilde{N}$	particle number density of nuclei (in m^{-3})
M	mass of nucleus
Ψ	wavefunction of sample system
ν_0, ν_1	quantum number of sample system before, after scattering event
E_0, E_1	energy of sample system before, after scattering event
$\boldsymbol{R}_i$	position vector of nucleus i

Part I

General part

1

Introduction

In the non-specialized scientific community the term neutron scattering is usually associated with Bragg peaks and structure determination and thus with neutron diffraction and neutron diffractograms, i.e. with plots of neutron scattering intensity versus the angle through which the beam is deflected. In comparison to this well-known technique, quasielastic neutron scattering, the topic of the present book, is clearly less commonly used. It concerns the small scattering intensity between the huge Bragg peaks. The coherent fraction of this '*Zwischenreflex*' scattering is known as diffuse scattering and contains, as a consequence of complex interference phenomena, information about short-range order and lattice distortion fields, like diffuse X-ray scattering. Neutrons, however, are more versatile in this respect: there is also incoherent scattering between Bragg peaks, in which case neutrons scattered at different atoms are not able to interfere with each other and thus carry information only about the single-particle behaviour. This single-particle behaviour in solids is of little scientific interest as long as the particles are immobile; but at sufficiently high temperatures the particles perform stochastic motions—diffusion or rotation—to such an extent that they transfer energy to or from the neutrons during the individual scattering event. For cold and thermal neutrons this energy gain or loss can amount to an appreciable fraction of their kinetic energy, and then the energy transfer is easily measurable; the resulting neutron spectrum (plots of neutron scattering intensity versus energy transfer $\hbar\omega$) exhibits a maximum centred at the elastic position $\hbar\omega = 0$, which is energetically broadened compared to the instrumental linewidth. This phenomenon is called quasielastic neutron scattering, QENS. From the previous arguments it is clear that QENS can be used to study diffusion phenomena in solids.

Diffusion in this connection means solid state diffusion in thermodynamic equilibrium; a gradient in concentration or, more generally, in chemical potential is *not* applied. Self-diffusion is easily comprehensible: an individual particle performs a random walk over its sublattice, analogously to Brownian motion of colloidal particles in liquids. This means single-particle behaviour and is studied by incoherent QENS with regard to jump rates, jump lengths, and jump directions. Less obvious is collective diffusion in thermodynamic equilibrium: this means the temporary and spatial development of density fluctuations, to be studied by coherent QENS.

The present book is intended to serve as a text and reference book on how quasielastic neutron scattering can be and has been used to elucidate these diffusion processes in solids. Both fields of science, neutron scattering and solid state diffusion, have benefitted essentially from military nuclear developments. Nuclear reactors and accelerator-based sources produce neutrons for neutron scattering and most of the radioactive tracer atoms, which were and are essential for quantitative solid state diffusion experiments and for an atomistic approach of material transport in solids. But also nuclear magnetic resonance and microwave techniques, both essential experimental methods of solid state diffusion, have derived profit from developments in the 1940s.

The history of neutron scattering starts in 1932, with the discovery of the neutron (Chadwick 1932); α particles from the radioactive decay of radium encounter beryllium (radium–beryllium source):

$$ {}^{9}_{4}\mathrm{Be} + {}^{4}_{2}\mathrm{He} \rightarrow \left({}^{13}_{6}\mathrm{C}\right) \rightarrow {}^{12}_{6}\mathrm{C} + {}^{1}_{0}\mathrm{n} \tag{1.1} $$

or in brief:

$$ {}^{9}\mathrm{Be}(\alpha, \mathrm{n})^{12}\mathrm{C}. $$

Such a source with a typical activity of 1 mCi produces $1.3 \cdot 10^4$ neutrons per s, by no means sufficient for neutron scattering investigations on materials.

But it did turn out to be sufficient to demonstrate the wave character of the neutrons and the effect of elastic neutron scattering, i.e. neutron diffraction (Halban and Preiswerk 1936, as well as Mitchell and Powers 1936). Probably the first 'real' neutron scattering experiment was the single-crystal diffraction study on NaCl by Shull in 1945 at the Oak Ridge Graphite Reactor, USA. A several orders of magnitude increase in neutron flux, achieved in the research reactor at Chalk River, Canada, enabled routine powder diffraction studies to be carried out; particularly remarkable are the determinations of magnetic structures where neutron scattering is unique: it cannot be done by any other method. A further two orders of magnitude increase in flux in the 1950s enabled the first inelastic neutron scattering experiments to be performed. An important milestone for high-resolution quasielastic neutron scattering is the invention and installation of the cold source (liquid H_2 moderator) in 1954 by Egelstaff at Harwell, UK. The final stage of research reactor development, at least up to today, was reached in 1972 with the commissioning of the high-flux reactor of the Institute Laue–Langevin in Grenoble, France; its flux of $1.5 \cdot 10^{15}\,\mathrm{cm}^{-2}\,\mathrm{s}^{-1}$, similar to that of the two high-flux reactors in the United States, has not been matched by newer reactors (medium-flux research reactors), which is amazing for such high-tech machinery.

Progress in neutron source development, however, has been made since then with a different approach, i.e. with accelerator-based pulsed sources. Neutrons are produced by the bombardment of a target by high-energy electrons or protons. Proton spallation—in which the proton knocks bits off the heavy nuclei in the target—is the most efficient neutron-producing process. The prototype spallation source was the ZIN-P in Argonne, which in the early 1980s was replaced by IPNS, Argonne's Intense Pulsed Neutron Source. A spallation neutron source offers experimental access to epithermal neutron scattering, and many pioneer experiments have been performed at IPNS and at the later-built British source ISIS at the Rutherford Appleton Laboratory near Oxford. The topic of this book, i.e. high-resolution cold neutron quasielastic scattering, is also possible because of the potency of pulsed neutron sources.

The newest neutron source at present is the Swiss Spallation Neutron Source SINQ, which produced its first neutron in December 1996 (Bauer 1991; Fischer 1995; Wagner *et al.* 1998). Designed as a continuous neutron source, it is driven by a 590 MeV accelerator with a design power of 1 MW on the SINQ target; from its performance it corresponds to a medium-flux research reactor.

To the same extent, since it is connected to the development of more powerful neutron sources, progress in neutron scattering is due to improvements in neutron scattering instrumentation. Important milestones for quasielastic neutron scattering are the triple axis spectrometer (Brockhouse 1958), the time-of-flight spectrometers working with a set of choppers (Lechner *et al.* 1973; Douchin *et al.* 1973) or, in a time-focusing arrangement, with one chopper and a set of monochromator crystals (Scherm *et al.* 1976; Blanc 1983; Janssen *et al.* 1997), the backscattering spectrometer (Alefeld *et al.* 1969; Birr *et al.* 1971), and the spin echo spectrometer (Mezei 1972; Hayter 1980; Monkenbusch 1992). Time-of-flight and backscattering spectrometers are particularly useful in the present context and therefore will be explained in detail in Chapter 7.

Scientific work in the field of neutron scattering is nowadays not restricted to the scientists of the (rather few) neutron scattering facilities: on the one hand, all neutron scattering centres—most of them officially operate as user facilities—offer beam time and use of their instruments to scientists from universities, other research laboratories, and industry; every interested scientist can have access to neutron scattering essentially for free (including sometimes even travel expenses). On the other hand, neutron scattering is, as will be outlined in the next two chapters, a very versatile and powerful tool for atomistic studies on condensed matter with applications in different branches of physics, chemistry, materials science, and biology. This versatility is due to the many aspects of elastic, quasielastic, inelastic, and deep inelastic neutron scattering, as well as coherent and incoherent neutron scattering. Consequently, neutron scattering is *not* 'big science' like elementary particle physics, where hundreds of scientists at dedicated huge machines work for years on just one problem, e.g. the detection of a new elementary particle. In contrast the practice of neutron scattering is characterized by a large variety of small groups working on many different scientific problems. For the majority of these groups neutron scattering is just one method among several other ones available at their respective home laboratory. The yearly neutron beam time at the neutron sources is between 2 and 20 days. Only the few national or international laboratories which operate neutron sources have groups of scientists devoted exclusively to neutron scattering. Therefore progress in neutron scattering usually occurs at a broad front and not in single famous experiments.

Neutron scattering and solid state diffusion have in common that they both can be separated into two problems: the single event and the cooperative combination (in space and time) of those events in the solid.

For *neutron scattering* the single event is the scattering of a neutron on a single nucleus. Since neutrons are also constituents of the nucleus, neutron–nucleus scattering is caused by and thus explores the nucleon potential within the nucleus and the nuclear forces (extremely short-range strong interactions). In condensed matter applications of neutron scattering, the single scattering event is a 'black box' characterized by just two empirical parameters, the absorption cross-section σ_a and the scattering length b which nowadays are listed in comprehensive tables. The proper aim of neutron scattering in condensed matter research is to deduce, from the spatial and temporal interference pattern of the elementary neutron waves emanating from each nucleus, the structure and dynamics of condensed matter, in our case the issue of the diffusive process in space and time within solids.

For *diffusion* the single event is the single hop, i.e. one change of site of a particle. The particles are located at the minima of the interparticle potential. Heavy particles are considered to perform, from time to time, a classical jump over the potential barrier separating two minima. Light particles, on the other hand, might be able to tunnel through the barrier. This tunnelling can be assisted by elementary excitations of the solid, e.g. by phonons. Again, this single event can be considered as a 'black box', which is characterized by one parameter, the jump rate Γ. The solid state diffusion problem is then the exploration of the diffusion mechanism; e.g. how many different jump rates exist, what is their temperature dependence, and how do these jumps cooperate in the space and time evolution of the diffusive process?

Throughout this book we restrict ourselves to solids with only one mobile component; problems of interdiffusion of different species are beyond our scope. The general part (Chapters 2–7) is intended to make interested scientists familiar with relevant theoretical and experimental aspects of neutron scattering and of solid state diffusion. Chapter 2 deals with the properties and the production of neutrons. Both Chapters 3 and 4, *The basic theory of neutron scattering* and *Diffusion in solids* are subdivided into two parts: one part for the single event—the neutron scattering on a single nucleus and the nature of the single hop, respectively—and the other part for the cooperative combination of these events in the solid. Then we combine these two branches of science in Chapter 5 and show what *incoherent quasielastic neutron scattering* can tell us about self-diffusion in solids, using models of increasing complexity. This is now, at least to a certain extent, a rather well-established field. In contrast, *coherent quasielastic neutron scattering* for the exploration of collective diffusion in solids is quite novel, and therefore our discussion in Chapter 6 must remain somewhat incomplete. Chapter 7 addresses experimental aspects of quasielastic neutron scattering which sometimes might look less fundamental but which are often decisive for the success of an experiment. This concludes the general part which, for that was our aim, is hopefully understandable and useful for non-experts in the field of neutron scattering and diffusion, i.e. also for people without special education in nuclear and theoretical physics.

The second part, on the other hand, is written more in the style of a review and summarizes the scientific applications of quasielastic neutron scattering to special solid state systems, i.e. to *hydrogen in metals*, to metal atom diffusion in *metals and alloys* and in *intercalation compounds* at elevated temperatures, and to diffusion in *solid ionic conductors* and in *solid state proton conductors*. These are all very active fields of research, and we hope that this book is also useful to experts in those fields. Some books and reviews for further reading, together with the special references, are listed at the end of the book.

The application of the technique of quasielastic neutron scattering is not, of course, restricted to solid state diffusion. Other important fields are rotational or conformational motions in plastic crystals (mostly organic chemical compounds) (Bée 1988) including rotational tunnelling phenomena (Press 1981; Prager 1991; Carlile *et al.* 1998), biological systems (Filabozzi *et al.* 1996), diffusion in supercooled water (DiCola *et al.* 1996) or in water in confined geometries (Cantù *et al.* 1997; Belissent-Funel *et al.* 1993), diffusion in molecular liquids (Bermejo *et al.* 1992), dynamic correlations near the liquid–glass transition (Frick and Richter 1995; Maurin *et al.* 1997) and, as a particular important topic, viscoelasticity and microscopic motions in polymer systems (Richter 1998).

2

The neutron

2.1 BASIC PROPERTIES

Neutrons, being constituents of nuclei, are liberated in nuclear reactions of neutron-rich nuclei, as in the fission of ^{235}U or the fusion of ^{3}H and ^{2}H. Furthermore, neutrons appear during the spallation process where heavy nuclei are bombarded by GeV-protons. Also high-energy γ quanta can produce neutrons by means of the nuclear photoeffect reaction. The properties of the neutron which are relevant for neutron scattering are listed in Table 2.1 together with the values of frequently used physical constants and energy conversion factors.

In the instant of generation neutrons have high kinetic energies (some MeV). For application in scattering experiments these neutrons are slowed down and

Table 2.1 Some properties of neutrons and some physical constants

Neutrons:	
mass:	$m = 1.675 \cdot 10^{-27}$ kg
electric charge:	0
spin:	$\frac{1}{2}$
magnetic moment:	$\mu_n = -1.913$ nuclear magnetons
life time:	10.6 min, $\mathrm{n} \rightarrow \mathrm{p} + \mathrm{e} + \bar{\nu}$, $\Delta E = 0.782$ MeV
Physical constants:	
elementary charge:	$e_0 = 1.602 \cdot 10^{-19}$ C
Boltzmann constant:	$k_B = 1.381 \cdot 10^{-23}$ J K^{-1},
	$k_B^{-1} = 11.6045$ K meV^{-1}
Planck constant:	$h = 6.626 \cdot 10^{-34}$ J s
	$\hbar = 1.0546 \cdot 10^{-34}$ J s $= 6.582 \cdot 10^{-10}$ μeV s
Avogadro constant:	$L = 6.022 \cdot 10^{23}$ mol^{-1}
Electron mass:	$m_e = 9.1095 \cdot 10^{-31}$ kg
Proton mass:	$m_p = 1.673 \cdot 10^{-27}$ kg
Energy conversion factors:	
1 meV $= 1.602\,19 \cdot 10^{-22}$ J	1 J $= 0.624\,14 \cdot 10^{22}$ meV
$= 11.6045$ K	1 K $= 0.086\,073$ meV
$= 0.241\,81 \cdot 10^{12}$ Hz	10^{12} Hz $= 4.135\,41$ meV
$= 8.0668$ cm^{-1}	1 cm$^{-1} = 0.123\,96$ meV
1 meV $= 0.096\,48$ kJ mol^{-1}	1 kJ mol$^{-1} = 10.364$ meV

finally thermalized in moderators. In a heavy water moderator, neutrons reach nearly thermal equilibrium because of the small D_2O absorption cross-section; also in H_2O, equilibrium is nearly achieved. So we can identify the temperature of the 'neutron gas' approximately with the water temperature. In a cold source moderator (see below) such as liquid H_2 or D_2, the distribution is not very close to Maxwellian, and one may chose an effective neutron temperature of 40 K. The mean neutron velocity can be quoted in several ways, calling T the neutron temperature:

$$\begin{aligned} &\text{the root mean square velocity} && \left(\overline{v^2}\right)^{1/2} = \left(\frac{3\,k_{\mathrm{B}}T}{m}\right)^{1/2} \\ &\text{the average velocity} && \bar{v} = \left(\frac{8k_{\mathrm{B}}T}{\pi m}\right)^{1/2} \\ &\text{the most probable velocity} && v_{\mathrm{T}} = \left(\frac{2\,k_{\mathrm{B}}T}{m}\right)^{1/2}. \end{aligned} \tag{2.1}$$

The energy of a single free neutron is simply kinetic:

$$E = \tfrac{1}{2}mv^2. \tag{2.2}$$

The mean energy of thermalized neutrons is hence given by

$$\overline{E} = \tfrac{1}{2}m\overline{v^2} = \tfrac{3}{2}k_{\mathrm{B}}T. \tag{2.3}$$

Conventionally, however, the most probable velocity is taken to connect the neutron energy and the moderator temperature, i.e.

$$E = k_{\mathrm{B}}T. \tag{2.4}$$

A 'neutron gas' at $T = 293.6$ K thus has an energy of 25 meV.

Neutrons can also be considered as particle waves with the de Broglie wavelength

$$\lambda = \frac{h}{mv}. \tag{2.5}$$

With h and m from Table 2.1 and $v = v^* = 2200\,\mathrm{m\,s^{-1}}$ for $T = 293.6$ K, a wavelength of 1.8 Å is obtained. Therefore, the energy of such a neutron is comparable to the kinetic energy and the interaction energies of atoms in solids (i.e. to the energy fluctuations in solids) and, simultaneously, the wavelength is comparable to typical interatomic spacings. These two properties combined make such slow neutrons a unique tool for studying, simultaneously, the spatial and temporal development of diffusive processes in solids.

As a particle wave, neutrons have a wavevector $\boldsymbol{k}$ with its modulus, the wavenumber

$$k = \frac{2\pi}{\lambda}. \tag{2.6}$$

Correspondingly, the neutron momentum is given by

$$\boldsymbol{p} = \hbar \boldsymbol{k}. \tag{2.7}$$

Combining Eqs. (2.2)–(2.7) we obtain

$$\varepsilon = k_{\rm B}T = \tfrac{1}{2}mv_{\rm T}^2 = \frac{h^2}{2m\lambda^2} = \frac{\hbar^2 k_{\rm T}^2}{2m} \tag{2.8}$$

where $k_{\rm T}$ is the wavenumber corresponding to $v_{\rm T}$. Note the λ^{-2} dependence of the energy, whereas electromagnetic radiation due to $E = hc/\lambda$ exhibits a λ^{-1} dependence.

With the units conventionally used in the field of neutron scattering, $[\lambda] = \text{Å}$, $[k] = \text{Å}^{-1}$, $[v] = \text{km s}^{-1} = \text{m ms}^{-1}$, $[E] = \text{meV}$, $[T] = \text{K}$, one obtains the following useful numerical transformations:

$$\lambda = 6.283\frac{1}{k} = 3.956\frac{1}{v} = 9.045\frac{1}{\sqrt{\varepsilon}} = 30.81\frac{1}{\sqrt{T}};$$

$$\varepsilon = \frac{T}{11.6045} = 5.227v^2 = 81.81\frac{1}{\lambda^2} = 2.072\,k^2.$$

A so-called thermal neutron with a velocity of $2200\,\text{m s}^{-1}$ (which is conventionally taken as the standard velocity) therefore has the following properties:

$$v = 2.20\,\text{m ms}^{-1} \qquad \frac{1}{v} = 455\frac{\mu\text{s}}{\text{m}} = \text{TOF (time-of-flight)}$$

$$\varepsilon = 25.3\,\text{meV} \qquad T = 293.6\,\text{K}$$

$$\lambda = 1.80\,\text{Å} \qquad k = 3.49 \cdot 10^{10}\,\text{m}^{-1} = 3.49\,\text{Å}^{-1}.$$

In condensed matter, neutrons can be scattered or absorbed. Both processes are characterized by cross-sections which are listed in tables in many neutron scattering books; a comprehensive compilation has been published by Sears (1992). Usually the absorption cross-section is proportional to $1/v$ and thus to λ. The value of the absorption cross-section is quoted for neutrons with the standard velocity $v = 2200\,\text{m s}^{-1}$ and $\lambda = 1.80\,\text{Å}$ and has to be transformed corresponding to the values of v or λ, respectively, of the actually applied neutrons:

$$\sigma_{\rm a}(\lambda) = \sigma_{\rm a}(\lambda = 1.8\,\text{Å}) \cdot \frac{\lambda}{1.8\,\text{Å}}. \tag{2.9}$$

Neutrons cannot be detected directly (since their interaction with matter is too weak), but only from ionizing products of nuclear reactions caused by the neutron. Three reactions are commonly used:

$$ {}^{3}_{2}\mathrm{He} + {}^{1}_{0}\mathrm{n} \rightarrow {}^{3}_{1}\mathrm{H} + {}^{1}_{1}\mathrm{H} + 0.76\,\mathrm{MeV} \tag{2.10} $$

$$ {}^{10}_{5}\mathrm{B} + {}^{1}_{0}\mathrm{n} \rightarrow {}^{7}_{3}\mathrm{Li} + {}^{4}_{2}\mathrm{He} + 2.8\,\mathrm{MeV} \tag{2.11} $$

$$ {}^{6}_{3}\mathrm{Li} + {}^{1}_{0}\mathrm{n} \rightarrow ({}^{7}_{3}\mathrm{Li}) \rightarrow {}^{3}_{1}\mathrm{He} + {}^{4}_{2}\mathrm{He} + 4.8\,\mathrm{MeV}. \tag{2.12} $$

Reactions (2.10) and (2.11) are utilized in gas counters, i.e. tubes filled with several bar of ^{3}He or BF_3 gas enriched in ^{10}B. The product nuclei have a high kinetic energy, ionize the counting gas, and thus produce some 10^3–10^4 ion pairs. The counter is operated in the proportional region; the electrons produced are accelerated to the anode (applied voltage 1–2.5 kV) whereby they produce numerous electrons by collision with gas molecules. Thus each neutron entering the counter generates an electrical current pulse. The pulses produced by (unwanted) γ radiation are very much weaker and can be well discriminated. Reaction (2.12) is utilized in scintillation counters where the primary α particle creates a light flash in a scintillator glass which is detected by photomultipliers.

Reaction (2.12) is the only neutron nucleus reaction without generation of a γ photon. Since secondary α particles can be shielded much more easily than secondary γ photons, the reaction (2.12) is sometimes used for neutron shielding at places where there is no space for the required massive γ shielding (^{6}Li in the chemical form of ^{6}LiF ceramics, or ^{6}LiF immersed in plastics, or ^{6}LiO-SiO_2 glass). Otherwise boron, cadmium, or gadolinium are used for neutron shielding (see the absorption cross-sections in the neutron cross-section tables of Sears (1992)).

2.2 NEUTRON SOURCES

Only two types of reactions are nowadays used in order to produce neutrons for research on a large scale: nuclear fission in research reactors, and nuclear spallation. Table 2.2 compares these methods. The key point is that neutron production is inseparably connected with heat production, and the amount of heat that can safely be removed ultimately limits the available neutron flux. This flux is low compared to the flux of sources for other radiation: a high-flux reactor delivers, after monochromatization, a neutron flux of the order of $10^8\,\mathrm{cm}^{-2}\,\mathrm{s}^{-1}$ at the sample position, whereas a laboratory X-ray rotating anode reaches $10^{17}\,\mathrm{cm}^{-2}\,\mathrm{s}^{-1}$, not to mention synchrotron gamma or laser light sources. Nuclear spallation is connected with less heat per neutron than nuclear fission, and spallation neutron sources are therefore considered to have a greater development potential, in particular since short neutron pulses can be generated. An alternative is the bremsstrahlung photoneutron effect, which, however, is an ineffective way to produce neutrons, probably without future. A list of nuclear reactors and spallation neutron sources with neutron scattering facilities can be found in *Neutron News* **6**, No. 4, 32

Table 2.2 Methods of neutron production

Primary particles	Reaction	Typical energy of primary particles	Target material	Number of neutrons per primary particle	Heat production per neutron
Electrons	Bremsstrahlung +photoneutron effect	100 MeV	Heavy nuclei	0.05*	2000 MeV*
Protons	Spallation	800 MeV	Heavy nuclei	30*	55 MeV*
Thermal neutrons	Fission	30 MeV	^{235}U	2.1**	175 MeV***
Fast neutrons	Fisson	1 MeV	^{239}Pu	3.0**	95 MeV***

*for a depleted ^{238}U target
**one of them is needed for sustaining the chain reaction
*** 192 MeV per fission event

(1995). A survey of the neutron scattering community and facilities in Europe has been published (1998) as a report of the European Science Foundation (see http://www.esf.org) and in an OECD study (Richter and Springer 1998). In the following we outline some essential features of neutron production. For comprehensive reviews on sources we refer to Windsor (1981), Carpenter and Yelon (1986), and Bauer (1992).

2.2.1 Steady state reactors

Fission is an autocatalytic exothermal nuclear reaction. It can be sustained as a chain reaction either by nearly unmoderated, i.e. fast neutrons, or by thermal neutrons, i.e. by neutrons after they have been slowed down and thermalized in the so-called moderator. The former case is only achieved in the pulsed research reactor IBR-2 in Dubna, Russia (and in fast breeder power reactors), whereas the latter case is used in all steady state research reactors (and conventional nuclear power plants). The control of the reactivity must be very precise and is accomplished with the help of the small fraction of neutrons that is released from some fission products with a sufficiently long time delay. Due to the average decay constant of the so-called delayed neutrons of $0.08\,s^{-1}$ of those fission products the reactor time constant is of the order of 10 s for a slightly overcritical reactor (delayed overcritical); it is only for this reason that safe and reliable control of the reactor is possible. Additionally, the temperature coefficient of the reactivity due to the moderator is negative, so during regular operation the reactor is self-stabilized (for more details on reactor physics and technique see Weinberg and Wigner (1958) and Smidt (1976)).

Since in most countries political acceptibility is a problem of research reactors, it seems appropriate to deal with certain safety aspects. The safety features of nuclear reactors can be subdivided into active, passive, and inherent categories. *Active safety* features are those which allow the operator to take action in case of possible hazards; these are fast emergency shut-down systems, emergency cooling water loops, with redundancy and spatial separation of the essential components. *Passive safety* features prevent or limit damage in the case of failure without the necessity to take action; for instance, in modern research reactors the cooling water tubes enter the reactor vessel from above, so the core by itself remains covered with water even in case of damage to the first cooling system; since the (thermal) power of research reactors is small (10–100 MW compared to typically 1 GW for nuclear power plants), natural convection is sufficient to remove the heat evolving after reactor shut-down (for a 23 MW research reactor the power decays from 1.5 MW immediately after shut-down to 0.3 MW within 20 minutes); therefore, core melting due to afterheat is prevented. An essential *inherent safety* feature is the negative temperature coefficient of the reactivity due to nuclear Doppler broadening, which implies the thermal broadening of the strong absorption resonances of ^{238}U and thus a decrease of self-shielding of neutron absorption in the fuel with increasing temperature. This effect acts immediately, much faster than mechanical shut-down systems, and makes the reactor inherently safe with respect to a power excursion, in the hypothetical case that the reactor should suddenly become prompt-overcritical (Smidt 1976). In research reactors working with highly enriched ^{235}U ($\geq 90\%$), however, this fast negative temperature coefficient is nearly missing because of the small ^{238}U content. However, the expulsion of coolant with increasing heat production reduces the reactivity and makes the core undercritical.

Research reactors contain a small amount of fuel (typically 3.5 kg ^{235}U) compared to nuclear power stations (typical core mass 100 t with a ^{235}U inventory of 3 t!), and are correspondingly less hazardous, with respect to the radioactive inventory. In high-flux research reactors the fuel is highly enriched, and the power density exceeds 1 MW litre^{-1}. This puts extreme requirements on the cooling, and heat removal problems eventually limit a sizable increase in flux beyond the level of today's high-flux reactors.

In a modern research reactor the reflector is D_2O. The core is strongly undermoderated: it relies on thermalized neutrons diffusing back into the core from the surrounding moderator to sustain the chain reaction. As a consequence of this 'drain', the thermal flux (the definition of this quantity follows later) peaks at about 15 cm from the core surface in the D_2O moderator and falls off slowly with increasing distance due to the small absorption cross-section and the large neutron diffusion length in D_2O. By contrast, the epithermal flux in the region of several eV and the fast flux (> 1 MeV) falls off much more rapidly with increasing distance from the core. This allows the beam tube noses to be placed in the region of the reflector maximum with good thermal flux and little fast neutron contamination. Fast neutron flux and γ radiation from the core into the beam tubes are reduced by a 'tangential' beam tube arrangement, i.e. by beam tubes looking past the core and not right onto it. Such an arrangement, which is typical for modern research reactors, is shown for the Grenoble high-flux reactor in Fig. 2.1.

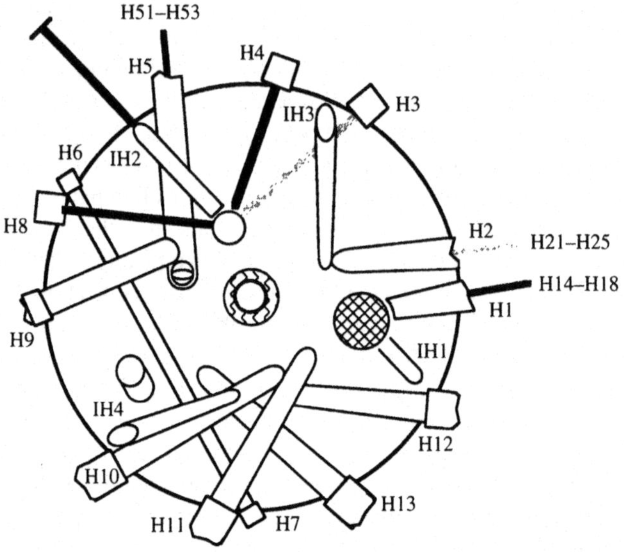

Fig. 2.1 Top view of the D_2O reflector vessel in the Institut Laue–Langevin, Grenoble. Centre: reactor core. The H beam holes bring the neutrons to the different spectrometers, diffractometers etc. Holes H14 bis H18 and IH1 are connected with the liquid D_2 cold source, 3 neutron guides conduct the cold neutrons to the instruments. H21 bis H25 contain a bundle of neutron guides for thermal neutrons. Beam holes H3, H4, H8, IH2 belong to the hot source. IH means inclined hole. V: vertical holes, in particular for irradiation purposes (from 'The yellow book' of the ILL 1994).

A research reactor is usually characterized by its neutron *thermal flux* ϕ_{th} and the corresponding neutron spectrum $\phi(\varepsilon)$ as a function of neutron energy ε in the water moderator surrounding the reactor core. However, what finally counts for the experimentalist is the *current density* $j(\varepsilon)$ extracted from the moderator, namely the number of neutrons with energy ε per unit area and second, which is provided essentially by neutron beam tubes or neutron guides whose nose enters the flux maximum in the reflector.

Assuming thermal equilibrium of the 'neutron gas', the number of the neutrons in a loss-free moderator per phase-space interval is then described by the *Maxwellian distribution*, namely (Maier-Leibnitz 1966; Beckurts and Wirtz 1964)

$$\mathrm{d}^6\tilde{n} = \tilde{n}\frac{\pi^{-3/2}}{k_{\mathrm{T}}^3}\mathrm{e}^{-(k_x^2+k_y^2+k_z^2)/k_{\mathrm{T}}^2}\,\mathrm{d}k_x\,\mathrm{d}k_y\,\mathrm{d}k_z\,\mathrm{d}x\,\mathrm{d}y\,\mathrm{d}z \tag{2.13}$$

where k_x, k_y, k_z are the components of the wave vector $\boldsymbol{k}$ of the neutron and $\boldsymbol{k}_{\mathrm{T}}$ denotes the wavevector corresponding to the most probable velocity v_{T}, see Eq. (2.8). $\tilde{n}$ is the number of neutrons per momentum-space element and per volume element in the moderator. $\mathrm{d}V_k = \mathrm{d}x\,\mathrm{d}y\,\mathrm{d}z$ is the volume element in space, and the volume element in momentum space is denoted by $dk_x\, dk_y\, dk_z$. This relation means that the majority of the $\tilde{n}$ neutrons is concentrated in a sphere whose radius is k_T.

We now derive the relation between the phase space density $\tilde{n}$ of the thermalized neutrons and the energy-dependent flux and current. The total or integrated *thermal flux* ϕ_{th} in the reflector is

$$\phi_{th} = \int \left(\frac{d\tilde{n}}{dV_k}\right) v\, dV_k = \tilde{n}\bar{v} = \tilde{n}\frac{2}{\sqrt{\pi}} v_T \tag{2.14}$$

where $\bar{v} = (2/\sqrt{\pi})v_T$ is the *average* velocity of the neutrons, see Eq. (2.1). We then derive the number of neutrons through an area element $df = dx\, dy$ emerging from a moderator volume $dx\, dy\, v\, dt = df\, v\, dt$ during a time dt in the z-direction. We introduce the energy interval $d\varepsilon$, the solid angle element $d\Omega$, and assume that the lateral components k_x, k_x of $\boldsymbol{k}$ are small ($k_z \simeq k$ or $v_z \cong v$). This yields

$$dk_x\, dk_y = k^2 d\Omega \tag{2.15}$$

and

$$dk_z = k\,\frac{d\varepsilon}{2\varepsilon}. \tag{2.16}$$

With $v^2/v_T^2 = k^2/k_T^2$ one finally obtains

$$d\dot{n}_z = \left(\frac{\phi_{th}}{4\pi}\right) e^{-\varepsilon/k_B T} \left(\frac{\varepsilon}{k_B T}\right)\left(\frac{d\varepsilon}{k_B T}\right) d\Omega\, df; \tag{2.17}$$

this is the number of neutrons per time unit entering a solid angle element $d\Omega$, from an area df, in an energy element $d\varepsilon$, which could be extracted by a narrow beam tube; we assume that the beam tube does not disturb the flux, and we neglect contributions from its side walls. One frequently uses the *luminosity* $L(\varepsilon)$ which we here define as the particle current per unit area, per energy interval and per solid angle, namely

$$L(\varepsilon) = \frac{d\dot{n}_z}{d\Omega\, df\, d\varepsilon} = \frac{\phi_{th}}{4\pi} e^{-\varepsilon/k_B T} \frac{\varepsilon}{(k_B T)^2}. \tag{2.18}$$

The total current density is then $\int L(\varepsilon)\, d\Omega\, d\varepsilon$. Integrated over all energies, one gets the total thermal flux $\phi_{th} = 4\pi L_{th}$. In general, Liouville's theorem states that the luminosity is constant everywhere along a beam path. Consequently, at any point we can calculate the *current density* in an energy interval $\delta\varepsilon$ from $L_{th}\delta\Omega\,\delta\varepsilon$ if $\delta\Omega$ is the solid angle accessible for the neutrons at this point in a plane across the beam axis, provided that there is no absorption or scattering in the beam (Maier-Leibnitz 1966; Schmatz 1973).

Although *flux* and *current density* have the same dimension $cm^{-2}\,s^{-1}$, they are, strictly speaking, different quantities. The current density is related to the directed flow of neutrons in a certain direction as described above, whereas the flux is the number of neutrons flowing in an undirected manner in the moderator. The flux ϕ_{th} is defined by the number of neutrons crossing the surface of a sphere from all sides which has a cross-section πR^2 equal to the unit area 1 cm^2, and which has a surface $4\pi R^2 = 4\,\text{cm}^2$. The flux ϕ_{th} can be measured by a thin and weakly absorbing foil, for instance a gold foil; if the activation cross-section of the nuclei in the foil is σ_a then the number of transmuted nuclei in a neutron beam with the flux ϕ_{th} is exactly $Z = \phi_{th}\, N_a\sigma_a$ after saturation if N_a is the number of nuclei in the foil (Beckurts and Wirtz 1964).

Beam holes in research reactors have cross-sections as big as 10 cm^2 or even more. This leads to a flux depression, and accurate calculations of the extracted current density by diffusion codes are difficult, in particular if adjacent beam holes interfere. Nevertheless, the value as calculated by our procedure is sufficient for an estimate and the flux depression is normally not larger than 10% or 20% in D_2O reflectors. For light-water reflected research reactors the situation is more complicated. For neutrons in H_2O the diffusion length is 2.8 cm as compared to values between 100 and 150 cm in D_2O, depending on purity. As a consequence gradients and flux depression are larger.

In order to measure small neutron energy transfers with high resolution as needed, e.g. for quasielastic neutron scattering experiments, it is advantageous to work with low incident energies in the meV region. The thermal neutron spectrum of the ambient temperature moderator has insufficient intensity in this energy range. Therefore part of the neutrons in the reflector can be cooled in a so-called *cold source*, thus increasing the flux in the meV region by reducing k_T and therefore the phase space volume filled by neutrons. The cold source is a vessel in front of the beam hole nose, sometimes located slightly outside the region of maximum thermal flux, compromising between neutron flux and heat removal requirements. Liquid H_2 and D_2 are the usual moderator materials. Also D_2O ice at 20–25 K can be used, but by radiolysis D and O radicals are formed which are immobile at that low temperature. In order to release this chemical energy by recombination, the solid D_2O moderator has to be warmed up periodically. Similiar problems occcur for solid methane. An advantage of an H_2 moderator is that its small volume reduces the total nuclear heat load (2–3 cm thickness of the vessel). On the other hand, a D_2 moderator has lower absorption and the vessel can be large. Consequently, the number of beam tubes originating on a D_2 source can be greater than on an H_2 source.

The flux profit achieved by a cold source is quantified by a gain factor $G(\varepsilon)$ which describes the increase of the extracted neutron current as compared to the value for the Maxwellian spectrum from the thermal moderator which may have a temperature of 320 K with $k_T = 3.7\,\text{Å}^{-1}$. The diagram calculated by Schmatz (1973) in Fig. 2.2 shows the momentum space density (see Eqs. (2.13) and (2.14)) for different sources, namely

$$p(k) = \frac{\phi_{th}}{2\pi} G(k) k_T^{-4} e^{-k^2/k_T^2} \tag{2.19}$$

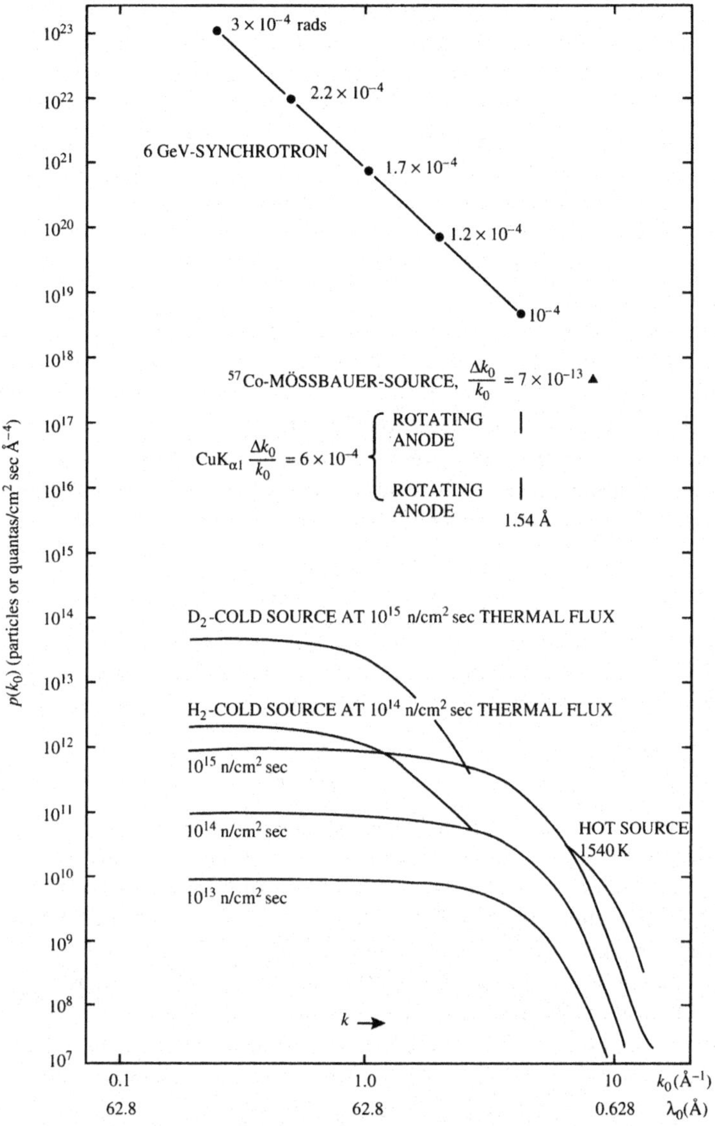

Fig. 2.2 Density in momentum space $p(k_0)$ of thermal, cold, and hot neutrons as a function of neutron wave number k_0 or wavelength λ, $p(k_0)$ (n/cm^2 Å^{-4}). The curves for cold and hot sources result from empirical data (from Schmatz 1973).

(in units of Å^{-4}). For a 320 K equilibrium spectrum one can choose $G = 1$. As can be seen, the gain from cold sources is of the order of 5–20, depending on wavelength. The curves for the cold and hot sources in the figure are representative results from existing installations. Obviously, the gain is much smaller than the theoretical value for a fully thermalized spectrum in the cold moderator. It is noteworthy that the momentum density for gamma quanta from a rotating anode X-ray source is about three orders of magnitude higher than the value obtained for a cold source in a high-flux reactor, choosing a width in wave number of $\Delta k/k = 10^{-3}$ corresponding to the K_α line of the X-ray source. For synchrotron light sources, the space density exceeds that of reactors by eight orders of magnitude or even more! The intensity drawback of neutron sources as compared to a synchrotron is partially compensated by the fact that the divergence and the beam size for a neutron source can be adapted to the problem, and is much larger than the values delivered from a synchrotron source.

So far a pure Maxwellian spectrum has been considered, not taking into account the moderation process. Fast neutrons from the reactor core penetrate into the D_2O reflector and are slowed down by elastic collisions, and then nearly reach thermal equilibrium before being captured. Consequently, the thermal spectrum merges into the so-called *epithermal spectrum* of the neutrons which are still in the process of moderation; for energies of above several $k_B T$, the epithermal flux Ψ follows the relation (Weinberg and Wigner 1958)

$$\phi(\varepsilon)\, d\varepsilon = \Psi \frac{d\varepsilon}{\varepsilon}. \tag{2.20}$$

The number of neutrons per unit time being slowed down to an energy ε, $\Psi = \varepsilon\phi(\varepsilon)$, is called the lethargy flux, which is obviously the flux per logarithmic energy interval. The ratio of thermal to epithermal flux can be calculated as

$$\frac{\phi_{th}}{\Psi} = \frac{\xi\sigma_s}{\sigma_a} \tag{2.21}$$

where σ_s and σ_a are the scattering and absorption cross-sections of the moderator nuclei, respectively. $\xi\sigma_s$ means the average logarithmic energy loss during scattering. It is largest for pure hydrogen ($\xi = 1$) and small for heavy nuclei $\xi \simeq 2/A$ with $A \gg 1$. The flux ratio ϕ_{th}/Ψ is 70 for light water, and about 2500 for heavy water which moderates less efficiently and has very low absorption. In steady state reactors, the epithermal flux is too weak to be used for spectroscopy, and it is a source of parasitic background. If a higher flux at energies above say 100 meV is needed, sometimes a hot source (e.g. a heated graphite block) is introduced into the reactor which shifts the Maxwell spectrum to higher energies, so increasing the 'heated' Maxwellian above the epithermal $1/\varepsilon$ flux.

Figure 2.3 summarizes typical thermal, cold and hot source fluxes per solid angle and wavelength interval of the Grenoble reactor at the beam hole noses in the D_2O reflector. The transition from the Maxwellian thermal behaviour to the epithermal $1/\varepsilon$ spectrum can be seen for the thermal spectrum.

2.2.2 Spallation neutron sources

Nuclear *spallation*, by contrast to nuclear fission, is a slightly endothermal process and relies on an external supply of protons with an energy of the order of 1 GeV

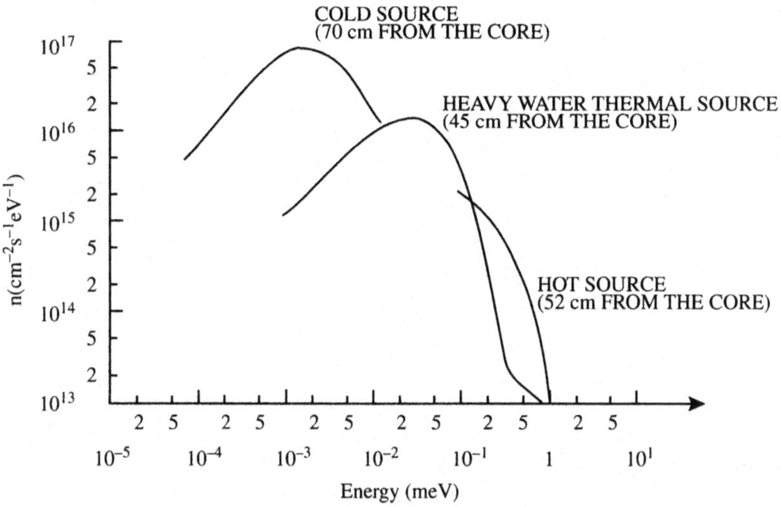

Fig. 2.3 Neutron spectra at typical thermal, cold, and hot beams at the high flux reactor of the Institut Laue–Langevin as a function of energy. For the thermal spectrum, the transition from the Maxwellian to the $1/\varepsilon$ spectrum can be seen (from 'The yellow book' of the ILL 1994).

(Bauer 1992). It is generally viewed as a two-stage process consisting of a nuclear cascade in which nucleons are directly knocked-on inside the nuclei, and a subsequent 'evaporation' phase where nucleons, preferably neutrons, are expelled from the excited nuclei to get rid of most of this energy from the cascade phase. Since the fast neutron release in a spallation target starts and stops immediately as the driving beam is switched off, almost any desired time structure in the fast neutron flux of a spallation source can be achieved. This ranges from a virtually continuous flux as produced by the 51 MHz cyclotron beam in SINQ, the medium-flux neutron source in Switzerland (Wagner *et al.* 1998), via fully thermalized pulses as in the SNQ project study (Bauer *et al.* 1985), to the usual microsecond regime for a small and 'decoupled' moderator. The time-structured beam comes from an accelerator, which is either a linear accelerator, a synchrotron, or a combination of both. The most prominent safety feature of a spallation neutron source is the circumstance that it can be switched off immediately, and no nuclear fuel and fission products come into play. For a detailed review on pulsed neutron spectroscopy and sources see Windsor (1981).

Figure 2.4 shows the floor plan of ISIS, the world's best first-generation pulsed spallation neutron source at the Rutherford Appleton Laboratory in Great Britain. It comprises three major parts: a fast cycling proton synchrotron (design current 200 μA at 800 MeV), a spallation target feeding moderators at three different temperatures, and an assembly of neutron scattering instruments in the experimental hall.

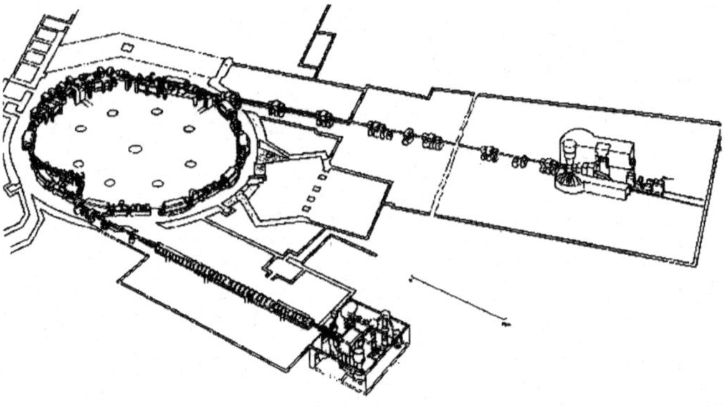

Fig. 2.4 Pulsed source ISIS at the Rutherford Appleton Laboratory. 70 MeV protons are injected into an accelerator ring. The 0.8 GeV protons hit a tantalum target which produces the neutrons. (With kind permission of the Rutherford Appleton Laboratory.)

Pulsed sources are characterized by the width of the primary proton current, the pulse frequency, the epithermal as well as the thermal neutron spectrum, and the energy-dependent duration and shape of the neutron pulses. This involves too many parameters for characterization, and the parameters separately do not relate in any simple way to those used to describe reactors. Typically, a pulsed source has several distinctly different moderators. The pulse length determines the resolution of the spectrometers or diffractometers at the source, whereas the repetition rate is related to the frame overlap, which means the intensity at the detectors during a given pulse, originating from the previous pulse or pulses.

In the existing pulsed sources, small moderators are used ('undermoderated sources'). These generate much shorter pulses and thus normally pulse-shaping devices are not necessary. Proton pulses from the accelerator of μs lengths are used, and the peak currents are 30–100 times larger than the average current. In the region of thermal energies, the spectrum emitted from a small moderator is still close to a Maxwellian-shaped distribution as described above. However, the 'neutron temperature' may be considerably higher (a factor 1.5 or so) than the temperature of the moderator, depending on the size and on the kind of the moderator. Typical pulse lengths in the neutron energy region between 10 and 20 meV are of the order of 30 μs for a 'decoupled' liquid H_2 or solid CH_4 target, where this figure depends strongly on size and, possibly, on 'poisoning' of the moderator with absorbing material; they are correspondingly shorter for a decoupled hydrogeneous moderator at room temperature. In the epithermal region, i.e. for energies sufficiently far from $k_B T$, the emitted spectrum can be calculated for an infinite moderator and the resulting flux is proportional to $1/\varepsilon$ as mentioned in the previous section. For the fast neutron pulse of a short pulse source, falling into the moderator within a narrow time interval at $t = 0$, the subsequent current density distribution emitted from the source surface as a function of time is (Windsor 1981)

$$J(\varepsilon, t) = \frac{K}{\varepsilon}\left[\xi \sum_{\mathrm{s}} \frac{v}{2\Gamma(2/\gamma)}\right]\left(\frac{t}{\tau}\right)^{2/\gamma} \mathrm{e}^{-t/\tau}; \tag{2.22}$$

Γ is the gamma function. K is a constant related to the source strength. The relaxation time which describes the decay of the intensity and thus roughly characterizes the pulse length is

$$\tau = \frac{\gamma}{\xi \tilde{N}_{\mathrm{s}} \sigma_{\mathrm{s}} v} \tag{2.23}$$

where σ_{s} is the scattering cross-section of the moderator atoms, $\tilde{N}_{\mathrm{s}}$ is their density, and $v = (2\varepsilon/m)^{1/2}$ is the neutron velocity. For a nuclear mass $A = 1$ the parameters are $\gamma = \xi = 1$; for $A \gg 1$ one gets $\gamma \approx 4/3\,A$ and $\xi \approx 2/A$. Obviously, τ is proportional to the mean time between moderating collisions. For a finite and small moderator the decay of $J(\varepsilon, t)$ with energy and with time is somewhat steeper than indicated by Eq. (2.22).

The pulse function $J(\varepsilon, t)$ depends on the dimensionless parameter t/τ in Eq. (22) such that the effective pulse width τ_{p} (half width at half maximum) is proportional to the reciprocal neutron velocity. For pulsed experiments the flight time from the source to the detector τ_{f} is also proportional to $1/v$, such that the resolution

$$\Delta\varepsilon = 2\frac{\Delta v}{v} = 2\frac{\tau_{\mathrm{p}}}{\tau_{\mathrm{f}}} \tag{2.24}$$

is energy independent.

Numerous (and sometimes misleading) discussions deal with the comparison between pulsed and steady state sources. We conclude this chapter with a few statements as a basis for such comparisons, taken from the study for a second-generation spallation source project, the ESS (European Spallation Source) (Richter and Springer 1998). For certain instruments it is generally believed that only the average neutron current of a source is relevant, and pulsing should not be profitable for such instruments. Triple axis spectrometers are supposed to belong to this class of instruments. Even there, however, pulsing may be advantageous, e.g. because background or higher-order suppression is possibly due to the pulse structure. On the other hand, instruments which inherently need short pulses have nearly the full benefit of pulsing. This benefit is roughly the peak flux divided by the average flux.

Taking the study for the European Spallation Source (ESS) (Kjems *et al.* 1997) as a reference we can quote certain quantitative statements. For 5 MW average proton beam power and 50 Hz pulse repetition rate, the thermal peak flux is about 300 times higher than the flux average; the latter is about $0.6 \cdot 10^{15}\,\mathrm{cm}^{-2}\,\mathrm{s}^{-1}$. This benefit can be more or less used in particular by time-of-flight diffractometers.

For pulsed spectrometers the benefit is often considerably smaller because, for a steady state reactor, the pulse repetition rate can be *adapted* for the corresponding experiments, e.g. to avoid frame overlap. For a pulsed source, however, there is normally only *one* repetition rate (perhaps two, in case of a second target). This leads to compromise and an intensity drawback. Certain high-resolution instruments do not need short pulses for resolution at all (e.g. backscattering, spin echo, SANS); in these cases, the pulsing of a source has nearly no or only a minor profit. So one concludes that a modern pulsed source with several MW target power is competitive with a good high-flux reactor for certain (non-pulsing) instruments, and is advantageous by factors between 10 and 100 for many instruments, which will be the majority.

2.3 NEUTRON GUIDES FOR BEAM TRANSPORT

Originating from the reflector or from the cold source of a reactor, the neutrons can be led to the instruments by beam tubes or by totally reflecting *neutron guides* (Maier-Leibnitz and Springer 1966). For a non-reflecting beam tube the accessible solid angle is entirely determined by its length and by its entrance and exit cross-sections. On the other hand, for neutron guides the solid angle is given by the critical angle γ_c for total reflection on the guide surface. For neutrons with the refractive index n given in Eq. (3.15) below one obtains

$$\gamma_c = \sqrt{2(1-n)} \tag{2.25}$$

or in terms of the lateral wavenumber spread of the beam,

$$\Delta k_x = 2k\gamma_c \tag{2.26}$$

where $2\gamma_c$ is the maximum divergence accepted by the neutron guide. A typical value is $\Delta k_x = 0.0214\,\text{Å}^{-1}$ for a nickel surface. The reflectivity of guides is much better than one would have expected after their discovery, i.e. less than 1% loss per reflection. Consequently very long neutron guides can be built without strongly reducing the transmitted intensity such that a great number of instruments can be located along one guide and, due to the large distance from the reactor source, the fast neutron and gamma background is correspondingly small. Furthermore, guides can be curved which moves the tube *off* the line of direct view which reduces the background. The transmitted intensity in a rectangular guide of width d, with a radius of curvature ρ is characterized by the accepted solid angle $\bar{\omega}$, averaged over the guide tube cross-section. This can easily be calculated. We only quote two limiting cases, namely

$$\frac{\bar{\omega}}{2\gamma^{*2}} = \begin{cases} 2\varepsilon^*/\varepsilon & \text{for } \varepsilon \ll \varepsilon^* \\ \frac{4}{3}(\varepsilon^*/\varepsilon)^2 & \text{for } \varepsilon \geq \varepsilon^* \end{cases} \tag{2.27}$$

where $\bar{\omega}$ is quoted in units of the so-called cut-off angle of the curved guide,

$$\gamma^* = \sqrt{2\frac{d}{\rho}}. \tag{2.28}$$

ε^* is the corresponding neutron energy, as obtained with $\gamma_c = \gamma^*$ in Eq. (2.25) from the energy dependence of the critical angle γ_c. For small ε, zigzag reflections dominate, where the neutron passes between the inner and outer wall of the curved guide. For higher energies garland reflections occur along the convex side of the guide. Obviously for small ε the solid angle compensates the linear energy dependence of the Maxwell–Boltzmann distribution in Eq. (2.18). The resulting energy dependence of the solid angle and the transmitted current, assuming a Maxwellian incident distribution, are presented in Fig. 2.5. Normally, bundles of curved guides are installed on cold sources, providing neutrons in a wavelength

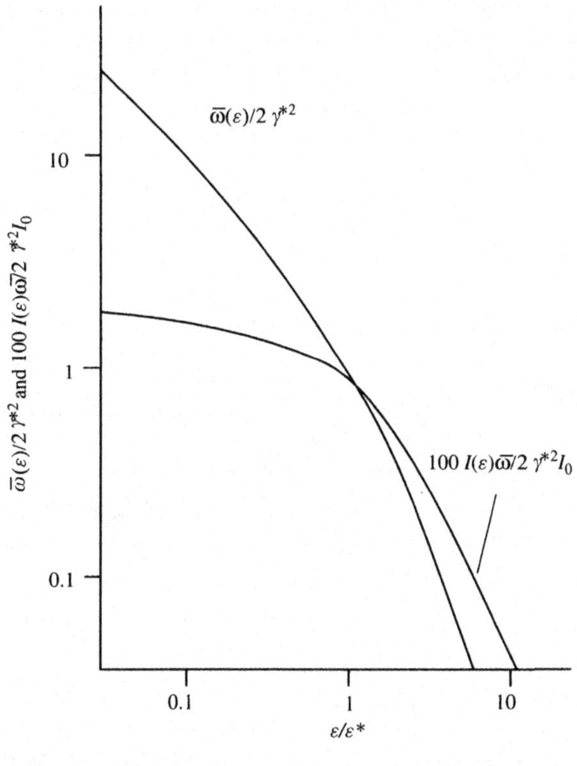

Fig. 2.5 Transmitted solid angle of a curved rectangular guide averaged over the guide cross-section in units of $2\gamma^*$ where $\gamma^{*2} = 2d/\rho$ is the cut-off angle, as a function of $\varepsilon/\varepsilon^*$ where ε^* is the energy corresponding to γ^*. Also the transmitted current is shown, assuming a Maxwellian incident current (from Maier-Leibnitz and Springer 1963, with permission from Elsevier Science).

range between 5 and 15 Å. The lateral width $2\Delta k$, for instance, for a ^{58}Ni guide, is sometimes smaller than the width allowed from the point of view of instrumental resolution. In particular, many instruments allow a large *vertical* divergence which enters the resolution but not to first order. In a sense, this has diminished the use of neutron guides for thermal neutrons where the divergence due to total reflection is relatively small.

Many years ago, Mezei (1976) and Mezei and Dagleish (1977) suggested extending the range of high reflection by a broadened Bragg reflection from a sequence of double layers with a gradient of d-spacing. Such coatings are called *supermirrors*. Using alternating layers of magnetic and non-magnetic materials, the angle of reflection depends also on the spin state of the neutron. Hence, supermirrors can also be used as efficient neutron polarizers, especially for cold neutrons. Multilayers for the transport of neutrons are usually composed of Ni and Ti, because the scattering contrast between these two materials is very large. Therefore, high reflectivities can be obtained with a relatively small number of layers. During the last few years supermirror coatings have been developed by several groups (see, for instance, Schärpf (1991); Majkrzak (1994) and Krist *et al.* (1995)). By applying some tricks to avoid Ti–Ni interdiffusion during the production process (which would result in contrast reduction and rough interfaces) Böni (1997) succeeded in producing supermirrors with a reflectivity $R \approx 94\%$ at two times Θ_c^{Ni} ('$m = 2$ mirror').

Figure 2.6 shows as an example of the reflectivity curve of a polarizing supermirror with a cut-off angle $m = 2.4$ times Θ_c. In addition to the excellent reflecting

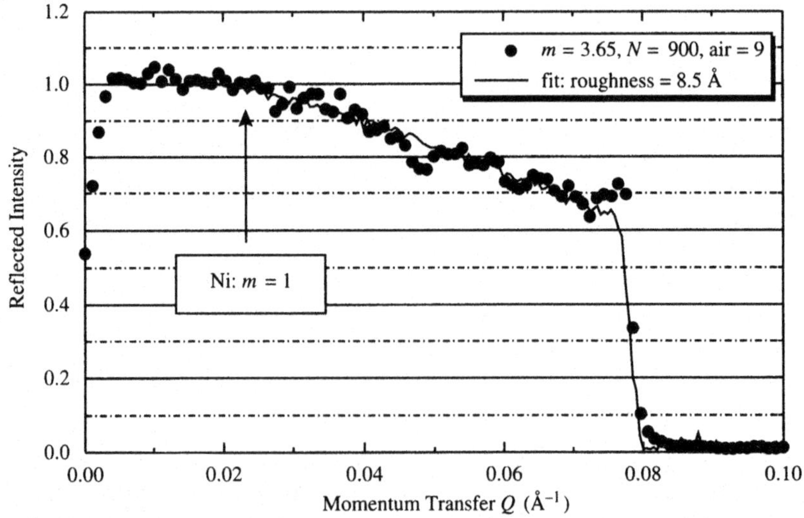

Fig. 2.6 Reflectivity of a polarizing supermirror. The polarization is of the order of 0.93 (from Böni 1997, with permission from Elsevier Science).

properties, these mirrors are remanent due to a magnetoelastic coupling to the (anisotropic) in-plain strain (Clemens *et al.* 1997). By extending the number of layers to 450, Böni and coworkers (1997) succeeded in producing supermirrors with $R > 70\%$ at $m \approx 3.4$. Using the advanced supermirrors in neutron guides results in an appreciable increase in the divergence of the neutron beams and thus in their neutron current density and intensity.

If these divergent beams subsequently can be focused onto the sample, appreciable intensity gain factors can be achieved. Two kinds of focusing devices have been developed, which are already operating in a number of instruments. One is a convergent guide covered with supermirrors (mostly convergent in one dimension, but sometimes even doubly focusing) which in laboratory jargon is called anti-trumpet. The other type of device is a focusing monochromator, consisting of a curved Bragg-reflecting single crystal, or a polygon of plane crystals on a curved substrate, or a polygon of plane crystals equipped with sophisticated mechanics which allows the curvature to be tuned vertically (Riste 1970), horizontally (Scherm *et al.* 1977) or both horizontally and vertically (Bührer *et al.* 1981; Bührer 1994; Lechner *et al.* 1994). Such devices on triple axis spectrometers lead to an increase of the neutron count rate by more than an order of magnitude without appreciable loss in energy resolution.

3

The basic theory of neutron scattering

To describe theoretically neutron scattering as a probe for condensed matter we first explain the scattering on a rigidly bound and isolated nucleus by the neutron–nucleus interaction; then we derive the corresponding matrix element in terms of perturbation theory which yields the differential scattering cross-section of the sample as a function of momentum and energy transfer of the scattered neutrons. With the concept of the van Hove correlation functions, this matrix formulation can finally be transformed into classical correlation functions which are convenient if the particle motion can be described by classical dynamical models. An introduction to the theory of thermal neutron scattering has been written by Squires (1978); an extensive representation is by Lovesey (1987)

3.1 THE SCATTERING LENGTH OF A BOUND NUCLEUS

We consider a rigidly *bound* nucleus exposed to a neutron plane wave $\exp(ikz)$ where $k = 2\pi/\lambda$ is the corresponding wavenumber, and λ the neutron wavelength. This plane wave interacts with the nucleus and one gets a total wavefunction consisting of the incident and the outgoing waves, namely

$$\psi_t(x, y, z) = e^{ikz} + \left(\frac{f}{r}\right)e^{ikr} \tag{3.1}$$

where $r(x, y, z)$ is the radial distance from the nucleus, and $f = (\delta_0/k)\exp(i\delta_0)$ where δ_0 is the phase shift caused by the scattering process. This holds for wavelengths λ large compared to the range of interaction or to the radius of the nucleus ($R_n \simeq 10^{-14}$ m). Under these conditions scattering is isotropic (s-wave scattering). From the quantity f we define the so-called *scattering length*, namely

$$b = -f = \frac{\delta_0}{k} \quad \text{for } \delta_0 \to 0. \tag{3.2}$$

Normally, this is a real quantity. It can be positive or negative and depends on the interaction potential between nucleus and neutron; the detailed shape of the potential does not matter. For the hypothetical case of a square potential of depth V_0 in a sphere of radius R_n one gets $b = (m/2\pi\hbar^2)(4\pi/3)(R_n^3)V_0$ where m is the neutron mass. Later we replace this potential simply by a delta function (*pseudopotential*) leading to the correct value of b. This will be discussed in Section 3.2.

The total wavefunction (3.1), including the incident plane wave, can be expanded in spherical Bessel functions. The first term of the expansion has spherical symmetry which allows us to write, instead of (3.1),

$$\psi_t(r) = (kr)^{-1} e^{i\delta_0} \sin(kr + \delta_0) \tag{3.3}$$

or, expanding the sine,

$$r\psi_t \simeq r + \left(\frac{\delta_0}{k}\right) = r - b. \tag{3.4}$$

For a 'hard sphere', i.e. an inpenetrable nucleus, one gets $R_n = b$. The function $r\Psi_t$ is shown in Fig. 3.1, where b is the abscissa intercept of $r\psi_t(r)$. One further defines the coherent bound scattering cross-section

$$\sigma_c = 4\pi b^2. \tag{3.5}$$

If a current of neutrons with density j_0 falls onto this nucleus, the total number of neutrons scattered isotropically into a sphere around this nucleus is $j_0\sigma_c$ per second. For a *free* nucleus at rest, going from the laboratory to the centre-of-gravity system, one obtains

$$\sigma_{\text{free}} = \sigma_s \frac{M^2}{(M+m)^2}; \tag{3.6}$$

M and m are the nucleus and the neutron mass, respectively. The reduced-mass factor is one-quarter for neutron–proton scattering and approaches unity for heavy nuclei.

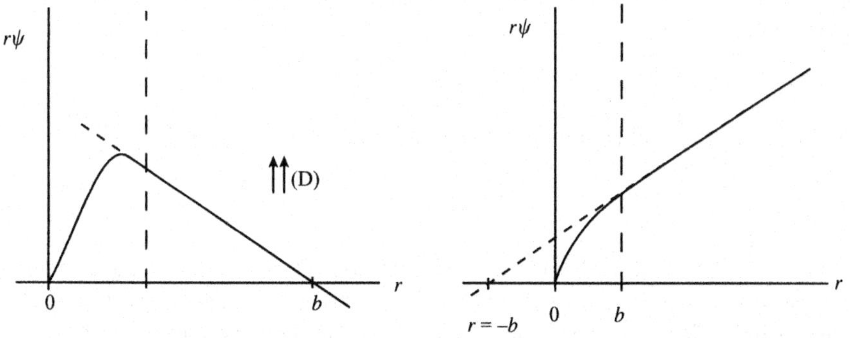

Fig. 3.1 Schematic behaviour of the total function $r\psi(r)$ for a neutron interaction with a nucleus consisting of the incident and the outgoing part. For scattering from a bound state the scattering length b is positive, and negative for scattering from unbound state. The amplitude of this scattered wave is entirely determined by b and not by the shape of the interaction potential inside the nucleus.

For neutrons the scattering length is a proportionality factor in the transition matrix elements to be discussed later, and it is only known experimentally. For comparison, in X-ray scattering the corresponding quantity is a function of the scattering angle θ with $f(\theta = 0) = r_0 Z$ where $r_0 = e^2/m_e c^2 = 0.2818 \cdot 10^{-14}$ m is the so-called classical electron radius and Z is the chemical number. The *irregularity* of the nuclear scattering length with Z and its relatively large value for light elements allows different atoms to be distinguished; the scattering contribution of such atoms could not be separated by X-ray diffraction. This plays an important role for hydrogen compounds or for investigations on light nuclei. It reflects an important aspect of neutron scattering for condensed matter physics in general.

There are many atoms where the nuclear species consists of several isotopes with scattering lengths $b_1, b_2, \ldots,$ and corresponding abundances $c_1, c_2, \ldots$. For the derivation of the scattering theory in the next section we assume a random distribution of nuclei on their sites, and two averages will be needed in the following. The average value of the scattering lengths is given by

$$\langle b \rangle = \sum_k c_k b_k. \tag{3.7}$$

The corresponding cross-section is called the coherent neutron scattering cross-section:

$$\sigma_c = 4\pi \langle b \rangle^2. \tag{3.8}$$

We point out that for the very light isotopes H, D, and T, pronounced isotope effects may appear even at room temperature (e.g. for certain metal hydrides), so that these isotopes are denoted by different chemical symbols.

Since the scattering lengths of the isotopes are different, there appears so-called *isotope incoherent* scattering, the meaning of which will be discussed in Section 3.2; it is defined by the so-called *incoherent* neutron scattering cross-section

$$\sigma_i = 4\pi \left(\langle b^2 \rangle - \langle b \rangle^2 \right) \tag{3.9}$$

where the average value of the squared scattering lengths is given by

$$\langle b^2 \rangle = \sum_k c_k b_k^2. \tag{3.10}$$

The cross-section σ_i describes the square fluctuation of b due to the isotope disorder. The sum $\sigma_s = \sigma_c + \sigma_i$ is called the total (bound) scattering cross-section. For nuclei with nuclear spin $I \neq 0$ there appears an additional scattering contribution, called *spin incoherent* scattering, if the scattering length depends on the relative orientation of neutron and nuclear spin. Since the neutron has spin $\frac{1}{2}$, we have two values b^+ and b^-, and one gets

$$\langle b \rangle = p^+ b^+ + p^- b^-. \tag{3.11}$$

Correspondingly, the *spin incoherent cross-section*, Eq. (3.9), can be expressed as

$$\sigma_{\rm i} = 4\pi\left(\langle b^2\rangle - \langle b\rangle^2\right) = 4\pi(b^+ - b^-)^2 p^+ p^-. \tag{3.12}$$

For a compound nucleus with $J_+ = I + \frac{1}{2}$ and $J^- = I - \frac{1}{2}$ the relative frequencies for both orientations are then

$$p^+ = \frac{2I+2}{4I+2} = \frac{I+1}{2I+1}$$

and

$$p^- = \frac{2I}{4I+2} = \frac{I}{2I+1} \tag{3.13}$$

respectively. This result can also be formulated more generally by the scattering-amplitude operator which is (Lovesey 1987)

$$\hat{b} = A + B\frac{\hat{\boldsymbol{\sigma}}}{2}\hat{\boldsymbol{I}} \tag{3.14}$$

with $A = p^+b^+ + p^-b^-$ and $B = 2(b^+ - b^-)/(2I+1)$. In this formula, $\hat{\boldsymbol{\sigma}}/2$ and $\hat{\boldsymbol{I}}$ are the spin operators of the neutron and nucleus, respectively. This formulation allows the calculation of scattering if the spins are not randomly distributed, normally at very low temperatures. Obviously, spin incoherent scattering occurs *as if* two species of different nuclei were randomly distributed over the atomic sites, having b^+ or b^- as scattering lengths with corresponding abundances p^+ and p^-. For the general case of both spin and isotope incoherence see Lovesey (1987).

Scattering lengths have been measured for most elements and for many isotopes, as well as several incoherent scattering cross-sections. Methods were used where the state of binding, the corresponding dynamics of the nucleus, and the mutual interference do not enter (Koester and Rauch 1990; Sears 1992). The most common method is the determination of the refractive index which is given by

$$n = 1 - \frac{\tilde{N}_0\langle b\rangle\lambda^2}{2\pi}; \tag{3.15}$$

$\langle b\rangle$ is the (average) bound coherent scattering length independent of the physical state of the sample material. $\tilde{N}_0$ is the atom number density. The refractive index can be determined from total reflection (Maier-Leibnitz 1962; Koester 1965) using the gravity constant for calibration or by neutron interferometry (Rauch *et al.* 1974; Rauch and Süda 1974). Typical accuracies are of the order of one per cent. An elegant method for powdered samples is the principle of the Christiansen Filter (Koester *et al.* 1984): a powder sample is mixed with a liquid which consists of two components with different refractive indices. By changing the composition one changes its refractive index until it is exactly matched with that of the immersed powder; multiple refraction and scattering on the powder disappears, or goes through a minimum. So the scattering length $\langle b \rangle$ can be measured relative to its value for the liquid which must be known *a priori.*

The determination of the free cross-section, i.e. the total scattering cross-section of a free nucleus at rest, σ_{free} (3.6), yields σ_c or $\sigma_s = \sigma_c + \sigma_i$, respectively, if there is an incoherent contribution. This quantity is obtained by a transmission experiment at higher energies since the sum over all coherent and incoherent, elastic and inelastic processes approaches the free cross-section as soon as the energy of the neutron is sufficiently high, but still small enough to have isotropic s-wave scattering. In many cases, the incoherent cross-section σ_i was simply obtained as the difference between the measured free total scattering cross-section and the coherent cross-section as determined from the refractive index. A direct measurement of σ_i is tedious. It can be obtained by a transmission experiment on a He-cooled sample at very small k_0 where phonon scattering is negligible and Bragg scattering does not occur.

For a table of the 'best values' for the coherent scattering lengths of elements and/or isotopes, for the total (bound) scattering cross-sections, and for the bound incoherent scattering cross-sections, we refer to Sears (1992).

An interesting case is hydrogen where the incoherent scattering cross-section is much larger than the coherent one. Since hydrogen plays a key role in many experiments to be described later we briefly explain the situation of the neutron–proton interaction. Since proton and neutron both have spin $\frac{1}{2}$, there is a triplet state of scattering with total spin $S = 1$, and singlet scattering with $S = 0$, having statistical weights $(2S + 1)$ equal to 1 or to 3, respectively. There is experimental evidence that the (p, n) triplet state is bound (the deuteron) and one gets $b^{(T)} > 0$. On the other hand, the singlet state is unbound with $b^{(S)} < 0$, where $|b^{(S)}|$ is about four times larger than $|b^{(T)}|$. This leads to a negative coherent scattering length $\langle b \rangle$, and, furthermore, to an unusually large incoherent scattering cross-section since σ_i is proportional to $(b^{(T)} - b^{(S)})^2$. We also draw attention to the case of deuterium where the coherent scattering length differs strongly in sign and magnitude from that of hydrogen. This plays a very important role in isotope *labelling* or in *index matching* methods for the investigation of hydrogeneous compounds.

In the following section, first-order perturbation theory or the Born approximation is used to derive the scattering matrix element. It is surprising that this approximation is successful since the neutron energies under consideration are in the range 10^{-3}–1 eV which is more than seven orders of magnitude smaller than the interaction potential V_0 between neutron and nucleus, so that the criterion for the approximation $R_n|[1 \pm (V_0/E_0)]^{1/2} - 1| \ll 1$ is not fulfilled at all. Nevertheless, as has been shown by Dietze and Nowak (1981), this approximation holds because the nucleus is not rigidly fixed. For instance, in a crystal it is smeared over a volume of about u^3 where $u \simeq 0.1$ Å is the mean vibrational amplitude. Dietze and Nowack have investigated the deviations from perturbation theory in calculating the reduced mass factor $\sigma_{\text{free}}/\sigma_s = M^2/(M + m)^2$ which has to be corrected by about $R_n/u \simeq 10^{-3}$. So far, nobody has been able to measure this ratio to such an accuracy and to investigate this intriguing effect. Another very small correction is caused by the presence of the nuclei surrounding a given nucleus; this leads to local field effects, which are related to the mutual atomic correlation in condensed matter (Sears 1985).

3.2 THE SCATTERING MATRIX ELEMENT; GENERAL PROPERTIES

The quantity to be calculated is $\mathrm{d}^2\sigma(\varepsilon_0, \varepsilon_1, \theta)/\mathrm{d}\Omega_1\,\mathrm{d}\varepsilon_1$, the differential cross-section for a process where neutrons with an incident energy ε_0 are scattered on a sample into a solid angle element $\mathrm{d}\Omega_1$ at an angle θ, and with an energy between ε_1 and $\varepsilon_1 + \mathrm{d}\varepsilon_1$ (see Fig. 3.2). In order to obtain the cross-section the transition matrix element will be derived by perturbation theory.

For the system consisting of the neutron beam and the sample, the Hamiltonian reads

$$H = H_0 + \frac{p_\mathrm{n}^2}{2m} + V. \tag{3.16}$$

The undisturbed system is characterized by the Hamiltonian H_0; the interaction potential between the neutron and the nuclei of the sample is V. $p_\mathrm{n}^2/2m = \hbar^2 k^2/2m$ is the kinetic energy of the neutron. Before and after scattering the neutron is described by ideal plane waves normalized in a cube of volume L^3, with wave vectors $\boldsymbol{k}_0$ and $\boldsymbol{k}_1$, namely

$$\psi_0 = L^{-3/2}\mathrm{e}^{\mathrm{i}\boldsymbol{k}_0\boldsymbol{r}} \quad \text{and} \quad \psi_1 = L^{-3/2}\mathrm{e}^{\mathrm{i}\boldsymbol{k}_1\boldsymbol{r}}. \tag{3.17}$$

From now on, the neutron wavefunctions ψ_0 and ψ_1 will be symbolized by $L^{-3/2}|\boldsymbol{k}_0\rangle$ and $L^{-3/2}|\boldsymbol{k}_1\rangle$, respectively. The unperturbed wavefunctions of the sample in quantum state n is $\Psi_n(R_1, R_2, \ldots, R_N)$ which depends on the coordinates $\boldsymbol{R}_i$ of all scattering nuclei. The corresponding quantum energies E_n are then given by

$$H_0\Psi_n = E_n\Psi_n. \tag{3.18}$$

First-order perturbation approximates the total wavefunction of the sample and the neutron by the product of their undisturbed wavefunctions, and one obtains the transition probability between the initial state $k_0 n_0$ to a final state $k_1 n_1$ as (Messiah 1962; Squires 1978; Lovesey 1987)

$$w(\boldsymbol{k}_0, n_0 \to \boldsymbol{k}_1, n_1) = 2\pi\hbar^{-1}L^{-6}|\langle\boldsymbol{k}_1 n_1|V|\boldsymbol{k}_0 n_0\rangle^2. \tag{3.19}$$

During the scattering process we conventionally consider the energy change of the *sample* which is related to the energy change of the neutrons during this process, $\varepsilon_0 - \varepsilon_1$, namely

$$\hbar\omega = E_{n_1} - E_{n_0} = \varepsilon_0 - \varepsilon_1. \tag{3.20}$$

This means energy conservation between neutron and sample; for energy gain (or loss) of the neutron $\hbar\omega$ is negative (or positive), respectively; $\hbar\omega = 0$ is a purely

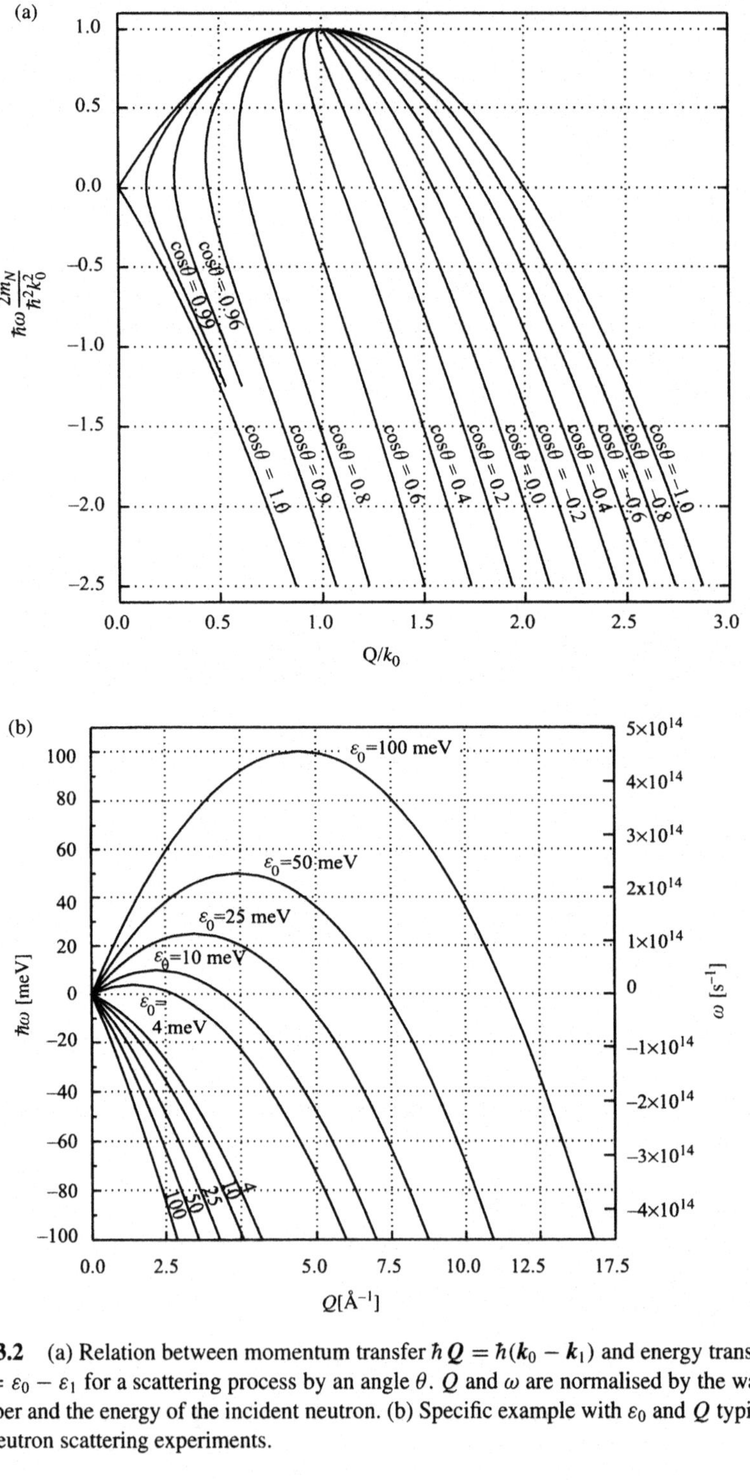

Fig. 3.2 (a) Relation between momentum transfer $\hbar \boldsymbol{Q} = \hbar(\boldsymbol{k}_0 - \boldsymbol{k}_1)$ and energy transfer $\hbar\omega = \varepsilon_0 - \varepsilon_1$ for a scattering process by an angle θ. Q and ω are normalised by the wave number and the energy of the incident neutron. (b) Specific example with ε_0 and Q typical for neutron scattering experiments.

elastic process. The scattering process is connected with a momentum transfer to the sample as a whole, namely

$$\hbar \boldsymbol{Q} = \hbar(\boldsymbol{k}_0 - \boldsymbol{k}_1). \tag{3.21}$$

This momentum is transferred to the sample as a whole. Q and ω are connected by the relation

$$Q^2 = k_0^2 + k_1^2 - 2k_0 k_1 \cos\theta; \tag{3.22}$$

θ is the scattering angle. This relation is sketched in Fig. 3.2 for different scattering angles and for a fixed incident energy ε. Only a restricted area in the Q, ω plane can be covered by a scattering experiment, the size of which increases with increasing energy ε_0.

To obtain the *total* transition or scattering probability for a given momentum transfer $\hbar\boldsymbol{Q}$, and an energy transfer $\hbar\omega$, we first take the sum over all initial states with the corresponding statistical weight of the thermal occupancy

$$p_{n_0} = \frac{\exp(-E_{n_0}/k_B T)}{\sum_{i_0} \exp(-E_{i_0}/k_B T)}. \tag{3.23}$$

Furthermore, we have to sum over *all final* states. A δ-function in $\hbar\omega$ will be introduced which *selects* those processes which correspond to a given energy change of the neutron $\hbar\omega = \varepsilon_0 - \varepsilon_1$. The total probability is then

$$w(\boldsymbol{k}_0 \to \boldsymbol{k}_1) = \sum_{n_0} p_{n_0}(T) \sum_{n_1} w(\boldsymbol{k}_0; n_0 \to \boldsymbol{k}_1; n_1)\, \delta(E_{n_0} - E_{n_1} + \hbar\omega). \tag{3.24}$$

The number of final states in momentum space is given by the phase-space density of final states, $(L/2\pi)^3$, times the size of the volume element $d\boldsymbol{k}_1$ in $\boldsymbol{k}$-space, namely $(L/2\pi)^3 \mathrm{d}\,\boldsymbol{k}_1$. With $\varepsilon_1 = \hbar^2 k_1^2/2m$ one gets

$$\mathrm{d}\boldsymbol{k}_1 = k_1^2\, \mathrm{d}\Omega\, \mathrm{d}k_1 = m k_1\, \mathrm{d}\varepsilon_1\, \mathrm{d}\Omega_1/\hbar^2. \tag{3.25}$$

This leads to the number of neutrons scattered per solid angle, per time unit, and per energy interval, if the nuclei are exposed to a neutron current of density j_0:

$$j_0\left(\frac{\mathrm{d}^2\sigma}{\mathrm{d}\Omega_1\, \mathrm{d}\varepsilon_1}\right) \mathrm{d}\Omega_1\, \mathrm{d}\varepsilon_1 = w(\boldsymbol{k}_0 \to \boldsymbol{k}_1)\left(\frac{L}{2\pi}\right)^3 \mathrm{d}\boldsymbol{k}_1. \tag{3.26}$$

The incident current density j_0 is the velocity $\hbar\boldsymbol{k}_0/m$ multiplied by the neutron density $1/L^3$, such that one finally obtains

$$\frac{\mathrm{d}^2\sigma}{\mathrm{d}\Omega_1\, \mathrm{d}\varepsilon_1} = \left(\frac{m}{2\pi\hbar^2}\right)^2 \frac{k_1}{k_0} \sum_{n_1} \sum_{n_0} p_{n_0} |\langle \boldsymbol{k}_1 n_1 | V | \boldsymbol{k}_0 n_0 \rangle|^2\, \delta(E_{n_0} - E_{n_1} + \hbar\omega). \tag{3.27}$$

The scattering process in real and in momentum space is shown in Fig. 3.3.

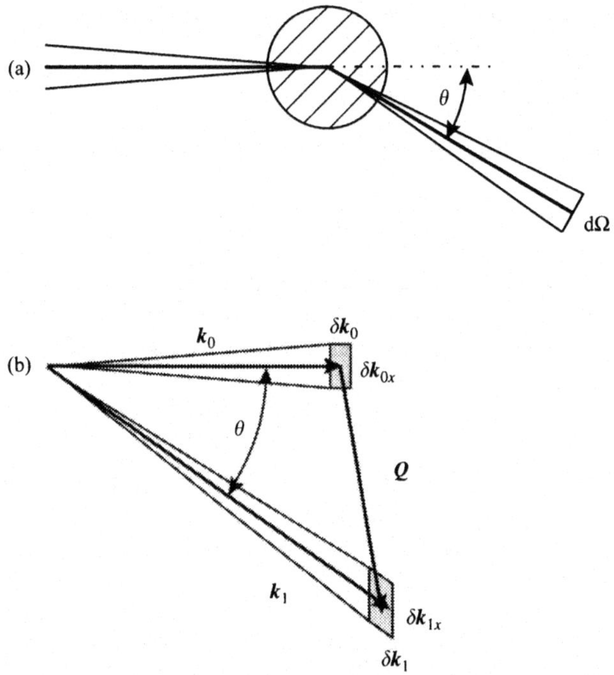

Fig. 3.3 The neutron scattering process in real space (a), and in reciprocal space (b). θ is the scattering angle, $\mathrm{d}\Omega$ is a solid angle element after scattering. In the theoretical treatment the incident beam has an exactly defined direction $\boldsymbol{k}_0$, whereas $\boldsymbol{k}_1$ is distributed over the resolution elements before and after scattering (shades). The longitudinal spread of the wave number $\Delta \boldsymbol{k}_{0x}$ and $\Delta \boldsymbol{k}_{1x}$ is determined by the angular divergency of the incident beam and of the analyser after scattering, respectively. The mean or 'nominal' directions are indicated by solid lines. The width of the $\boldsymbol{Q}$ vector results from a convolution of the contributions of the volume element, before and after scattering.

Now we introduce the interaction potential between a neutron and nuclei, $V(\boldsymbol{r}-\boldsymbol{r}_j)$. Fortunately, the range of this potential is small compared to the neutron wavelength λ in the experiments. This allows us to replace the interaction potential of all nuclei j by a sum of δ-functions of *zero range*, namely

$$V(\boldsymbol{r}) = \sum_j V(\boldsymbol{r}-\boldsymbol{R}_j) = \left(\frac{2\pi\hbar^2}{m}\right)\sum_j b_j\delta(\boldsymbol{r}-\boldsymbol{R}_j). \tag{3.28}$$

Obviously, the properties of the potential enter into one parameter, the scattering length b_j, already discussed in the previous section. It is an empirical quantity and normally of the order of the nuclear radius (10^{-14} m). With Eq. (3.28) the integration of the scattering cross-section over all energies and over the full solid angle just leads to the correct result value σ_c as given in Eq. (3.5). This formulation of the interaction is called the *Fermi pseudopotential*.

The introduction of the sum in the potential leads to

$$\frac{d^2\sigma}{d\Omega_1\, d\varepsilon_1} = \frac{k_1}{k_0} \sum_{n_1} \sum_{n_0} p_{n_0} \sum_{j=1}^{N} b_j |\langle n_1 | e^{i\boldsymbol{Q}\boldsymbol{R}_j} | n_0 \rangle|^2 \, \delta(E_{n_0} - E_{n_1} + \hbar\omega). \tag{3.29}$$

This is the cross-section for an ensemble of N nuclei. In many papers one uses cross-section *per atom* which introduces a factor $1/N$. Obviously, the short range of the interaction (δ-function in Eq. (3.28)) together with the plane waves $\exp(i\boldsymbol{k}_0\boldsymbol{r})$ and $\exp(i\boldsymbol{k}_1\boldsymbol{r})$ has led to the replacement of the potential sum by the phase factors $\exp(i\boldsymbol{Q}\boldsymbol{R}_j)$, where $\boldsymbol{Q} = \boldsymbol{k}_0 - \boldsymbol{k}_1$, which greatly simplifies the following formalism.

The squared sum can be rewritten as a double sum over all atoms:

$$\frac{d^2\sigma}{d\Omega_1\, d\varepsilon_1} = \frac{k_1}{k_0} \sum_{n_1} \sum_{n_0} p_{n_0} \sum_{i=1}^{N} \sum_{j=1}^{N} b_i b_j \times \langle n_0 | e^{i\boldsymbol{Q}\boldsymbol{R}_i} | n_1 \rangle \langle n_1 | e^{-i\boldsymbol{Q}\boldsymbol{R}_j} | n_0 \rangle \delta(E_{n_0} - E_{n_1} + \hbar\omega). \tag{3.30}$$

Now we introduce time-dependent space coordinates. First the δ-function with energy argument (required by energy conservation) is Fourier transformed:

$$\delta(E_{n_o} - E_{n_1} + \hbar\omega) = (2\pi\hbar)^{-1} \int_{-\infty}^{+\infty} e^{it(E_{n_0} - E_{n_1})/\hbar} e^{-i\omega t}\, dt. \tag{3.31}$$

Then we use the closure relation $\sum_{n_1} \langle n_0 | \boldsymbol{A} | n_1 \rangle \langle n_1 | \boldsymbol{B} | n_0 \rangle = \langle n_0 | \boldsymbol{A}\boldsymbol{B} | n_0 \rangle$ where $\boldsymbol{A}$ and $\boldsymbol{B}$ are operators, and introduce H instead of E_n by $\exp[-iHt/\hbar]|n_0\rangle = \exp[-iE_0 t/\hbar]|n_0\rangle$:

$$\frac{d^2\sigma}{d\Omega_1\, d\varepsilon_1} = \frac{k_1}{k_0} \frac{1}{2\pi\hbar} \sum_{n_0} p_{n_0} \sum_i \sum_j b_i b_j \times \int_{-\infty}^{\infty} dt\, e^{-i\omega t} \langle n_0 | e^{-i\boldsymbol{Q}\boldsymbol{R}_j} e^{iHt/\hbar} e^{i\boldsymbol{Q}\boldsymbol{R}_i} e^{-iHt/\hbar} | n_0 \rangle. \tag{3.32}$$

The next important reformulation concerns the time evolution of $\boldsymbol{R}$, as introduced by the operation

$$\exp[i\boldsymbol{Q}\boldsymbol{R}(t)] = \exp[iHt/\hbar] \exp[i\boldsymbol{Q}\boldsymbol{R}] \exp[-iHt/\hbar]$$

(the Heisenberg representation). This leads to the expression (Squires 1978)

$$\frac{d^2\sigma}{d\Omega_1\, d\varepsilon_1} = \left(\frac{k_1}{k_0}\right) \frac{1}{2\pi\hbar} \sum_i \sum_j b_i b_j \int_{-\infty}^{\infty} e^{-i\omega t} \langle e^{-i\boldsymbol{Q}\boldsymbol{R}_j(0)} e^{i\boldsymbol{Q}\boldsymbol{R}_i(t)} \rangle \, dt \tag{3.33}$$

where all sums go from 1 to N. The sum $\sum p_{n_0}$ has been omitted from here on and $\langle \cdots \rangle$ means the thermal average over the expectation value of the operator in the brackets.

In order to deal with the double sum over all nuclei a discussion of the *isotope* and *nuclear spin disorder* is needed. In many cases, an atomic species has several isotopes. For not too low temperatures these are randomly distributed over the atomic sites. Since the scattering averages over all possible distributions of the isotopes, we must introduce the averages as defined already in the previous section, and we use the relations

$$\langle b_i b_j \rangle = \begin{cases} \langle b^2 \rangle & \text{for } i = j \\ \langle b \rangle^2 & \text{for } i \neq j. \end{cases} \tag{3.34}$$

The sums in Eq. (3.33) can then be separated into a double sum over *all* pairs (including $i = j$), and a term which separates the sum over terms $i = j$ only. This now reads

$$\frac{\mathrm{d}^2\sigma}{\mathrm{d}\Omega_1\,\mathrm{d}\varepsilon_1} = \left(\frac{k_1}{k_0}\right)\frac{1}{2\pi\hbar}\Bigg[\langle b^2\rangle \sum_i \sum_j \int_{-\infty}^{\infty} \mathrm{e}^{-\mathrm{i}\omega t}\,\mathrm{d}t \left\langle \mathrm{e}^{-\mathrm{i}\boldsymbol{Q}\boldsymbol{R}_j(0)}\mathrm{e}^{\mathrm{i}\boldsymbol{Q}\boldsymbol{R}_i(t)}\right\rangle$$

$$+ \left(\langle b^2\rangle - \langle b\rangle^2\right) \int_{-\infty}^{\infty} \mathrm{e}^{-\mathrm{i}\omega t}\,\mathrm{d}t \sum_j \left\langle \mathrm{e}^{-\mathrm{i}\boldsymbol{Q}\boldsymbol{R}_j(0)}\mathrm{e}^{\mathrm{i}\boldsymbol{Q}\boldsymbol{R}_j(t)}\right\rangle\Bigg]. \tag{3.35}$$

The first term describes coherent scattering. The second term behaves *as if scattering were incoherent*: no phase factors enter, and the scattering of the N nuclei is just N times the scattering of a single one and $\sum_j$ is to be replaced by N. We point out that the word 'incoherent' is misleading since there is always a coherent superposition of the outgoing neutron waves (or, in other words, a superposition of the amplitudes). Only the averaging process (over random spins or isotopes) leads to a cancellation of the phase factors. In order to illustrate this point let us assume that we had two *different atomic species* randomly distributed over the sites, for instance in a disordered alloy. Also here a scattering term could appear without phase factors, called 'Laue scattering' in X-ray diffraction; see Eq. (6.8). However, in the case of neutron scattering we use the word 'incoherent' by convention for disordered *isotopic* species or for *spin* disorder only, whereas *chemical disorder* is conventionally taken into account in the 'coherent' scattering part.

In this sense we separate the double differential cross-section into two parts, introducing the so-called (coherent) *intermediate scattering function*

$$I(\boldsymbol{Q}, t) = N^{-1} \sum_{i=1}^{N} \sum_{j=1}^{N} \left\langle e^{-\mathrm{i}\boldsymbol{Q}\boldsymbol{R}_j(0)}\mathrm{e}^{\mathrm{i}\boldsymbol{Q}\boldsymbol{R}_i(t)}\right\rangle \tag{3.36}$$

which again includes all terms ($i \neq j$ and $i = j$), and the contribution $I_s(\boldsymbol{Q}, t)$ from the N self-terms only

$$I_s(\boldsymbol{Q}, t) = N^{-1} \sum_{j=1}^{N} \left\langle e^{-i\boldsymbol{Q}\boldsymbol{R}_j(0)} e^{i\boldsymbol{Q}\boldsymbol{R}_j(t)} \right\rangle \equiv \left\langle e^{-i\boldsymbol{Q}\boldsymbol{R}(0)} e^{i\boldsymbol{Q}\boldsymbol{R}(t)} \right\rangle \tag{3.37}$$

$\left(\text{since } N^{-1} \sum = 1\right)$. This leads to the coherent double differential cross-section

$$\left(\frac{d^2\sigma}{d\Omega_1\, d\varepsilon_1}\right)_c = \left(\frac{\sigma_c}{4\pi}\right)\left(\frac{k_1}{k_0}\right) N\, S(\boldsymbol{Q}, \omega) \tag{3.38}$$

where

$$S(\boldsymbol{Q}, \omega) = \frac{1}{2\pi\hbar} \int I(\boldsymbol{Q}, t) e^{-i\omega t}\, dt. \tag{3.39}$$

We recognize that $S(\boldsymbol{Q}, \omega)$ is defined per nucleus, dimension eV^{-1}. The corresponding incoherent cross-section is then

$$\left(\frac{d^2\sigma}{d\Omega_1\, d\varepsilon_1}\right)_i = \left(\frac{\sigma_i}{4\pi}\right)\left(\frac{k_1}{k_0}\right) N\, S_i(\boldsymbol{Q}, \omega) \tag{3.40}$$

with

$$S_i(\boldsymbol{Q}, \omega) = \frac{1}{2\pi\hbar} \int I_s(\boldsymbol{Q}, t) e^{-i\omega t}\, dt. \tag{3.41}$$

σ_c and σ_i are the *bound* coherent and incoherent cross-sections, respectively (Eqs. (3.8) and (3.9)). Here we have introduced the usual *scattering functions* $S(\boldsymbol{Q}, \omega)$ and $S_i(\boldsymbol{Q}, \omega)$, which are also called *coherent* and *incoherent dynamical structure factors* or *scattering functions*, respectively.

The intermediate scattering function can be reformulated with density–density correlation functions. We replace exponentials by particle density operators, such that

$$\rho(\boldsymbol{Q}, t) = \int \rho(\boldsymbol{r}, t) e^{-i\boldsymbol{Q}\boldsymbol{r}}\, d^3r = \sum_j e^{-i\boldsymbol{Q}\boldsymbol{R}_j(t)} \tag{3.42}$$

with

$$\rho(\boldsymbol{r}, t) = \sum_j \delta\left[\boldsymbol{r} - \boldsymbol{R}_j(t)\right] \tag{3.43}$$

which describes the particle coordinates at points $\boldsymbol{r}$ and at times t.

This leads to

$$I(\boldsymbol{Q}, t) = N^{-1} \langle \rho(-\boldsymbol{Q}, 0) \rho(\boldsymbol{Q}, t) \rangle. \tag{3.44}$$

In the classical approximation the density operators can be understood as the particle density functions. Integrating the scattering function over all energy transfers (which means measuring the scattering intensity without energy analysis under the condition that $\boldsymbol{Q}$ is fixed) the exponential $\exp(-\mathrm{i}\omega t)$ in Eq. (3.35) transforms into a delta function such that only contributions with $t = 0$ remain, or: R_i and R_j are taken at the same time. This yields the instantaneous scattering function or the structure factor, namely $S(Q) = I(Q, 0)$, where

$$S(\boldsymbol{Q}) = N^{-1}\sum_{i=1}^{N}\sum_{j=1}^{N}\langle \mathrm{e}^{-\mathrm{i}\boldsymbol{Q}\boldsymbol{R}_j}\mathrm{e}^{\mathrm{i}\boldsymbol{Q}\boldsymbol{R}_i}\rangle = N^{-1}\sum_{i=1}^{N}\sum_{j=1}^{N}\big\langle \exp\big[\mathrm{i}\boldsymbol{Q}\cdot(\boldsymbol{R}_i - \boldsymbol{R}_j)\big]\big\rangle, \tag{3.45}$$

since operators commute at $t = 0$. In this formalism, the hydrogen nucleus plays an exceptional role since σ_i is much larger than σ_c (see the previous section) so that hydrogen scattering is dominated by the self-term in the above equations. On the other hand, many nuclei have both coherent and incoherent contributions which complicates the interpretation of experimental results: one measures the sum of coherent and incoherent double differential cross-sections, where the coherent part includes *both* distinct and self-terms, whereas the incoherent part contains the self-terms only.

In the cross-sections, Eqs. (3.38) and (3.40), we recognize features which are typical consequences of perturbation theory. First, the scattering functions depend on $\boldsymbol{Q}$ and ω only, and not individually on $\boldsymbol{k}_0$, $\boldsymbol{k}_1$, ε_0, and ε_1. Secondly, the cross-section includes a typical prefactor $k_1/k_0 = (\varepsilon_1/\varepsilon_0)^{1/2}$; the factor k_0 results from the relation between current and density (small speed leads to a high density and enhances scattering); the factor k_1 is related to the size of the volume element in $\boldsymbol{k}$-space, $\mathrm{d}\boldsymbol{k}_1$, which yields the number of accessible states. Finally, we notice that the differential neutron scattering cross-section is factorized into quantities $\langle \mathrm{b}\rangle^2$ or $\langle \mathrm{b}^2\rangle - \langle \mathrm{b}\rangle^2$ which depend only on the interaction potential of the nuclei, and into functions $S_\mathrm{i}(\boldsymbol{Q}, \omega)$ or $S(\boldsymbol{Q}, \omega)$ which are related to the condensed matter properties, i.e. the atomic structure and dynamics of the material system to be investigated.

So far, the scattering function has been calculated for one atomic species. For a mixture of several atomic species the cross-section is written as a superposition of products with partial scattering functions. For two nuclear species A, B with numbers N_A, N_B the cross-section can be written as

$$\frac{\mathrm{d}^2\sigma}{\mathrm{d}\Omega_1\,\mathrm{d}\varepsilon_1} = \frac{k_1}{k_0}\left[\frac{\sigma_\mathrm{A}}{4\pi}N_\mathrm{A}S^\mathrm{A} + \frac{\sigma_\mathrm{B}}{4\pi}N_\mathrm{B}S^\mathrm{B} + \frac{(\sigma_\mathrm{A}\sigma_\mathrm{B})^{1/2}}{4\pi}(N_\mathrm{A}N_\mathrm{B})^{1/2}S^\mathrm{AB}\right]. \tag{3.46}$$

For incoherent scattering there is no cross-term. $\mathrm{S^A}$, $\mathrm{S^B}$, and $\mathrm{S^{AB}}$ have the form as in Eq. (3.45) counting atomic pairs AA, BB, and AB, respectively (Squires 1978).

There are several general properties of the scattering functions which are frequently used and will be briefly summarized here (Squires 1978). The *sum rules* concern the moments of the scattering functions. The moments of $S(\boldsymbol{Q}, \omega)$ are defined as

$$\langle \omega^n \rangle = \int_{-\infty}^{\infty} S(\boldsymbol{Q}, \omega)\omega^n \, \mathrm{d}\omega. \tag{3.47}$$

We easily get the zeroth moments which are

$$\int_{-\infty}^{\infty} S(\boldsymbol{Q}, \omega) \, \mathrm{d}\omega = S(\boldsymbol{Q}) \quad \text{and} \quad \int_{-\infty}^{\infty} S_i(\boldsymbol{Q}, \omega) \, \mathrm{d}\omega = 1. \tag{3.48}$$

Also the higher moments can be calculated. One obtains the first moments for velocity-independent forces

$$\hbar \langle \omega \rangle = \frac{\hbar^2 Q^2}{2M} \tag{3.49}$$

both for coherent and incoherent scattering. This means that the *average* energy transfer $\hbar\langle\omega\rangle$ for a neutron scattered on an atom of mass M is that of a free particle, i.e. as for an ideal gas. For the classical approximation to be discussed in the subsequent section, $S_\mathrm{i}^{(\mathrm{cl})}(\boldsymbol{Q}, \omega)$ is symmetric in ω such that $\langle\omega\rangle^{\mathrm{cl}} = 0$ which obviously contradicts the sum rule. For $n = 2$ one obtains

$$\hbar^2 \langle \omega^2 \rangle_\mathrm{i} = \frac{\hbar^2 Q^2 k_\mathrm{B} T}{M} + \left(\frac{\hbar^2 Q^2}{2M}\right)^2 \tag{3.50}$$

which is the square width of the scattering spectrum for an ideal gas of temperature T. The fourth moment also depends on the atomic interaction potential, as displayed in Fig. 3.4, namely (de Gennes 1959)

$$\frac{\langle \omega^4 \rangle_\mathrm{i}}{3\langle \omega^2 \rangle_\mathrm{i}^2} = 1 + \frac{\Omega_0^2}{\langle \omega^2 \rangle_\mathrm{i}}. \tag{3.51}$$

For an isotropic system Ω_0 can be related to the pair potential $V(r)$, namely

$$\Omega_0^2 = \frac{4\pi}{2M} \int g(r) \frac{\partial^2 V(r)}{\partial x^2} \, \mathrm{d}\boldsymbol{r}. \tag{3.52}$$

Ω_0 has the meaning of an average vibrational frequency; $g(r)$ is the pair correlation function. For an ideal gas Eq. (3.52) corresponds to a Gausssian distribution. For interacting systems of particles, the wings of the spectrum are stronger than for a Gaussian such that the spectrum is closer to a Lorentzian function.

Finally, the scattering functions $S(\boldsymbol{Q}, \omega)$ and $S_\mathrm{i}(\boldsymbol{Q}, \omega)$ fulfil the condition of detailed balance

$$S(\boldsymbol{Q}, \omega) = \mathrm{e}^{\hbar\omega/k_\mathrm{B}T} S(-\boldsymbol{Q}, -\omega). \tag{3.53}$$

If there is no magnetic field, the invariance of the system with respect to the sign of the time means that $S(\boldsymbol{Q}, \omega)$ is independent of the sign of $\boldsymbol{Q}$. The incident neutrons should reach thermal equilibrium by many collisions in a very large and non-absorbing sample, namely a Maxwellian with temperature T. This implies that the number of neutrons gaining a certain energy has to be equal to the number losing this energy. Eq. (3.52) takes care that this general condition applies.

As has been shown in this section, the double differential cross-section is a function of $\boldsymbol{Q}$ and ω, apart from the phase-space factor (k_1/k_0) (Eqs. (3.38) and (3.40)). Experimentally, however, the cross-section is measured for a fixed scattering angle θ, and $\boldsymbol{Q}$ is connected with the energy after scattering ε_1 or with $\hbar\omega$ as shown in Fig. 3.2. As can be seen, only for a spectral distribution which is narrow compared to the incident energy ε_0 will the experiment at fixed θ values approximately yield $S(\boldsymbol{Q}, \omega)$ for fixed $\boldsymbol{Q}$. For the rather narrow quasielastic spectra due to atomic diffusion this criterion is practically fulfilled in most cases. However, care must be taken in measuring $S(\boldsymbol{Q})$ where an integration over ω for fixed $\boldsymbol{Q}$ (and not for fixed θ) is required. This is only possible if ε_0 is high, compared to the quasielastic spectra but also with respect to phonon scattering processes.

3.3 THE VAN HOVE CORRELATION FUNCTIONS

In the previous section we have treated neutron scattering as a quantum-mechanical process expressed in terms of wavefunctions, transition probabilities, time-dependent perturbation theory, etc. and, eventually, ensemble averages. In Chapter 4 the statistical aspects of (classical) particle diffusion in condensed matter will be outlined. The 'interface' between these quite different types of theories is given by thermal averages which can be expressed in terms of so-called correlation functions, $G(\boldsymbol{r}, t)$ and $G_s(\boldsymbol{r}, t)$. The correlation function $G(\boldsymbol{r}, t)\,\mathrm{d}\boldsymbol{r}$ is the conditional probability that, given a particle was at time $t = 0$ at the origin $\boldsymbol{r} = 0$, *any* particle is found at time t at the position $\boldsymbol{r}$ in a volume element $\mathrm{d}\boldsymbol{r}$. 'Any particle' means the same or a different particle, i.e. the correlation function contains a self and a distinct part. The self-part is separately registered as the self-correlation function $G_s(\boldsymbol{r}, t)\,\mathrm{d}\boldsymbol{r}$ which is the conditional probability that, given a particle was at time $t = 0$ at the origin $\boldsymbol{r} = 0$, the *same* particle is found at time t at the position $\boldsymbol{r}$ in a volume element $\mathrm{d}\boldsymbol{r}$.

The formulation of the correlation functions in connection to neutron scattering theory is due primarily to van Hove (1954; 1958); the van Hove correlation function—as it is therefore frequently called—is defined as the Fourier transforms in space and time of the intermediate scattering function Eq. (3.36) (van Hove 1958; Squires 1978; Lovesey 1987):

$$G(\boldsymbol{r}, t) = (2\pi)^{-3} \int I(\boldsymbol{Q}, t) \mathrm{e}^{-\mathrm{i}\boldsymbol{Q}\boldsymbol{r}}\, \mathrm{d}\boldsymbol{Q}$$
$$= (2\pi)^{-3} N^{-1} \sum_i \sum_j \int \left\langle \mathrm{e}^{-\mathrm{i}\boldsymbol{Q}\boldsymbol{R}_j(0)} \mathrm{e}^{\mathrm{i}\boldsymbol{Q}\boldsymbol{R}_i(t)} \right\rangle \mathrm{e}^{-\mathrm{i}\boldsymbol{Q}\boldsymbol{r}}\, \mathrm{d}\boldsymbol{Q} \qquad (3.54)$$

Since $\boldsymbol{R}_i(0)$ and $\boldsymbol{R}_j(t)$ do not commute except when $t = 0$, the factor with $\boldsymbol{R}_i(t)$ must be kept to the right of the factor with $\boldsymbol{R}_j(0)$. By algebraic manipulation (Squires 1978) one can write

$$G(\boldsymbol{r}, t) = N^{-1} \sum_i \sum_j \int \langle \delta[\boldsymbol{r}' - \boldsymbol{R}_j(0)]\, \delta[\boldsymbol{r}' + \boldsymbol{r} - \boldsymbol{R}_i(t)] \rangle \, \mathrm{d}\boldsymbol{r}'. \tag{3.55}$$

Obviously, contributions from the double sum over the product of δ-functions appear only if the conditions $\boldsymbol{r}' - \boldsymbol{R}_j(t) = 0$ and $\boldsymbol{r}' + \boldsymbol{r} - \boldsymbol{R}_j = 0$ coincide. With Eq. (3.43),

$$\rho(\boldsymbol{r}, t) = \sum_j \delta[\boldsymbol{r} - \boldsymbol{R}_j(t)], \tag{3.56}$$

one can express $G(\boldsymbol{r}, t)$ in terms of a density–density correlation function:

$$G(\boldsymbol{r}, t) = N^{-1} \int \langle \rho(\boldsymbol{r}', 0)\, \rho(\boldsymbol{r}' + \boldsymbol{r}, t) \rangle \, \mathrm{d}\boldsymbol{r}' \tag{3.57}$$

where the integral over the whole sample yields

$$\int G(\boldsymbol{r}, t)\, \mathrm{d}\boldsymbol{r} = N. \tag{3.58}$$

Before discussing the meaning of $G(\boldsymbol{r}, t)$ we relate this function to the scattering function introduced earlier, namely, with Eq. (3.39):

$$S(\boldsymbol{Q}, \omega) = (2\pi\hbar)^{-1} \int_{-\infty}^{\infty} \int_{-\infty}^{\infty} \mathrm{e}^{\mathrm{i}(\boldsymbol{Q}\boldsymbol{r} - \omega t)} G(\boldsymbol{r}, t)\, \mathrm{d}\boldsymbol{r}\, \mathrm{d}t. \tag{3.59}$$

Inversely one obtains

$$G(\boldsymbol{r}, t) = \frac{\hbar}{(2\pi)^3} \iint \mathrm{e}^{-\mathrm{i}(\boldsymbol{Q}\boldsymbol{r} - \omega t)} S(\boldsymbol{Q}, \omega)\, \mathrm{d}\boldsymbol{Q}\, \mathrm{d}\omega. \tag{3.60}$$

If one considers only the self-terms $i = j$ one yields the incoherent scattering function, namely

$$S_{\mathrm{i}}(\boldsymbol{Q}, \omega) = (2\pi\hbar)^{-1} \iint \mathrm{e}^{\mathrm{i}(\boldsymbol{Q}\boldsymbol{r} - \omega t)} G_{\mathrm{s}}(\boldsymbol{r}, t)\, \mathrm{d}\boldsymbol{r}\, \mathrm{d}t. \tag{3.61}$$

and, inversely,

$$G_{\mathrm{s}}(\boldsymbol{r}, t) = \frac{\hbar}{(2\pi)^3} \iint \mathrm{e}^{-\mathrm{i}(\boldsymbol{Q}\boldsymbol{r} - \omega t)} S_{\mathrm{i}}(\boldsymbol{Q}, \omega)\, \mathrm{d}\boldsymbol{Q}\, \mathrm{d}\omega. \tag{3.62}$$

We note here that S and S_{i} have the dimension of an inverse energy, whereas G and G_{s} have the dimension of an inverse volume.

We first consider the case $t = 0$, where the operators commute (in Eq. (3.32), $\exp[\mathrm{i}Ht/\hbar] = 1$); the integral over the product of the two delta functions in Eq. (3.55) can be carried out, and we get

$$G(\boldsymbol{r}, 0) = N^{-1} \sum_i \sum_j \langle \delta[\boldsymbol{r} - \boldsymbol{R}_i(0) + \boldsymbol{R}_j(0)] \rangle. \tag{3.63}$$

Then we take the sum over the coordinates $\boldsymbol{R}_i$ assuming translational invariance, i.e. that all nuclei are equivalent ($N^{-1} \sum_i = 1$). Furthermore, we separate the self-terms $i = j$, thus

$$G(\boldsymbol{r}, 0) = \delta(\boldsymbol{r}) + \sum_{i \neq 0} \langle \delta[\boldsymbol{r} - \boldsymbol{R}_0(0) + \boldsymbol{R}_i(0)] \rangle. \tag{3.64}$$

We introduce the static pair correlation function $g(\boldsymbol{r})$ and the average particle number density ρ, and write Eq. (3.64) as

$$G(\boldsymbol{r}, 0) = \delta(\boldsymbol{r}) + \rho g(\boldsymbol{r}). \tag{3.65}$$

For $t \to \infty$, the particle number densities at different space coordinates are uncorrelated such that in Eq. (3.57) the density–density correlation function transforms into

$$G(\boldsymbol{r}, \infty) = N^{-1} \int \langle \rho(\boldsymbol{r}') \rangle \langle \rho(\boldsymbol{r}' + \boldsymbol{r}) \rangle \, \mathrm{d}\boldsymbol{r}'. \tag{3.66}$$

For a uniform system, i.e. a monoatomic gas or a liquid, this yields $\boldsymbol{G}(\boldsymbol{r}, \infty) = N^{-1} \int \rho^2 \, \mathrm{d}r' = \rho$.

The meaning of the pair correlation function can be explained as follows: We consider a vector $\boldsymbol{r}$ which originates at a particle coordinate $\boldsymbol{r}_i$ and which ends in a small volume element $\mathrm{d}\boldsymbol{r}$ around $\boldsymbol{r}_j$. Obviously, the double sum in Eq. (3.55) integrated over this volume element counts the number of such particle pairs at $\boldsymbol{R}_i$ and $\boldsymbol{R}_j$ which are simultaneously separated by $\boldsymbol{r} = \boldsymbol{R}_i - \boldsymbol{R}_j$. Consequently, $g(\boldsymbol{r})\,\mathrm{d}r$ is the ensemble averaged probability of finding a particle in a volume element at $\boldsymbol{R}_i$ if another one is, simultaneously, at $\boldsymbol{R}_j$. We arbitrarily place particle $\boldsymbol{R}_j = 0$ at the origin.

Integrating the coherent scattering function $S(\boldsymbol{Q}, \omega)$ over all energies we obtain again the structure factor already introduced in Eq. (3.45), namely

$$S(\boldsymbol{Q}) = \int_{-\infty}^{\infty} S(\boldsymbol{Q}, \omega) \, \mathrm{d}\omega = \int G(\boldsymbol{r}, 0) \mathrm{e}^{\mathrm{i}\boldsymbol{Q}\boldsymbol{r}} \, \mathrm{d}\boldsymbol{r}$$

$$= 1 + \rho \int [g(\boldsymbol{r}) - 1] \mathrm{e}^{\mathrm{i}\boldsymbol{Q}\boldsymbol{r}} \, \mathrm{d}\boldsymbol{r}. \tag{3.67}$$

The -1 in the square brackets removes the delta function at $Q = 0$ which originates from diffraction on the sample as a whole. For $t > 0$, $G(\boldsymbol{r}, t)$ loses its classical meaning; then it is a complex function of its variables. If we assume that the system

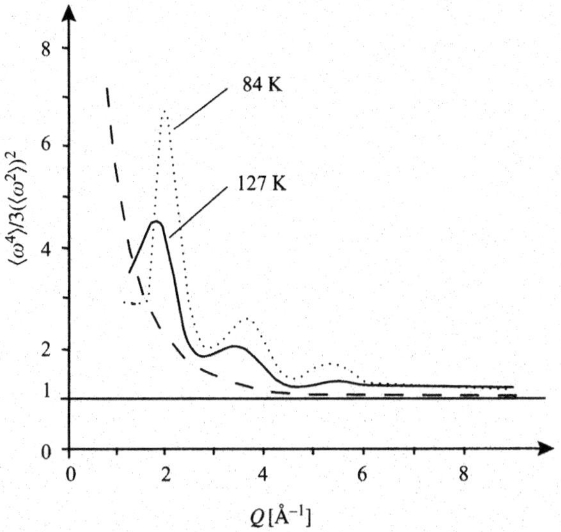

Fig. 3.4 The ratio of the 4th and the square of the 2nd moment of the structure factor for coherent scattering, calculated for a typical liquid noble gas, as a function of momentum transfer (from de Gennes 1959, with permission from Elsevier Science). The maxima indicate the *line narrowing effect*. The dashed line shows the behaviour of incoherent scattering.

behaves *as if* it were classical, the operators commute in Eq. (3.55) and in analogy to Eq. (3.63) we obtain the time-dependent and *classical correlation function*

$$G^{(\mathrm{cl})}(\boldsymbol{r}, t) = N^{-1} \sum_i \sum_j \langle \delta[\boldsymbol{r} - \boldsymbol{R}_i(t) + \boldsymbol{R}_j(0)] \rangle. \tag{3.68}$$

For simplicity we assume that all nuclei are equivalent. Then for fixed j, the sum over i in Eq. (3.68) gives the same value whatever the value of j. So the sum over j is N times the term with $j = 0$. Thus

$$G^{\mathrm{cl}}(\boldsymbol{r}, t) = \sum_i \langle \delta[\boldsymbol{r} - \boldsymbol{R}_i(t) + \boldsymbol{R}_0(0)] \rangle. \tag{3.69}$$

We conclude from this equation that $G^{\mathrm{cl}}(\boldsymbol{r}, t)\,\mathrm{d}\boldsymbol{r}$ is the probability that, given a particle at the origin at time $t = 0$, any particle (including the original particle) is in the voulume $\mathrm{d}\boldsymbol{r}$ at position $\boldsymbol{r}$ at time t.

Again separating the terms $i = j$ we get the *classical self-correlation function*

$$G_{\mathrm{s}}^{(\mathrm{cl})}(\boldsymbol{r}, t) = \langle \delta[\boldsymbol{r} - \boldsymbol{R}_0(t) + \boldsymbol{R}_0(0)] \rangle \tag{3.70}$$

with $G(\boldsymbol{r}, 0) = \delta(\boldsymbol{r})$, which means for $t = 0$ the scattering particle is certainly at the origin. Thus $G_{\mathrm{s}}^{\mathrm{cl}}(\boldsymbol{r}, t)\,\mathrm{d}\boldsymbol{r}$ is the probability that, given a particle at the origin at time $t = 0$, the same particle is in the volume $\mathrm{d}\boldsymbol{r}$ at position $\boldsymbol{r}$ at time t.

The introduction of the classical correlation functions, $G_{\mathrm{s}}^{\mathrm{cl}}(\boldsymbol{r}, t)$ or $\mathrm{G}^{\mathrm{cl}}(\boldsymbol{r}, t)$, for $t > 0$ is an approximation. In order to get a quantitative argument for the conditions under which this holds we calculate the de Broglie wavelength λ_M of a scattering nucleus with mass M, moving with thermal velocity. This is $\lambda_M \approx \hbar(2Mk_BT)^{-1/2} \simeq 10^{-11}$–$10^{-10}$ m at room temperature. Obviously, a classical description of the motion makes no sense as long as the distance travelled by the particle during time t, namely $t(k_BT/2M)^{1/2}$, is smaller than λ_M. This means that the classical interpretation of the self-correlation functions fails if

$$t < t_0 = \frac{\hbar}{2k_BT} \simeq 10^{-13}\,\text{s}. \tag{3.71}$$

The duration of the diffusive step, namely the residence time of a particle on a lattice site, is longer than 10^{-12} s in most cases to be considered, and this allows us to use the classical form. We mention, however, that the time which a particle needs to *move* from one lattice site to another is much shorter. Assume that the corresponding speed is the thermal velocity. Then the transition time is of the order of $10^{-10}\,\text{m}/10^3\text{m}\,\text{s}^{-1} \cong 10^{-13}$ s for a hydrogen atom. This causes scattering processes which cannot be described by the classical formalism. The corresponding contributions of the spectrum are in an energy range beyond the quasielastic spectrum, say for $\hbar\omega >$ several 10^{-3} eV (cf. the case of vanadium hydride in Section 8.2.1). Also lattice vibrations at higher frequencies cannot be described by the classical correlation function.

The real-valued classical correlation function, $G_s^{cl}(\boldsymbol{r}, t)$, leads to an incoherent scattering function, $S_i^{cl}(\boldsymbol{Q}, \omega)$, which is symmetric in $\hbar\omega$. This violates the principle of detailed balance in Eq. (3.53). Comparison with experimental data, therefore, may necessitate a correction of the scattering function, calculated by classical models, with the detailed balance factor such that

$$S(\boldsymbol{Q}, \omega) = e^{\hbar\omega/2k_BT}\, S^{cl}(\boldsymbol{Q}, \omega). \tag{3.72}$$

This 'improves' the scattering function, without introducing new physics as concerns the particle dynamics.

The classical correlation function can be approximated in the following way. We assume an atom A at time $t = 0$ at the origin. $g(\boldsymbol{r}')\,d\boldsymbol{r}'$ is the probability of finding simultaneously another atom B in $d\boldsymbol{r}'$ at $\boldsymbol{r}'$. We introduce a *conditioned probability* $H(\boldsymbol{r}, \boldsymbol{r}', t)$ which describes the probability that atom B moves from $\boldsymbol{r}'$ to $\boldsymbol{r}$ during a time t *if* an atom A was at $\boldsymbol{r} = 0$ for $t = 0$. Under these circumstances, $g(\boldsymbol{r}')\,d\boldsymbol{r} H(\boldsymbol{r}, \boldsymbol{r}', t)$ is the probability of finding atom B in $d\boldsymbol{r}'$ at $\boldsymbol{r}'$ for $t = 0$, and migrating to $\boldsymbol{r}$ during the time t. The total probability of finding atom B at $\boldsymbol{r}$ for a time t, wherever the atom was for $t = 0$, is then

$$G_d^{(cl)}(\boldsymbol{r}, t) = \int H(\boldsymbol{r}, \boldsymbol{r}', t)\, g(\boldsymbol{r}')\, d\boldsymbol{r}'. \tag{3.73}$$

This equation is a starting point for approximations. The *convolution approximation* by Vineyard (1958) replaces the conditioned probability by the self-correlation function, namely

$$H(\boldsymbol{r}, \boldsymbol{r}', t) \simeq G_s(\boldsymbol{r} - \boldsymbol{r}', t) \tag{3.74}$$

where G_s is the probability that an atom moves from $\boldsymbol{r}'$ to $\boldsymbol{r}$ in time t, without taking into account the position of other atoms. Consequently, this approximation will certainly fail for small mutual distances. However, it also fails for large distances, i.e. in the hydrodynamic limit where $1/Q$ is small compared to the interatomic distance. After Fourier transformation the convolution approximation leads to a factorization of the intermediate scattering function

$$I^{(\text{conv})}(Q, t) \approx I_s(Q, t)\, I(Q, 0)$$

or

$$S^{(\text{conv})}(Q, \omega) \approx S(Q)\, S_i(Q, \omega). \tag{3.75}$$

Obviously, the convolution approximation contradicts the sum rule (Eq. 3.50). Several attempts have been made to find a relation between the self-correlation function and the pair correlation function which, however, are of limited use (Singwi 1965; Glass and Rice 1968; Singwi *et al.* 1970).

3.4 THE ASYMPTOTIC BEHAVIOUR OF $G_s(\boldsymbol{r}, t)$ AND OF $S(Q)$

In the classical limit the van Hove theory analyses the motion of particles by a Fourier transformation of correlation functions in space and in time. Experimentally, the obtainable information is limited by the energy and momentum resolution of the spectrometer. Let us assume that the spectrometer has a rectangular resolution function of width $\Delta\hbar\omega$ such that the measured intensity is an integral of Eq. (3.61) over energy transfers between $-\Delta\hbar\omega/2$ and $+\Delta\hbar\omega/2$, where we set the spectrometer at a nominal value $\hbar\omega = \varepsilon_0 - \varepsilon_1 \simeq 0$. Integration of Eq. (3.61) over ω yields the intensity in this interval, namely

$$\begin{aligned} I_s(Q, \Delta\omega) &= \int_{-\Delta\omega/2}^{\Delta\omega/2} S_i(Q, \omega)\, d\omega \\ &= \frac{1}{2\pi\hbar}\Delta\omega \iint e^{iQr} G_s(\boldsymbol{r}, t) \frac{\sin(t\Delta\omega/2)}{t\Delta\omega/2}\, d\boldsymbol{r}\, dt. \end{aligned} \tag{3.76}$$

The window function $(\sin x)/x$ introduces a time interval of width $1/\Delta\omega$, and we obtain a time average $G_s(\boldsymbol{r}, t)$ over $1/\Delta\omega$. Details of $G_s(\boldsymbol{r}, t)$ *finer* than this interval remain unresolved.

Assume, for instance, that $G_s(\boldsymbol{r}, t)$ is a function decaying exponentially to zero as time increases, as for a diffusing particle in an infinite medium (see Fig. 3.5). Then the resulting spectrum is a Lorentzian curve, centred at $\hbar\omega = 0$. Reducing the resolution width $\Delta\omega$, the time window in Eq. (3.76) increases such that the average of $G_s(\boldsymbol{r}, t)$ goes to zero if t goes to infinity.

Assume, on the other hand, the particle carries out confined motion, for instance in a restricted volume, such as a moving particle bound to other atoms which are

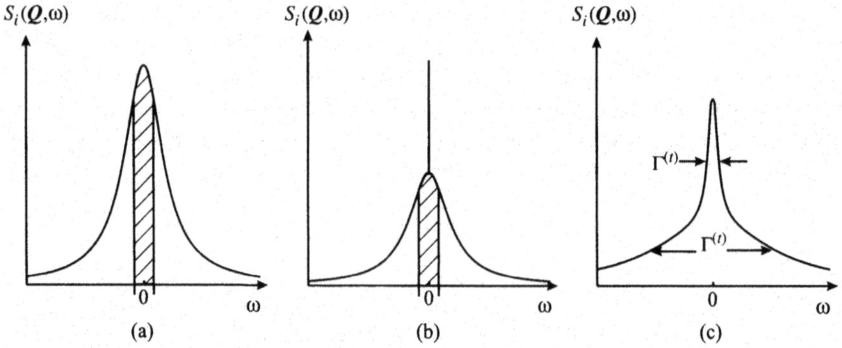

Fig. 3.5 (a) Qualitative shape of the incoherent scattering function for neutron scattered on a particle diffusing freely in a large volume. The intensity in a resolution interval $\hbar\Delta\omega$ goes to zero for $\Delta\omega \rightarrow 0$. (b) For a particle which carries out a random motion in a finite or restricted region of space, the quasielastic line is superimposed by a delta function whose intensity remains finite for $\Delta\omega \rightarrow 0$. (c) The particle carries out a combined motion, namely a diffusion, and during one diffusive step it stays a finite time interval in a restricted volume. This leads to the superposition of two quasielastic lines; the width of the broad one is given by the confined motion, and the width of the narrow component is determined by the diffusion process.

fixed. Under these circumstances, the intensity $I(Q, \Delta\omega)$ first decreases as $\Delta\omega$ is reduced, and then approaches a finite value. $G_s(\boldsymbol{r}, t \rightarrow \infty) = G^{\infty}(\boldsymbol{r})$ remains finite, even if $\Delta\omega$ approaches zero. This means that the spectrum consists of a quasielastic part superimposed on infinitively sharp line or δ-function (Fig. 3.5).

We see that the scattering on a bound or locally 'confined' particle leads to a δ-function in the energy distribution of scattered neutrons, i.e. to a finite intensity of an exactly elastic process for a finite momentum transfer $\hbar\boldsymbol{Q}$ which goes to the macroscopic sample *as a whole*. This corresponds to the elastic γ line observed in Mössbauer spectroscopy; see Section 7.8.

As has been shown the frequency width $\Delta\omega$ of the scattered neutron leads to an average over a time interval:

$$\tau_{\text{coh}} = 1/\Delta\omega \qquad (3.77)$$

which has the meaning of the *coherence time* of the neutron. It is determined by the resolution of the spectrometer and it is not an 'intrinsic' property of the neutron wave (except if $1/\Delta\omega$ was comparable with the neutronic life time). For presently existing spectrometers this coherence time is between 10^{-12} and 10^{-8} s. If the energy window covers *all* energies from $-\infty$ to $+\infty$ (i.e. if there is no energy analysis), the resulting window shrinks to zero and one obtains $S(\boldsymbol{Q})$ and the *instantaneous* correlation function $g(\boldsymbol{r})$ as usual in X-ray diffraction.

Analogous arguments hold for the scattering vector $\boldsymbol{Q}$. Here we consider the value of $\boldsymbol{Q}$, and not its corresponding width. For small $\boldsymbol{Q}$, the factor in the Fourier transform $\exp(i\boldsymbol{Q}\boldsymbol{r})$ approaches unity as long as $\boldsymbol{Q}\boldsymbol{r}$ is sufficiently small. Under this condition, the contributions to the integral results from distances $r < 2\pi/Q$. For larger $\boldsymbol{r}$, the exponential oscillates rapidly such that contributions of the correlation function cancel out. This is true with the exception that $\boldsymbol{Q}$ coincides with the periodicity of the correlation function; i.e. with the reciprocal lattice vector $\boldsymbol{G}$. This leads to the Bragg reflections of a lattice for the coherent scattering. Also for *incoherent* scattering, an anomaly appears in $S_i(\boldsymbol{Q}, \omega)$ for $\boldsymbol{Q} = \boldsymbol{G}$ if the *self-correlation* function has a periodicity which is caused by the diffusion of a single scattering particle over a periodic lattice. It leads to *incoherent line narrowing* (see Section 5.3.2).

For instance, let us consider a hydrogen atom, i.e. a proton, which is diffusing over an interstitial lattice. The incoherent narrowing is then understood by the diffraction of the neutron on *a single proton spread over a finite range of the lattice* by diffusion during the neutronic coherence time defined above. For, say, $\tau_{coh} = 10^{-9}$ s and for a diffusion constant of $D = 10^{-5}\,\mathrm{cm^2\,s^{-1}}$, the size of the diffracting region is $(D\tau_{coh})^{1/2} \simeq 10\,\text{Å}$ and contains up to 1000 interstitial sites! Our arguments imply that for the Q-range covered by existing spectrometers (between 10^{-2} and $10\,\text{Å}^{-1}$), the accessible 'observation range' is between 0.1 and several 100 Å, whereas the observations times go from 10^{-8} to 10^{-13} s.

Finally we calculate the structure factor $S(\boldsymbol{Q})$ in the limiting case $Q \to 0$. In a large sample, with an average particle number density ρ, we consider a small volume V surrounded by transparent walls. Because of the thermal motion of the particles, the number N inside this volume fluctuates around the average value $\langle N \rangle = \rho\,V$. One understands that the mean square fluctuation of particles in this volume is obtained by the relation

$$\langle N^2 \rangle = \int_V \int_V G(\boldsymbol{r}'' - \boldsymbol{r}', 0)\,\mathrm{d}\boldsymbol{r}''\,\mathrm{d}\boldsymbol{r}'. \tag{3.78}$$

Translational invariance allows us to replace one integral by the average particle number:

$$\langle N \rangle^2 = \langle N \rangle \int_V G(\boldsymbol{r}, 0)\,\mathrm{d}\boldsymbol{r}. \tag{3.79}$$

With $G(\boldsymbol{r}, 0) = \delta(\boldsymbol{r}) + \rho g(\boldsymbol{r})$ (Eq. (3.65)) and with $Q = 0$ in Eq. (3.67), i.e. with

$$S(0) = 1 + \rho \int (g(\boldsymbol{r}) - 1)\,\mathrm{d}\boldsymbol{r}, \tag{3.80}$$

this leads to

$$\begin{aligned}\frac{\langle N^2 \rangle}{\langle N \rangle} &= \int \delta(\boldsymbol{r})\,\mathrm{d}\boldsymbol{r} + \rho \int (g(\boldsymbol{r}) - 1)\,\mathrm{d}\boldsymbol{r} + \rho \int \mathrm{d}\boldsymbol{r} \\ &\approx S(0) + \langle N \rangle\end{aligned} \tag{3.81}$$

which can be rearranged to

$$S(0) = \frac{\langle N^2 \rangle - \langle N \rangle^2}{\langle N \rangle}. \tag{3.82}$$

In Chapter 4 a relation will be derived which relates these particle number fluctuations to the so-called thermodynamic factor and thus to the thermodynamic properties of the sample. For an ideal gas of non-interacting particles the mean square fluctuation $\langle N^2 \rangle - \langle N^2 \rangle^2$ is equal to $\langle N \rangle$ such that $S(0) = 1$. This means that the scattering behaves as if there were no mutual interference from the different particles. In Eq. (3.82) and with $Q \to 0$ we tacitly assume that the sample size is infinite. For a finite sample of dimension $d = 1$ cm this means that Q has to be greater than 10^{-8} Å^{-1}, otherwise *all* partial waves from the sample would constructively interfere such that the total scattering intensity would be N_0^2 where N_0 is the total number of particles in the sample, and $S(Q < 1/d) = N_0$. This is the interference on the sample as a whole, unresolved for samples of macroscopic size. But in the field of small-angle neutron scattering on polymers, i.e. on mesoscopic objects, this effect is utilized for the determination of the molar mass of the macromolecules.

3.5 DEBYE–WALLER FACTOR; LATTICE VIBRATIONS

The topic of this book is the diffusive motion of atoms or ions in solids. However, when sitting on a lattice site, the diffusing atom joins the vibrations of the host lattice or performs localized vibrations, provided that the residence time is sufficiently long. For this reason this section briefly deals with the influence of these motions on the scattering spectrum. To simplify the problem it is assumed that the vibrational motions are completely decoupled from the diffusion.

The lattice vibrations influence the intensity $I_0(\boldsymbol{Q})$ of the elastic or quasielastic line which appears in the incoherent spectrum: $I_0(\boldsymbol{Q})$ is reduced with increasing scattering vector $\boldsymbol{Q}$, and this dependence is described by the so-called Debye–Waller factor well known in crystallography. This reduction of the elastic or quasielastic intensity is counterbalanced by the increase of the inelastic scattering intensity due to phonon or multiphonon processes such that, in agreement with the first sum rule, the total intensity remains constant. We deal with the calculation of the Debye–Waller factor and then also with the intensity of inelastic processes, in both cases assuming a simple cubic Bravais lattice.

We first describe an isolated harmonic oscillator which approximates the localized vibrations of hydrogen dissolved in a metal. A hydrogen atom on an interstitial site in a metal (see Chapter 8) has three vibrational degrees of freedom; thus it can be considered as a three-dimensional Einstein oscillator which can be factorized into three one-dimensional harmonic oscillators. We prepare the system in the ground state, simply by lowering the temperature such that $T \ll \hbar\omega_0/k_B$ is the Einstein frequency. In the typical case of a bcc metal hydride, $\hbar\omega_0 \approx 100$ meV for the localized hydrogen vibration, as compared to the acoustic (band) vibrations which are in the region of 10 or 20 meV. The ground state and first excited state wavefunctions for the harmonic oscillator are:

$$|0\rangle = \Psi_0(x) = \sqrt{\frac{\alpha}{\sqrt{\pi}}}\mathrm{e}^{-\alpha^2x^2/2} \tag{3.83}$$

$$|1\rangle = \Psi_1(x) = \sqrt{\frac{2\alpha^3}{\sqrt{\pi}}}x\mathrm{e}^{-\alpha^2x^2/2} \tag{3.84}$$

with the abbreviation $\alpha^2 = M\omega_0/\hbar$. The first five functions are displayed in Fig. 3.6. All wavefunctions are normalized, for instance $\langle 0|0\rangle = 1$. The mean square displacement in the ground state (zero-point vibration) is then

$$\langle x^2\rangle = \langle 0|x^2|0\rangle = \frac{\alpha}{\sqrt{\pi}}\int_{-\infty}^{\infty} x^2\mathrm{e}^{-\alpha^2x^2}\,\mathrm{d}x = \frac{1}{2\alpha^2}. \tag{3.85}$$

For hydrogen in metals, with mass M and $\hbar\omega_0 \approx 100$ meV we obtain $\langle x^2\rangle = \hbar/2m\omega_0 \simeq 0.02\text{Å}^2$, and the mean displacement of a hydrogen atom vibrating on its interstitial site is typically of the order of 0.1 Å.

In order to calculate the double differential neutron scattering cross-section we start from Eq. (3.29) which, however, simplifies if the system is in the ground state:

$$\left(\frac{\mathrm{d}^2\sigma}{\mathrm{d}\Omega\,\mathrm{d}\varepsilon}\right)_{0\to n} = \left(\frac{k_1}{k_0}\right)\left(\frac{\sigma_\mathrm{s}}{4\pi}\right)\sum_{n=0}^{\infty} S_{0\to n}(\boldsymbol{Q},\omega) \tag{3.86}$$

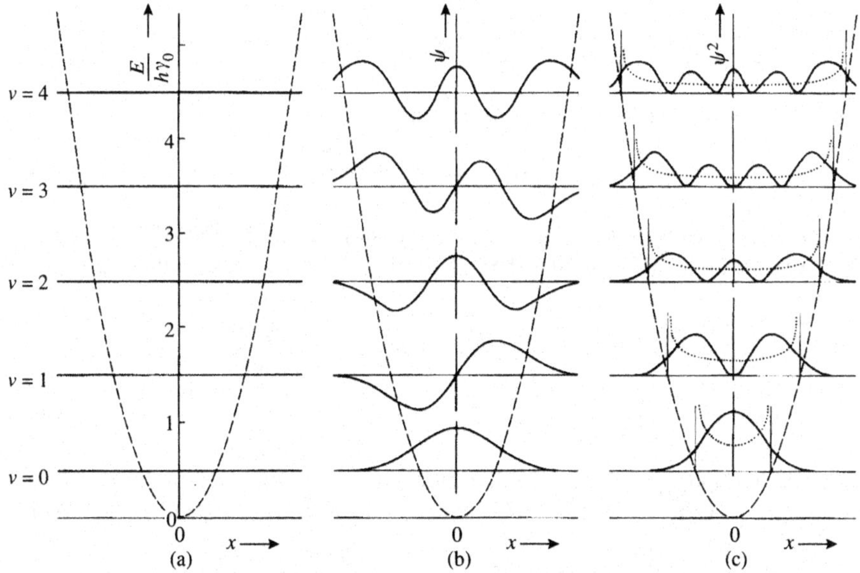

Fig. 3.6 Harmonic potential (a) with five harmonic oscillator wave functions (b) and their probability densities (c).

with the scattering function for the transition from $|0\rangle$ to $|n\rangle$:

$$S_{0\to n}(\boldsymbol{Q},\omega) = |\langle n|\mathrm{e}^{\mathrm{i}Qx}|0\rangle|^2\,\delta(\hbar\omega - n\hbar\omega_0); \tag{3.87}$$

$n\hbar\omega_0$ is the energy difference between the nth level and the ground state; it is positive since we only consider neutron energy loss processes. The system consists of independent oscillators without interactions which vibrate without any mutual phase relationship. In that case the coherent scattering function contains no distinct terms; consequently, coherent and incoherent scattering functions are identical. That is the reason why the total scattering cross-section $\sigma_\mathrm{s} = \sigma_\mathrm{i} + \sigma_\mathrm{c}$ enters into Eq. (3.86).

For elastic neutron scattering with the transition from $|0\rangle$ to $|0\rangle$, the transition matrix element in Eq. (3.87) is calculated as

$$\begin{aligned}\langle 0|e^{\mathrm{i}Qx}|0\rangle &= \int_{-\infty}^{\infty} \Psi_0^*(x)\mathrm{e}^{\mathrm{i}Qx}\Psi_0(x)\,\mathrm{d}x \\ &= \frac{\alpha}{\sqrt{\pi}}\mathrm{e}^{-Q^2/4\alpha^2}\int_{-\infty}^{\infty}\mathrm{e}^{-(\alpha^2x^2-\mathrm{i}Qx-Q^2/4\alpha^2)}\,\mathrm{d}x \\ &= \mathrm{e}^{-Q^2/4\alpha^2}\end{aligned} \tag{3.88}$$

or

$$\langle 0|e^{\mathrm{i}Qx}|0\rangle = \mathrm{e}^{(-1/2)Q^2\langle x^2\rangle} = \mathrm{e}^{-W_\mathrm{LM}}, \tag{3.89}$$

and inserting this into Eq. (3.87), one obtains

$$S_{0\to 0}(\boldsymbol{Q},\omega) = \mathrm{e}^{-2W_\mathrm{LM}}\,\delta(\omega), \tag{3.90}$$

i.e. the elastic scattering decreases with increasing Q on account of the Debye–Waller factor for localized modes (LM). For a three-dimensional oscillator we have

$$2W_\mathrm{LM} = \langle(\boldsymbol{Q}\cdot\boldsymbol{u})^2\rangle = \frac{1}{3}Q^2(u_x^2+u_y^2+u_z^2) = \frac{\hbar^2Q^2}{6M}\sum_i\frac{1}{\hbar\omega_i}$$
$$i = x, y, z. \tag{3.91}$$

In crystallography, this Debye–Waller factor is usually included in the structure factor F_{hkl} via the so-called atomic temperature factor e^{-W} (Willis and Pryor 1975).

At ambient temperatures, $2W_\mathrm{LM}$ is temperature dependent because of the temperature-dependent occupation of the excited states. In other words, at higher temperatures, the scattering functions $S_{1\to 1}, S_{2\to 2}, \ldots,$ also contribute to the elastic scattering in a proper thermal average, such that $2W_{\mathrm{LM}}$ in Eqs. (3.90) and (3.91) has to be replaced by (Lovesey 1987)

$$2W_{\mathrm{LM}} = \frac{\hbar^2 Q^2}{6M_{\mathrm{H}}} \sum_{i=1}^{3} \frac{1}{\hbar\omega_i} \coth\left(\frac{\hbar\omega_i}{2k_{\mathrm{B}}T}\right). \tag{3.92}$$

For low temperatures ($\hbar\omega_{\mathrm{i}} \gg 2k_{\mathrm{B}}T$) the coth function approaches unity and Eq. (3.91) is recovered. Eqs. (3.91)–(3.93) show that the elastic line (δ-function in Eq. 3.91) or the quasielastic line (in the case of diffusive motions), respectively, in the neutron spectrum does not change its shape but only its intensity due to the vibrations. Vibrational transitions from $|0\rangle$ to $|n\rangle, n = 1, 2, \ldots$, give rise to inelastic scattering intensity outside the energy window of quasielastic neutron scattering, and are outside the scope of this book.

The resulting total transition matrix element in Eq. (3.89) can also be considered as the Fourier transform of the probability density $\Psi_0^* \Psi_0$ which in the case of a harmonic oscillator is a Gaussian distribution. Thus the Debye–Waller factor is just the square of the Fourier transformed probability density of the particle or, in other words, the Debye–Waller factor is effectively the 'form factor' of the nuclear density.

We now consider the acoustic vibrations (band modes, BM) of the scattering atoms and characterize these phonons by a density of states $Z(\omega)$. For a simple cubic Bravais lattice, the formulation is simple and shows the characteristic features which appear also for more complicated cases. The incoherent scattering cross-section on an atom in a lattice is obtained as a sum of multiphonon terms ($n = 1, 2, \ldots$). This expansion reads (Sjölander 1958)

$$S_{\mathrm{i}}(\boldsymbol{Q}, \omega) = \mathrm{e}^{-Q^2u^2}\Big[\delta(\omega) + (Q^2u^2)\, G_1(\omega) + \frac{(Q^2u^2)^2}{2!}\, G_2(\omega) + \cdots + \frac{(Q^2u^2)^n}{n!}\, G_n(\omega) + \cdots\Big]. \tag{3.93}$$

The δ-function again corresponds to the purely elastic (zero-phonon) processes, which appear for rigidly bound atoms, and which, in the case of diffusion, are energetically broadened. The one-phonon function $G_1(\omega)$ is proportional to the normalized phonon density of states, $Z(\omega)$, namely

$$G_1(\omega) = Z(\omega)\frac{\mathrm{e}^{-\hbar\omega/k_{\mathrm{B}}T}}{\omega \sinh(\hbar\omega/2k_{\mathrm{B}}T)}. \tag{3.94}$$

For $\omega \to 0$, the scattering spectrum approaches a constant value because $Z(\omega)$ is proportional to ω^2 which cancels against the amplitude u^2, see Eq. (3.93), and the temperature factor which are both propotional to $1/\omega$, as long as $\hbar\omega/k_{\mathrm{B}}T \ll 1$. Thus $G_1 = \text{const.}$ for small ω. The two-phonon term is given by

$$G_2(\omega) = \int_{-\infty}^{+\infty} G_1(\omega - \omega')\, G_1(\omega')\, \mathrm{d}\omega'. \tag{3.95}$$

Correspondingly, $G_n(\omega)$ is an n-fold self-convolution of the single phonon function $G_1(\omega)$. All functions $G_n(\omega)$ are normalized to unity. Despite the irregular form of the density of states and of $G_1(\omega)$, the multiple convolution smears out these irregularities and the spectral distribution soon approaches a Gaussian-shaped curve. Equation (3.93) is consistent with the sum rule $\int S(\boldsymbol{Q}, \omega)\,\mathrm{d}\omega = 1$ since, integrating the Sjölander formula (3.93) over all ω, the resulting sum in the square bracket $\sum_0^\infty (Q^2u^2)^n/n!$ is just the inverse of the Debye–Waller factor. For a cubic Bravais lattice the Debye–Waller exponent for the band modes can be written as

$$2W_{\mathrm{BM}} = \frac{\hbar Q^2}{2M}\int \frac{Z(\omega)}{\omega}\coth\left(\frac{\hbar\omega}{2k_{\mathrm{B}}T}\right)\mathrm{d}\omega. \tag{3.96}$$

This equation is analogous to Eqs. (3.91) and (3.92), but without the factor $\frac{1}{3}$ for the approximate spatial average and with the sum over discrete vibrational excitation replaced by an integral over a continuous density of states $Z(\omega)$. In the limit $T \to 0$ K, the Debye–Waller exponent does not vanish because of the zero-point vibrations of the nuclei. The coth function in Eq. (3.96) approaches unity, and with the Debye spectrum,

$$Z(\omega) = \begin{cases} 3\omega^2/\omega_{\mathrm{D}} & \text{for } \omega \le \omega_{\mathrm{D}} \\ 0 & \text{for } \omega > \omega_{\mathrm{D}}, \end{cases} \tag{3.97}$$

we obtain, with the use of Eq. (3.96) and with $\hbar\omega_{\mathrm{D}} = k_{\mathrm{B}}\Theta_{\mathrm{D}}$

$$2W_0 = \frac{3}{2}\left(\frac{\hbar^2Q^2}{2M}\right)\frac{1}{k_{\mathrm{B}}\Theta_{\mathrm{D}}} \quad \text{for } T \ll \Theta_{\mathrm{D}} \tag{3.98}$$

where Θ_{D} is the Debye temperature.

In the high-temperature limit the coth function in Eq. (3.96) has a small argument, and because $\coth(x) \approx 1/x$ for small x we obtain

$$2W_{\mathrm{HT}} = 6\left(\frac{\hbar^2Q^2}{2M}\right)\frac{T}{k_{\mathrm{B}}\Theta_{\mathrm{D}}} \quad \text{for } T \gg \Theta_{\mathrm{D}}. \tag{3.99}$$

For practical purposes a more detailed power expansion is useful which yields (Lovesey 1987)

$$2W_{\mathrm{BM}} = \frac{\hbar^2Q^2}{M}\frac{3k_{\mathrm{B}}T}{(k_{\mathrm{B}}\Theta_{\mathrm{D}})^2}\left\{1 + \frac{1}{36}\left(\frac{\Theta_{\mathrm{D}}}{T}\right)^2 - \frac{1}{3600}\left(\frac{\Theta_{\mathrm{D}}}{T}\right)^4\right\}$$
$$\text{for } T \ge 0.4\Theta_{\mathrm{D}}. \tag{3.100}$$

For hydrogen dissolved in a metal, the situation is complex. As an approximation we can calculate $\langle u^2\rangle$ as an independent superposition of the square amplitude caused by the localized vibrations $\langle u^2 \rangle_{\rm LM}$, Eq. (3.92), and the contribution from the acoustic band of phonons of the host lattice $\langle u^2 \rangle_{\rm BM}$, Eq. (3.100), namely

$$2W = 2W_{\rm LM} + 2W_{\rm BM}. \tag{3.101}$$

The lattice term caused by the acoustic phonons can easily be calculated under the assumption that the hydrogen atom exactly follows the motion of the host lattice atoms and mirrors the host lattice density of states.

Actually the situation is more complicated. One effect is that in particular the hydrogen amplitude can strongly exceed the host lattice amplitude at specific resonance frequencies (see Kley 1966; Lottner, Schober, and Fitzgerald 1979b; Dawber and Elliott 1963; Springer 1978). The situation has been discussed and investigated experimentally for the case of hydrogen in niobium (see for instance Springer and Richter 1987). Another effect is that the wavelength of lattice phonons with frequencies comparable with the diffusive jump rate $1/\tau$ is much larger than the jump distance. Consequently, even for frequencies $\omega < 1/\tau$ both motions coexist. This implies that the quasielastic line and the scattering on the phonons of the host lattice can be superimposed such that the phonon spectrum behaves as a frequency independent 'background'. This means that the phonon intensity can be extrapolated from $G_1(\omega \to 0)$ and subtracted from the quasielastic spectrum. As concerns the calculation of the Debye–Waller factor, the contributions of the modes in $\langle u^2 \rangle_{\rm BM}$ down to $\omega = 0$ have to be taken into account in Eq.(3.96). On the other hand, optic or acoustic zone boundary modes have wavelengths comparable with the jump distance; however, the frequency is much higher than the jump rate. This means that we can assume that the independent superposition of the jump motion and the lattice vibrations may be a relatively good approximation at least for bcc hydrides with high-energy optic modes. Egelstaff and Schofield (1962a, b), Rahman *et al.* (1962) and Cocking (1992) have dealt with the problem of the combined diffusive and oscillatory motion of atoms in liquids. Assuming that both motions are independent, the intermediate scattering function can formally be written as

$$I(Q,t)/I(Q,0) = {\rm e}^{-Q^2[f(t)+\langle u^2(t)\rangle]}. \tag{3.102}$$

$f(t)$ is the relaxation function for the (purely) diffusive motion and $\langle u^2(t) \rangle$ describes the harmonic oscillator motion. The artificial separation between the two kinds of motion can be avoided by molecular dynamic simulations of the motion, as will be shown for H in Pd in Fig. 8.10. Under these circumstances the quasielastic spectrum, merging into the vibrational spectrum, could be fully obtained. For a given energy window such a calculation could then be compared directly with experiment.

4

Diffusion in solids

Ideal crystalline solids comprise a periodic sequence of atoms on so-called lattice sites which are the minima of the mutual interaction potential. If this statement is taken literally, i.e. if all structural sites are occupied and if anti-structure atoms do not occur, then diffusion is not imaginable. Therefore the lemma '*corpora non agunt nisi fluida*' dominated scientific thinking for a long time, in such a way that numerous indications of the mobility of atoms in solids were disregarded or misinterpreted. The hardening of iron in carbon-containing material, where carbon diffuses into the iron, for example, had been applied for many centuries without realizing the mechanism of this process. The first to express doubt about '*corpora non agunt nisi fluida*' was Gay-Lussac (1846). The first papers about the observation of diffusion and reaction processes in solids appear in the literature in the 1880s. Spring (1880) reports that a compressed mixture of two metal powders starts to melt at the melting temperature of the corresponding alloy well below the melting temperature of the individual metals: at the contact points of the metal particles, thin layers of the alloy are formed by solid state diffusion.

The first quantitative investigation of diffusion in solid metals is due to Roberts-Austen (1896), who determined the diffusion coefficient of Pb in Au. A larger number of diffusion studies have been undertaken since the 1920s, which can be considered as the basis of the now developing systematic investigation of atomic site exchanges in solid metals and ionic crystals.

Of fundamental importance for an atomistic understanding of diffusion is the work by Einstein (1905) about the relationship between the diffusion coefficient and the atomic mobility which at that time, however, was not recognized as being relevant for solid state diffusion. Frenkel (1926) was the first to point out that lattice defects are necessary for diffusion and postulated what is now called a *Frenkel defect* consisting of an atom in the interstitial lattice and a vacancy on a regular lattice site. Another typical defect type is the Schottky defect which occurs in ionic crystals and is regarded as a pair comprising a cation and an anion vacancy (Wagner and Schottky 1930). Important contribution to the thermodynamics of those defects are due to Wagner (1938, for example).

The first measurement of self-diffusion was performed by Groh and v. Hevesey (1920) who let radioactive Pb diffuse into natural Pb. However, only after World War II were radioactive isotopes of essentially all elements generally available. This initiated a drastic increase in the number and, also due to the improvements in counting electronics, in the precision of tracer diffusion studies.

Furthermore, important experimental techniques, like neutron scattering, radio-frequency techniques (NMR and ESR) and microwave techniques (impedance spectroscopy), were invented during or briefly after the war and since then have been continuously improved. Consequently, the progress in the field of solid state diffusion was dramatic. Scattering techniques like quasielastic neutron scattering or dynamic light scattering are particularly powerful tools for diffusion research because they enable one to study the diffusive process simultaneously in space and time.

The first quasielastic neutron scattering study of diffusion is due to Brockhouse (1958; 1959) who investigated atomic motions in liquid water. The first application of this techniques to a solid state diffusion problem is by Sköld and Nelin (1967), namely to hydrogen diffusion in α-PdH$_x$. However, only after the commissioning of the high-flux reactor of the Institute Laue–Langevin (1972) with its cold source and its high-resolution time-of-flight and backscattering spectrometers could solid state diffusion routinely be studied on a wide variety of systems; since then quasielastic neutron scattering has contributed a great deal to the microscopic understanding of solid state diffusion.

As will be described in detail in the next chapter, quasielastic neutron scattering on the one hand enables one to determine diffusion coefficients and on the other hand allows one to study elementary atomic jump processes. Due to the limited resolution, however, there are lower limits for the diffusion coefficient, $D > 10^{-7}\,\mathrm{cm}^2\,\mathrm{s}^{-1}$, and for the jump rates, $\tau^{-1} > 10^8\,\mathrm{s}^{-1}$, respectively.

Solid state diffusion comprises a sequence of site changes (hops) of the atoms or molecules in the solid on well-defined lattice sites. In the following we will emphasize four different aspects of solid state diffusion. In Section 4.4 we consider the nature or the *mechanism of a single event*, i.e. how does an atom manage to get from its present lattice site to a neighbouring site, and how does it perform a site exchange? For a heavy atoms this is done by a classical, thermally activated jump over the barrier provided by the interaction potential. Theoretically, this is described in the framework of classical transition state theory (TST) originally developed for chemical reactions (Eyring 1935; 1936) and later applied to diffusion (see, e.g. Stearn and Eyring 1940). The interaction potential, however, is not static due to lattice vibrations. Certain vibrational modes, so to speak, open the door between two sites and thus considerably diminish the potential barrier in this direction. Thus there is a close connection between diffusion and lattice vibrations. For light atoms like the different hydrogen isotopes, including the positive muon, diffusion is a quantum mechanical process involving tunnelling. Keywords are *phonon-assisted tunnelling* or *electron-restricted tunnelling* due to particle–phonon or particle–electron interactions, respectively. These are both incoherent tunnelling types, also called *hopping*. Incoherent in this connection means that there is no phase relationship between subsequent tunnelling events. Whatever the mechanism of the site exchange, it can—as a black box so to speak—be characterized just by one parameter, the jump rate Γ (dimension s^{-1}). In Section 4.3 we consider the *stochastic aspects* of solid state (jump) diffusion.

Statistical methods are applied to study the outcome of a sequence of jumps, i.e. the spatial/temporal development of the diffusive process. We will consider Markov and non-Markov processes, Poisson processes, introduce the principle of detailed balance and describe jump diffusion quantitatively by a complicated rate equation, the so-called master equation.

Microscopically (Section 4.2) solid state diffusion is characterized by the following parameters:

- the jump rate Γ; if the lattice sites are energetically or crystallographically inequivalent, so are the jump rates; then Γ_{mn} is the jump rate from site m to site n; implicitly the jump rate is connected to the site energy;
- the jump vector from site m to site n, $\mathbf{s}_{mn}$,
- the coordination number z indicating the number of neighbouring sites.

If all jump rates are equal, $\Gamma_{mn} = \Gamma$, then the residence time of a particle on its site is given by

$$\Gamma = \frac{1}{z\tau}. \tag{4.1}$$

Thus generally τ^{-1} is *not* the jump rate (except for a dumb-bell-like arrangement of two sites where $z = 1$) which is emphasized here in order to avoid confusion. If all the jump vectors have the same modulus it makes sense to introduce the jump length $l = |\mathbf{s}_{mn}|$; otherwise l represents an average and sometimes an effective jump length.

Macroscopically (Section 4.1) diffusion is characterized by various diffusion coefficients which will be introduced and defined in due course. The central connection between macroscopic and microscopic diffusion is the famous Einstein–Smoluchowski relation:

$$D_{\mathrm{E}} = \frac{l^2}{2d\tau} \tag{4.2}$$

where D_{E} is the Einstein diffusion coefficient and $d = \{1, 2, 3\}$ the dimensionality of the diffusive process. Equation (4.2) considers only the simplest case: all sites are energetically equivalent and only diffusional jumps to nearest neighbour sites are allowed.

The residence time depends on temperature and usually obeys the Arrhenius law, which for classical diffusion will be derived in Section 4.4.1:

$$\tau = \tau_0 \exp(E_{\mathrm{a}}/k_{\mathrm{B}}T) \tag{4.3}$$

where τ_0 is called the pre-exponential factor and E_{a} the activation energy. Correspondingly,

$$D_{\mathrm{E}} = D_{\mathrm{E},0} \exp(-E_{\mathrm{a}}/k_{\mathrm{B}}T). \tag{4.4}$$

The fastest imaginable diffusion process would be a free flight of the particles between sites with an upper limit $D_{\mathrm{E}} \approx D_{\mathrm{E},0}$ which is given by the expression of the diffusion coefficient of a perfect gas, $D_{\text{gas}} = \lambda\bar{v}/3$. The mean free path, λ, equals the jump length and is about 3 Å. The mean speed is given by Eq. (2.1) and yields $2500\,\text{m}\,\text{s}^{-1}$ for hydrogen at 300 K. Thus $D_{\text{E},0} < 2.5 \cdot 10^{-7}\ \text{m}^2\,\text{s}^{-1}$ at room temperature. This value is reached for H in V at 600 °C (Völkl and Alefeld 1978). Subsequent sections are restricted to solids which contain only one mobile component. Interdiffusion and cross-effects in the diffusion (characterized by Onsager coefficients) are beyond the scope of this book. We refer to textbooks on diffusion, e.g. to Philibert (1985; 1991), Heumann (1992), Kärger *et al.* (1998).

4.1 COLLECTIVE DIFFUSION

We start with a particle flux j which is also called the particle current density and means particles crossing unit area per unit time. It is given as the product of the particle velocity v and the particle density (particles per unit volume):

$$j = v\frac{N}{V} = vcL \tag{4.5}$$

(c is the molar concentration, L is the Loschmidt or Avogadro constant). The velocity is expressed by the product of the mobility B (dimension: time per mass) and driving force. In the case of macroscopic diffusion the driving force is the gradient of the chemical potential; thus it is essentially an entropic force: a system with a concentration inhomogeneity strives for homogeneity, i.e. for a state of maximum entropy (enthalpy terms, of course, can in special cases be responsible for ordering phenomena including short-range order):

$$v = -\frac{1}{L}B\frac{\partial\mu}{\partial x}. \tag{4.6}$$

(The Avogadro constant shows up in the denominator because μ is a molar quantity.) Experimentally the gradient of the chemical potential is often thought of as a concentration gradient:

$$\frac{\partial\mu}{\partial x} = \frac{\partial\mu}{\partial c}\frac{\partial c}{\partial x}; \tag{4.7}$$

in summary,

$$j = -Bc\frac{\partial\mu}{\partial c}\frac{\partial c}{\partial x}. \tag{4.8}$$

Now we utilize the Einstein relation between the mobility B and the Einstein diffusion coefficient

$$D_{\text{E}} = Bk_{\text{B}}T \tag{4.9}$$

which can be derived by solving the equation of motion with a random force, known as the Langevin equation. This can be considered as an expression of the fluctuation dissipation theorem which, qualitatively, says: 'the stronger a systems reacts on an external disturbance, the stronger are its spontaneous fluctuations', and which thus mutually relates spontaneous and externally forced deviations from thermodynamic equilibrium; here B is a dissipative and D a fluctuative quantity. With the molar flux $J = j/L$ we obtain

$$J = -D_{\mathrm{E}} \left(\frac{c}{RT} \frac{\partial \mu}{\partial c} \right) \frac{\partial c}{\partial x} \tag{4.10}$$

or, in other words, we obtain Fick's first law

$$J = -D_{\mathrm{chem}} \frac{\partial c}{\partial x}; \tag{4.11}$$

the chemical (or Fick's) diffusion coefficient is thus given by

$$D_{\mathrm{chem}} = \phi D_{\mathrm{E}} \tag{4.12}$$

with the so-called thermodynamic factor

$$\phi = \frac{c}{RT} \frac{\partial \mu}{\partial c}. \tag{4.13}$$

(The notation $\tilde{D}$ is often used (Mehrer 1991). For mnemotechnical reasons we prefer the notation D_{chem}, since too many different kinds of diffusion coefficients have to be carefully distinguished. Since only systems with one mobile component are considered in this book, chemical and intrinsic diffusion coefficients are identical.)

Let us consider some relevant examples. We start with an *ideal lattice gas*: the particles are distributed over discrete points in space (lattice sites) and exhibit no interactions except self-site blocking and no correlations. An ideal lattice gas can be considered as an ideal mixture of particles and vacancies on a lattice. The hop of a particle to an adjacent site is only possible if this site is available for occupation and not blocked by another atom, the probability of which is given by the site availability (or blocking) factor $V(c)$; this is supposed to be a function of the average occupancy c. Sometimes this factor is called the vacancy wind factor (Mehrer 1991). For lattice gases the occupancy or concentration c is the number of particles divided by the number of sites and is thus dimensionless. In the case of self-site blocking the vacancy availability factor is given by

$$V(c) = 1 - c \tag{4.14}$$

and approaches zero for $c \rightarrow 1$; for multiple site blocking (see below) zero is approached at lower concentration. This factor has to be included into the expression for the chemical diffusion coefficient of a lattice gas:

$$D_{\mathrm{chem}} = (1 - c)\phi D_{\mathrm{E}}. \tag{4.15}$$

An ideal mixture means that the Gibbs energy function ΔG contains no enthalpy term, but only the entropy of mixing:

$$\Delta G = RT\{c \ln c + (1 - c)\ln(1 - c)\}. \tag{4.16}$$

The derivative with respect to concentration is the chemical potential

$$\mu = \frac{\partial \Delta G}{\partial c} = RT \ln\left(\frac{c}{1 - c}\right) \tag{4.17}$$

and a further derivative yields the thermodynamic factor

$$\phi = \frac{c}{RT}\frac{\partial \mu}{\partial c} = \frac{1}{1 - c}. \tag{4.18}$$

We insert this result into Eq. (4.15) and see that the vacancy availability factor and the thermodynamic factor just compensate; hence in the case of an ideal lattice gas

$$D_{\text{chem}} = D_{\text{E}}, \tag{4.19}$$

i.e. again (like in the previous example) the chemical diffusion coefficient does not depend on concentration.

In the second example we allow for *interactions* and treat the lattice gas as a regular mixture (Bragg and Williams 1934; 1935). In this approximation for the entropy term in the Gibbs energy function the expression for ideal mixing is used, and additionally an enthalpy term is taken into account in terms of pair interactions:

$$\begin{aligned}\Delta G &= -RT\Delta S_{\text{m}} + \Delta H_{\text{m}} \\ &= RT\{c \ln c + (1 - c)\ln(1 - c)\} + Lc(1 - c)w.\end{aligned} \tag{4.20}$$

Note that here, in spite of its local character (only nearest neighbours are involved), the interaction is described in terms of the average concentration. In this way we have implicitly introduced a mean field approximation. Without reducing generality, we consider a mixture of hydrogen atoms H and vacancies v on an interstitial lattice; then the pair interaction parameter w is given by

$$w = w_{\text{Hv}} - \tfrac{1}{2}w_{\text{HH}} - \tfrac{1}{2}w_{\text{vv}}; \tag{4.21}$$

$w > 0$ means: H-v pairs are energetically unfavourable with the consequence of H-H attraction; $w < 0$ means: H-v pairs are favourable relative to H-H repulsion. From Eq. (4.20) we obtain the chemical potential

$$\mu = RT\{\ln c - \ln(1 - c)\} - (2c - 1)Lw \tag{4.22}$$

and in a further step the thermodynamic factor

$$\phi = \frac{1}{1-c}\left[1 - 2c(1-c)\frac{w}{k_B T}\right]. \tag{4.23}$$

Thus via Eq. (4.15) the chemical diffusion coefficient is given by

$$D_{\text{chem}} = \left[1 - 2c(1-c)\frac{w}{k_B T}\right] D_E. \tag{4.24}$$

The concentration dependence of the chemical diffusion coefficient in the Bragg–Williams approximation is displayed in Fig. 4.1; since the effect of lattice expansion with increasing H content is neglected in this approximation, D_{chem} in the full lattice is equal to the value in the empty lattice, D_E. In between, $D_{\text{chem}} > D_E$ for H-H repulsion and $D_{\text{chem}} < D_E$ for H-H attraction.

If the interaction parameter w is positive (H-H attraction), then at sufficiently low temperatures the system will demix into a lattice gas at low concentration and a lattice liquid at large concentration separated by a two-phase region. The two-phase region is enclosed by the binodal or phase boundary. The spinodal curve $T_s(c)$

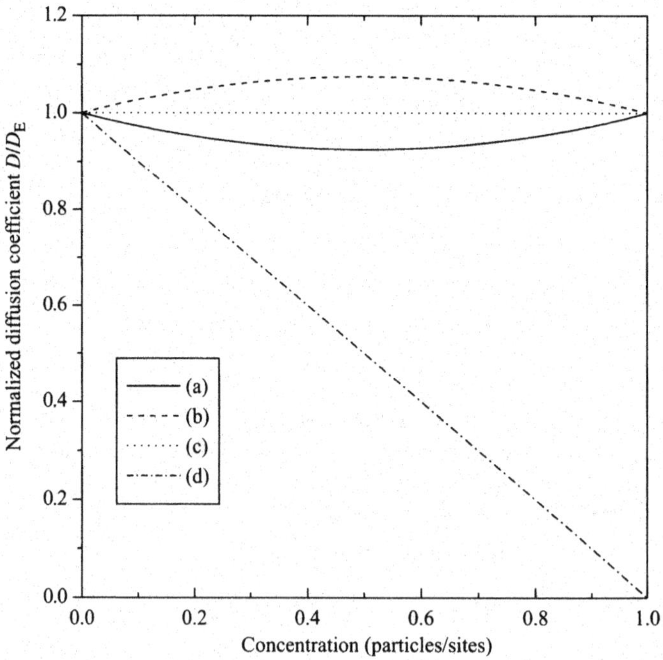

Fig. 4.1 Concentration dependence of chemical diffusion coefficient in the Bragg–Williams approximation; for comparison also the tracer diffusion coefficient is displayed. (a) D_{chem} with attractive interactions, (b) D_{chem} with repulsive interactions, (c) D_{chem} without interactions, (d) D_t.

corresponds to those temperatures where $\partial\mu/\partial c$ and thus the thermodynamic factor Eq. (4.24) vanishes, whereas within the spinodal region it is negative resulting in a negative chemical diffusion coefficient (up-hill diffusion). Thus Eq. (4.23) can be expressed as

$$\phi = \frac{1}{1-c}\frac{T - T_s(c)}{T} = \frac{\varepsilon}{1-c} \tag{4.25}$$

with $T_s(c) = 2c(1-c)w/k_B$. Again, c is the average concentration, i.e. Eq. (4.25) implies a mean field approximation. The maximum spinodal temperature $T_s(c)$ is found at the critical concentration, c_{crit}, which in the mean field approximation equals $\frac{1}{2}$; here the binodal and spinodal regions coincide ($\partial^2\mu/\partial c^2 = 0$) and $T_s(c_{crit})$ is called the critical temperature T_c. Even at temperatures well above T_c (the region of the lattice fluid) the thermodynamic factor $\phi(c)$ is still reminiscent of the two-phase region and exhibits a deep minimum at $c = c_{crit}$ thus slowing down the chemical diffusion coefficient which disapears at $T = T_c$ (critical slowing down); see Section 8.1.3.

Finally we allow for multiple site blocking. This means a hydrogen atom, in addition to blocking its own site for occupation by another H atom, also strongly influences the occupancy of neighbouring sites. The case of complete blocking of neighbouring sites has been considered by Faux and Ross (1987) using Monte Carlo simulation for interstitial diffusion on the tetrahedral sites of bcc metals. Their result for the site availability factor is shown in Fig. 4.2. We point out that for an *ordered* and stoichiometric hydride the closest distance between two hydrogen atoms turned out to be 2.1 Å (Switendick 1979; Westlake 1983).

If we finally include *correlation effects* by means of the mobility correlation factor $f_m(c)$, the chemical diffusion coefficient is given as

$$D_{chem}(c) = V(c)\,\phi(c)\,f_m(c)D_E. \tag{4.26}$$

The mobility correlation factor is displayed in Fig. 4.3, but its physical meaning will only later be discussed in connection with the tracer correlation factor $f_t(c)$ in Section 4.2.3. In the limit of no interactions except self-site blocking, $V(c) = 1 - c$ and $f_m(c) = 1$ (curves 0 in Figs. 4.2 and 4.3); then the chemical diffusion coefficient equals Einstein's diffusion coefficient.

It is interesting to note that the thermodynamic factor ϕ of a lattice liquid can be obtained from an elastic neutron scattering experiment in the limit $Q \to 0$. In Section 3.3 it was shown that

$$S(0) = \frac{\langle(\Delta N)^2\rangle}{\langle N\rangle} \tag{4.27}$$

where $\langle N\rangle$ is the average particle number and $\langle(\Delta N)^2\rangle = \langle N^2\rangle - \langle N\rangle^2$ the average square fluctuation of N in a certain volume. Since this square fluctuation is of the order of $\langle N\rangle$, $S(0)$ is of the order of unity. We consider an open subsystem at constant volume V^* and constant temperature T^*, with particle number fluctuations. At equilibrium the Helmholtz free energy function $A = U - TS$ (U denotes

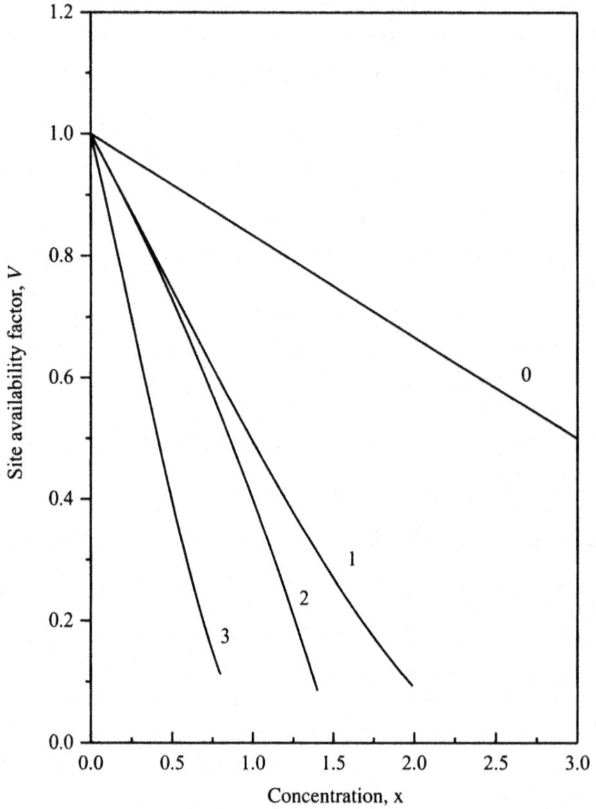

Fig. 4.2 Site availability factor (Faux and Ross 1987, with permission from IOP Publishing Ltd) 0: self-site blocking only, 1, 2, 3: blocking of one/two/three coordination spheres.

internal energy, S means entropy) is a minimum, and in its neighbourhood it can be expressed by a Taylor expansion as

$$A(N) = A(\langle N\rangle) + \left(\frac{\partial A}{\partial N}\right)_{T,V} \Delta N + \frac{1}{2}\left(\frac{\partial^2 A}{\partial N^2}\right)_{T,V} (\Delta N)^2 + \cdots . \qquad (4.28)$$

The derivatives with respect to N are evaluated at $N = \langle N\rangle$. At equilibrium the first derivative vanishes (minimum condition). The second derivative contains the thermodynamic factor:

$$\left(\frac{\partial^2 A}{\partial N^2}\right)_{T,V} = \frac{\partial \mu}{\partial N} = \frac{RT}{\langle N\rangle}\left\{\frac{\langle N\rangle}{RT}\frac{\partial \mu}{\partial N}\right\} = \frac{RT}{\langle N\rangle}\phi_N. \qquad (4.29)$$

Here the Helmholtz energy is formulated as function of particle number which, however, is equivalent to the formulation as a function of concentration (particles per sites, $c = N/N_s$); then $\partial A/\partial N$ represents the chemical potential.

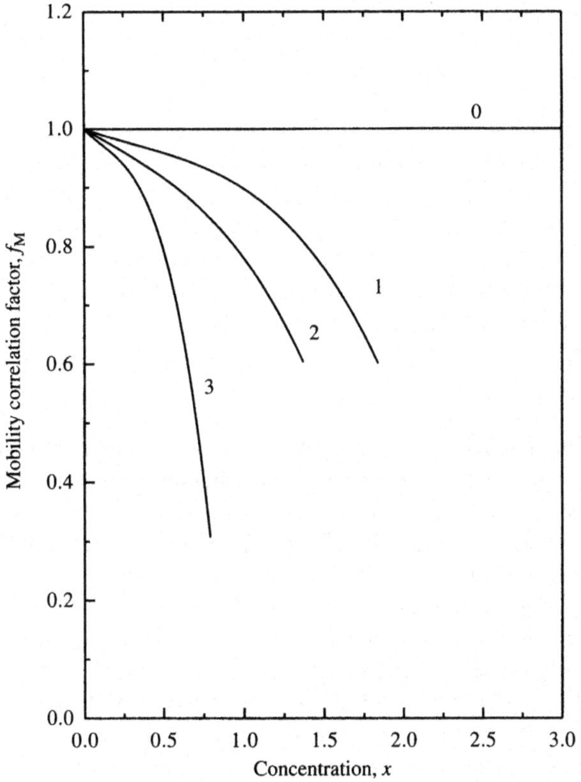

Fig. 4.3 Mobility correlation factor (Faux and Ross 1987, with permission from IOP Publishing Ltd) 0: self-site blocking only, 1, 2, 3: blocking of one/two/three coordination spheres.

The probability that the particle number in our volume V^* lies between N and $N + \mathrm{d}N$ is

$$f(N)\,\mathrm{d}N \propto \exp\left(-\frac{A(N)}{RT}\right)\mathrm{d}N. \tag{4.30}$$

Thus from Eqs. (4.28) and (4.29)

$$f(N)\,\mathrm{d}N \propto \exp\left\{-\frac{\phi}{2}\frac{(\Delta N)^2}{\langle N\rangle}\right\} \tag{4.31}$$

which, in view of $\Delta N = N - \langle N\rangle$, is a Gaussian centred at the mean particle number $\langle N\rangle$. The formula for the standard deviation of a Gaussian gives

$$\langle(\Delta N)^2\rangle = \langle N\rangle/\phi. \tag{4.32}$$

Therefore, from Eq. (4.27) it follows that

$$S(0) = 1/\phi. \tag{4.33}$$

Generally, a scattering experiment at $Q \to 0$ averages over a very large volume and thus yields thermodynamic quantities (the thermodynamic limit). For a normal liquid usually volume fluctuations are considered and $S(0)$ is proportional to the isothermal compressibility; a lattice liquid or lattice gas, however, is essentially incompressible, and it is more appropriate to consider particle number fluctuations. For an *ideal lattice gas*, the thermodynamic factor equals $1/(1-c)$ (Eq. 4.18); therefore, the scattering function at $Q = 0$ equals

$$S(Q) = c(1-c) \tag{4.34}$$

where the additional factor c appears since for a lattice gas (mixture of particles and vacancies on an interstitial lattice) the scattering function is quoted per particle and not per site. Without short-range order effects the result (4.34) is valid at all Q values and represents the well-known Laue scattering for a random binary mixture (see Eq. 6.8). On the other hand, for a perfect gas, $\phi = 1$ and thus $S(Q) = 1$. This indicates a state of total disorder, whereas a lattice gas at finite concentration is correlated in so far as double occupancy is forbidden.

We now briefly consider *diffusion of ions in ionic solids*. It can conveniently be determined from electrical conductivity measurements if there is no electronic contribution or if the electronic contribution can be separated, e.g. by measurements of the transport numbers. In most cases only one kind of ion is mobile. For the particle flux we again have

$$j = v\tilde{N}, \tag{4.35}$$

where $\tilde{N}$ is the number of ions per volume; v is now the drift velocity in an electric field E:

$$v = uE = \frac{u}{ez} \cdot ezE. \tag{4.36}$$

u means ionic mobility (dimension: $\mathrm{cm^2\,s^{-1}\,V^{-1}}$), e is the elementary charge, $F = eL$ is known as Faraday's constant, and z indicates the number of charges per ion (e.g. $z = 2$ for O^{2-}). In Eq. (4.35) the factor ezE is the force, which acts on the ion; thus the other factor

$$\frac{u}{ez} = B \tag{4.37}$$

is the mobility. The mobility B is a kinetic property which does not depend on the type of driving force. Thus also here the Einstein relation (Eq. 4.9), $B = D_\mathrm{E}/k_\mathrm{B}T$, is valid; hence

$$u = \frac{D_{\mathrm{E}}ez}{k_{\mathrm{B}}T} = \frac{D_{\mathrm{E}}Fz}{RT}. \tag{4.38}$$

If we multiply the particle flux (Eqs. (4.36) and (4.35)),

$$j = uEcL, \tag{4.39}$$

by the charge ze, we get the charge flux which equals the current density i (in A cm^{-2}):

$$i = zeuEcL. \tag{4.40}$$

The current density divided by the electric field yields, by definition, the specific conductivity σ. Hence, with Eq. (4.37), we end up with the well-known Nernst–Einstein equation (Nernst 1888; Einstein 1905):

$$\sigma = \frac{F^2 z^2}{RT} D_\sigma c. \tag{4.41}$$

In this expression, in view of the large concentration of mobile ions which are usually involved, D_{E} has to be replaced by the conductivity diffusion coefficient

$$D_\sigma = V(c) f_{\mathrm{m}}(c) D_{\mathrm{E}} \tag{4.42}$$

in order to include blocking and correlation effects. It is noteworthy that in conductivity measurements the driving force for the particle motion is the electric field and not the gradient of the chemical potential; therefore, other than the chemical diffusion coefficient D_{chem}, the conductivity diffusion coefficient D_σ does not contain a thermodynamic factor. Like the chemical diffusion coefficient it contains the mobility correlation factor which, therefore, is sometimes also called the conductivity correlation factor.

Fick's first law, Eq. (4.11), looks straightforward, but for an actual determination of diffusion coefficients it is not useful, because it is very difficult to experimentally establish a constant concentration gradient and measure the diffusing amount of material. It is much easier to probe the distance dependence of the concentration along the diffusional path (concentration profile) after a certain time or the time dependence of the concentration at a certain distance. The diffusion coefficient can be evaluated from those data by means of *Fick's second law*:

$$\frac{\partial c}{\partial t} = D_{\mathrm{chem}} \frac{\partial^2 c}{\partial x^2}. \tag{4.43}$$

This is the one-dimensional form of Fick's second law provided D_{chem} does not depend on concentration. For isotropic three-dimensional diffusion it reads

$$\frac{\partial c}{\partial t} = D_{\mathrm{chem}} \left(\frac{\partial^2 c}{\partial x^2} + \frac{\partial^2 c}{\partial y^2} + \frac{\partial^2 c}{\partial z^2} \right) = D_{\mathrm{chem}} \nabla^2 c. \tag{4.44}$$

For anisotropic solids the diffusion coefficient is a tensor; in the case of axial symmetry the two quantities $D_\perp$ and $D_\parallel$ determine the diffusion process:

$$\mathbf{D} = \begin{pmatrix} D_\perp & 0 & 0 \\ 0 & D_\perp & 0 \\ 0 & 0 & D_\parallel \end{pmatrix}. \tag{4.45}$$

Correspondingly, Fick's laws for the flow of atoms in anisotropic single crystals are formulated as

$$J_i = -D_{ij} \frac{\partial c}{\partial x_j}, \quad i, j = \{1, 2, 3\} \tag{4.46}$$

and

$$\frac{\partial c}{\partial t} = \nabla \mathbf{D} \, \nabla c. \tag{4.47}$$

Fick's second law is a partial differential equation of second order and can be solved for experimentally relevant boundary and initial conditions. For a complete treatment see Crank (1975).

In connection with incoherent quasielastic neutron scattering (Chapter 5) which measures single-particle diffusion we need to consider three-dimensional diffusion starting from a point source. The diffusion equation (4.44) is also valid in this case, as will be derived in Eqs. (4.89)–(4.91), but with D_{chem} replaced by the self-diffusion coefficient D_s. This equation has to be solved with the initial condition $c(r, 0) = \delta(r) n_{\text{mol}}$ where n_{mol} is the number of moles. It can easily be verified that

$$c(\boldsymbol{r}, t) = \frac{n_{\text{mol}}}{(4\pi D_s t)^{3/2}} \exp\left(-\frac{r^2}{4 D_s t}\right) \tag{4.48}$$

is a solution. Then it is straightforward to calculate the mean square displacement

$$\langle r^2(t) \rangle = 6 D_s t. \tag{4.49}$$

These results establish a direct link to the Einstein–Smoluchowski relation (4.2). The linear increase with time of the mean square displacement is direct evidence for a diffusive process.

4.2 SINGLE-PARTICLE DIFFUSION

This section deals with diffusion processes which occur in thermodynamic equilibrium and do not involve mass transport, i.e. with self-diffusion or single-particle diffusion. If self-diffusion is studied by using isotopes as tracers it is called tracer diffusion. Except for light elements the difference between self and tracer diffusion is negligible. It is characterized by the self-diffusion coefficient D_s (or, in the case of isotope effects, by the tracer diffusion coefficient D_t). (Often the notation D^* is used (Mehrer 1991). For mnemotechnical reasons we prefer the notation D_s or D_t (Mehrer does distinguish between D_s and D_t) since too many kinds of diffusion coefficients have to be carefully distinguished.) The microscopic diffusion measurement techniques of *nuclear magnetic resonance* (*NMR*) and *incoherent quasielastic neutron scattering* (*IQENS*) use the particle's spin in order to label it; thus these techniques directly yield the self-diffusion coefficient. Before we deal with solid state diffusion we briefly mention a few features of gas and liquid self-diffusion for comparison.

4.2.1 Gas, liquid-like, and solid state diffusion

It is characteristic for diffusion in gases that the time of free flight of a particle is much larger than the time of collision, i.e. the total time is essentially the sum of the flight times between collision events. A perfect gas consists of a swarm of atoms or molecules in permanent random motion. The particle size is much smaller than the average particle distance. There is no interaction except elastic collisions (elastic means conservation of kinetic energy of translation of the collision pair, i.e. no excitation of internal degrees of freedom). The primary microscopic parameters are the collision cross-section σ and the mass M. For instance for Ar these quantities are $\sigma = 0.26\,\mathrm{nm}^2$, $M = 39.95 \cdot 10^{-3}\,\mathrm{kg\,mol^{-1}\,L^{-1}}$. Derived quantities are the Maxwellian velocity distribution with the root mean square speeds $(\langle v_x^2\rangle)^{1/2} = (k_B T/M)^{1/2}$ and $(\langle v^2\rangle)^{1/2} = (3\,k_B T/M)^{1/2}$ and the mean speed $\bar{v} = (8\,k_B T/\pi M)^{1/2}$ as well as the collision frequency $\nu = \sqrt{2}\,\sigma\,\bar{v}N/V$ and the mean free path

$$\lambda = \frac{\bar{v}}{\nu} = \frac{1}{\sqrt{2}\sigma}\frac{k_B T}{p}. \tag{4.50}$$

For argon, at $T = 300$ K and $p = 105$ Pa $= 1$ bar, the mean free path according to Eq. (4.50) equals 112 nm. This is much larger than the observation range of quasielastic neutron scattering of not more than about 4 nm, see Section 3.4. Thus, in the measuring regime of QENS, the gas atoms perform a free flight with a constant velocity v. This holds also for a single atom in condensed matter in the limit $t \to 0$ (short-time behaviour, free-particle limit). The probability that such a free particle is displaced by the vector $\boldsymbol{r}$ within the time period t is proportional to the probability that its velocity is between $\boldsymbol{u}$ and $\boldsymbol{u} + \mathrm{d}\boldsymbol{u}$, with $\boldsymbol{u} = \boldsymbol{r}/t$. If the last identity is inserted in the Maxwellian velocity distribution function, we obtain for the perfect gas:

$$P(r,t) = \left(\frac{M}{2\pi k_B T t^2}\right)^{3/2} \exp\left(-\frac{M r^2}{2 k_B T t^2}\right). \tag{4.51}$$

This quantity indicates the probability of finding a particle at time t in the distance r if it has been at $t = 0$ at the origin $r = 0$, i.e. it represents the self-correlation function, see Section 3.3.

Thus we are able to calculate the mean square displacement for a perfect gas:

$$\langle r^2(t)\rangle = \int_{-\infty}^{+\infty} r^2 P(r,t)\,\mathrm{d}^3 r = \frac{3k_\mathrm{B}T}{M}t^2 = \langle v^2\rangle t^2 \tag{4.52}$$

which, other than for diffusion, is proportional to t^2. At very long times, of course, the mean square displacement is dominated by collision processes and exhibits the diffusional linear time dependence $\langle r^2(t)\rangle = 6D_\mathrm{s}t$, with the self-diffusion coefficient

$$D_\mathrm{s} = \tfrac{1}{3}\lambda\bar{v}. \tag{4.53}$$

The mean square displacement for gas diffusion is schematically shown in Fig. 4.4c.

The corresponding curve for self-diffusion in a liquid is displayed in Fig. 4.4b. At very short times and very low displacements the particles move freely within the cage formed by the surrounding particles. For displacements larger than the cage the mutual collisions slow down the particles, resulting in a slow long-range diffusion. Features of liquid-like diffusion in solids occur in certain ionic conductors, which for that reason are called superionic conductors; in these systems, the

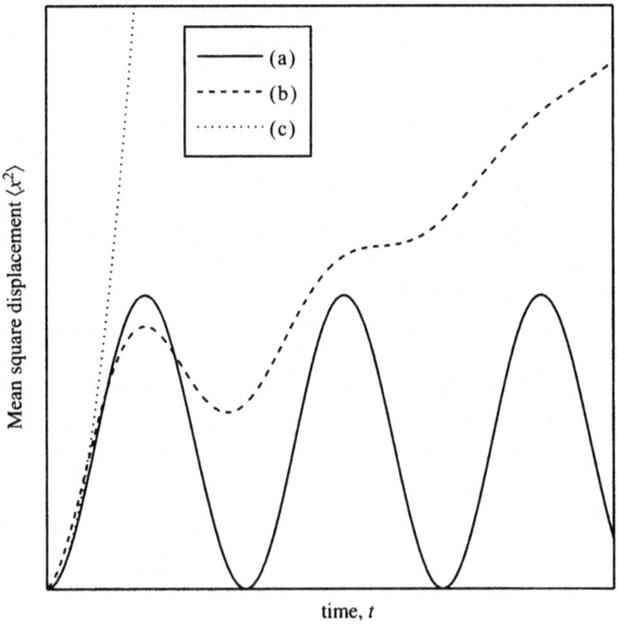

Fig. 4.4 Mean square displacement versus time for (a) single crystal in one specific direction, (b) liquid, (c) perfect gas; short-time behaviour: free flight; long-time behaviour: diffusion.

sublattice with the mobile ions exhibits far more sites than atoms, so that vacancies on this sublattice are abundant. The mobile ions are dynamically disordered, or in other words and in a certain sense, the corresponding sublattice is molten.

For solid state diffusion, opposite to gas and liquid-like diffusion, the time of flight, or better the jump time, is much shorter than the residence time, i.e. the total time is essentially the sum of the residence times. This statement is the basis of all jump diffusion models to be discussed in the next section. We verify this statement by considering, e.g., hydrogen in palladium. The residence time τ amounts to 10^{-8} s – 10^{-10} s, depending on temperature. The time of flight can be estimated from the jump length $l \approx 3$ Å (one lattice constant) and the mean thermal velocity $\bar{v} = (8\,k_B T/\pi M_H)^{1/2}$:

$$t = \frac{l}{v} = 3.8 \cdot 10^{-12}\ \text{s}. \tag{4.54}$$

For systems with very fast diffusion and thus very short residence times, like hydrogen in vanadium at elevated temperatures, jump models become questionable. The hydrogen atoms are no more properly localized which is in accordance with the broadening of the localized vibrational peaks. Therefore, the lower the temperature, the better jump models are applicable. The mean square displacement for a solid is displayed in Fig. 4.4a. Again at very short times we notice the free-particle behaviour. The subsequent oscillations are due to the vibrations in the solid. Since there is a broad spectrum of vibrational frequencies and since the vibrations in different directions are out of phase, the oscillations are damped and the time-dependent translational mean square displacement $\langle x^2(t)\rangle$ converges to $\langle u^2\rangle$, the mean square displacement of the vibrations. In the following we consider four atomistic mechanisms of diffusion: the vacancy mechanism, the divacancy mechanism, the direct interstitial mechanism, and the interstitialcy (or indirect interstitial) mechanism (Mehrer 1991).

4.2.2 Atomistic mechanisms of diffusion

In the *vacancy mechanism* vacant lattice sites enable diffusion to take place. A substitutional foreign atom or a self-atom diffuses by jumping into a neighbouring vacancy (see Fig. 4.5). In solids this is the most common mechanism. The diffusion coefficient is proportional to the concentration of vacancies, and if many vacancies are available then diffusion is fast. In metals the formation of a vacancy requires an energy of about 1 eV; in thermodynamic equilibrium the number of vacancies, N_v, is strongly temperature dependent:

$$N_v/N = \exp(-\Delta G_v/RT) \tag{4.55}$$

where $\Delta G_v = \Delta H_v - T\Delta S_v$ is the Gibbs free energy of vacancy formation; thus even at elevated temperatures the vacancy concentration is small, e.g. at 1000 K $N_v/N \approx 10^{-5}$. Therefore quasielastic neutron scattering experiments

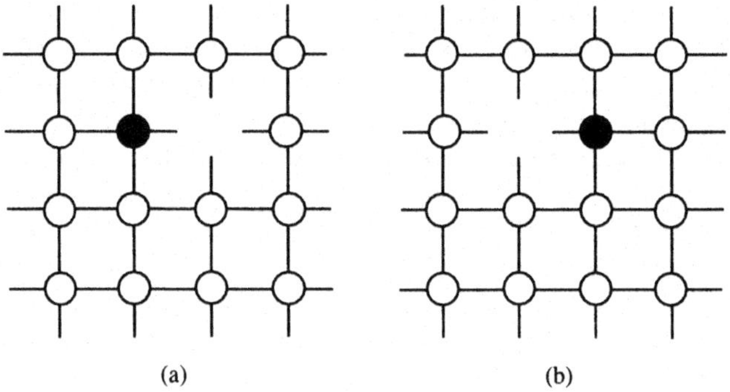

Fig. 4.5 Illustration of the vacancy mechanism: the tagged self-atom (full circle) moves, by jumping into the vacancy on its right-hand side (a), to the right (b) by one nearest neighbour distance of the regular lattice atom.

are possible only close to the melting point. The self-diffusion coefficient of a lattice atom caused by the presence of thermally activated vacancies is temperature dependent due to the temperature dependence of the jump rate (Eq. 4.3) and due to the temperature dependence of the vacancy concentration:

$$D_{\mathrm{s}}^{\mathrm{v}}(T) = f_{\mathrm{t}} D_{\mathrm{E},0} \exp\left(\frac{\Delta S_{\mathrm{v}}}{R}\right) \exp\left(\frac{-\Delta H_{\mathrm{v}}}{RT}\right) \exp\left(\frac{-E_{\mathrm{a}}}{k_{\mathrm{B}} T}\right). \tag{4.56}$$

This leads to an effective activation energy which is the sum of E_{a} and ΔH_{v}. The correlation factor f_{t} takes into account the enhanced back-jump probability and will be explained in Section 4.2.3. Correlated back-jumps are fast because the jump rate of the vacancies and correspondingly their diffusion coefficient is 10^4–10^5 times larger than those of the lattice atoms. In ionic crystals vacancies can be artificially introduced by doping with aliovalent ions. ZrO_2 (zirconia), for example consists of Zr^{4+} cations and O^{2-} anions. If part of the Zr^{4+} is replaced by a cation of lower valency, for instance by Y^{3+}, then vacancies in the O^{2-} sublattice establish electroneutrality: $Zr_{1-x}Y_xO_{2-x/2}$. This so-called Y-stabilized zirconia (YSZ) is a well-known oxygen conductor with a high oxygen diffusivity.

In the case of the *divacancy mechanism* bound pairs of vacancies on neighbouring lattice sites enable diffusion to take place. A host atom or a substitutional foreign atom on a regular site diffuses by jumping into one vacancy of the neighbouring pair. The divacancy mechanism has been proposed to contribute to self-diffusion in face-centred cubic metals above $\frac{2}{3}T_{\mathrm{m}}$ (T_{m} is the melting temperature) besides the vacancy mechanism (see, the example, Na self-diffusion in solid Na in Chapter 9).

In the *interstitialcy* (or *indirect interstitial*) *mechanism* self-interstitials allow diffusion: a self-interstitial replaces a substitutional atom which then in turn replaces a neighbouring lattice atom (Fig. 4.6). A self-interstitial is also named an anti-structure atom. The resulting self-diffusion coefficient, $D_{\mathrm{s}}^{\mathrm{i}}$, can be expressed

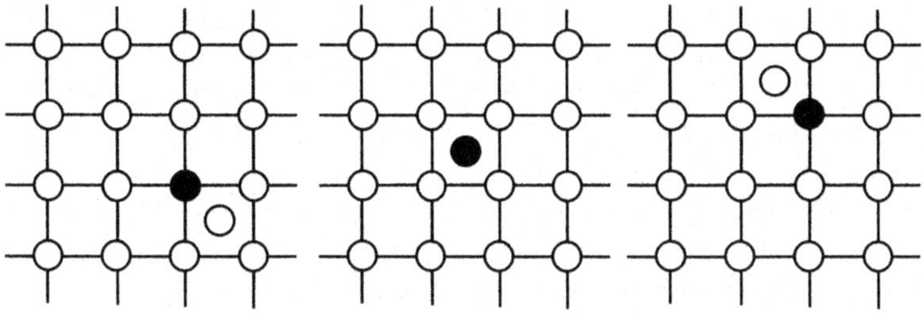

Fig. 4.6 Illustration of the interstitialcy mechanism: in (a) a self-interstitial (open circle in the center of the lattice cell) has approached a tagged self-atom (full circle); in (b) the tagged atom has exchanged its original position with the self-interstitial. In this way the tagged atom has temporarily become an interstitial, whereas the original self-interstitial has disappeared by occupying a regular lattice site. In (c) the tagged atom has jumped into a regular site by pushing a self-atom into an interstitial site.

in terms of the Einstein diffusion coefficient in the interstitial lattice, D_E, of the mean residence time on a substitutional (regular) site, τ_0, and of the mean time of one jump sequence in the interstitial lattice, τ_i:

$$D_\mathrm{s}^\mathrm{i} = \frac{\tau_\mathrm{i}}{\tau_\mathrm{i} + \tau_0} D_\mathrm{E}. \tag{4.57}$$

There are no correlations in the interstitial lattice, or in other words, the expected correlation factor equals unity, because the interstitial lattice is essentially empty and, except in special cases close to phase transitions, $\tau_\mathrm{i} \ll \tau_0$. The ratio τ_i/τ_0 can thus be replaced by N_i/N where N_i is the number of self-interstitial (anti-structure) atoms. Since the ratio N_i/N depends on the Gibbs energy of formation of a self-interstitial, ΔH_i, again in a Boltzmann way, the temperature dependence of the self-diffusion coefficient is given by

$$D_\mathrm{s}^\mathrm{i}(T) = D_{\mathrm{E},0} \exp\left(\frac{\Delta S_\mathrm{i}}{R}\right) \exp\left(\frac{-\Delta H_\mathrm{i}}{RT}\right) \exp\left(\frac{-E_\mathrm{a}}{k_\mathrm{B}T}\right). \tag{4.58}$$

This kind of diffusion mechanism cannot usually be investigated by means of a bulk microscopic method like NMR or QENS, because the fraction of self-interstitials of the order of ppb or ppm is far too low and cannot be separated from the contribution of the static atoms. The only possibility for a microscopic method is to measure in the proximity of order–disorder phase transitions where the anti-structure sites are well populated (see H diffusion in V_2H, Section 8.2.3).

Finally we consider the *(direct) interstitial* mechanism as it is called, e.g., by Mehrer (1991). Foreign atoms that are located exclusively in interstitial sites of an otherwise perfect crystal may diffuse simply by jumping from interstitial site to interstitial site. Prominent representatives are the different hydrogen isotopes, carbon, nitrogen, and oxygen. The host lattice atoms are not involved; the host lattice is, however, indirectly involved via lattice distortions, small polaron effects, etc. Then we are dealing with a lattice gas, lattice liquid, or lattice fluid, depending on concentration and temperature. Contrary to other one-component systems in metal hydrides the concentration (number of H atoms/number of available sites) can be varied from virtually zero to virtually one, a scientifically very attractive feature. In the limit $c \to 1$ the interstitial mechanism is equivalent to the vacancy mechanism, and generally the interstitial mechanism can be considered as a vacancy mechanism on the interstitial lattice.

4.2.3 Concentration dependence of the self-diffusion coefficient

The concentration dependence of the self-diffusion coefficient is due to site blocking and to correlation effects:

$$D_s(c) = V(c) f_t(c) D_E. \tag{4.59}$$

Blocking can involve self-site blocking only, see Eq. (4.14), or multiple site blocking; it is taken into account by the same site availability factor $V(c)$ which also appears in the expression of the chemical diffusion coefficient, Eq. (4.26). For hydrogen diffusion in bcc metals this factor is displayed in Fig. 4.2. The Einstein diffusion coefficient $D_E = l^2/6\tau$ contains the mean residence time in the empty lattice, τ; some authors combine this quantity with the site availability factor such that

$$\tau_{\text{eff}}(c) = \tau / V(c) \tag{4.60}$$

is the mean residence time at the lattice site for a given concentration. Looking back we recognize that Eq. (4.55) represents an expression of the site availability factor for the vacancy mechanism of metal atom diffusion in metals, and thus the corresponding diffusion coefficient $D_s^v(T)$ (Eq. 4.56) contains such a site availability factor in accordance with Eq. (4.59). Considering the limiting cases we notice that $\lim_{c \to 0} V(c) = 1$, i.e. site blocking is irrelevant for low concentrations (hydrogen diffusion in dilute α phase metal hydrides, for example). On the other hand, in the full lattice $V(c)$ and thus the diffusion coefficient would vanish. In reality this does not happen: actually for entropy reasons a lattice cannot be completely full; in addition, another diffusion process can become dominant, for instance an interstitial mechanism on the interstitial lattice.

Correlations between subsequent diffusional jumps are a common phenomenon. We consider the vacancy mechanism in a cubic lattice. Provided that an atom or ion jumps once per lapse of time τ_{eff} and that the distance and direction of the ith jump is given by the vector $\boldsymbol{r}_i$, its location $\boldsymbol{R}$ after n jumps ($n = t/\tau_{\text{eff}}$) from the original position may be given by

$$\boldsymbol{R} = \boldsymbol{r}_1 + \boldsymbol{r}_2 + \cdots + \boldsymbol{r}_n; \tag{4.61}$$

$\langle \boldsymbol{R}^2 \rangle$, the mean square displacement, may be expressed by

$$\langle \boldsymbol{R}^2 \rangle = (\boldsymbol{r}_1 + \boldsymbol{r}_2 + \cdots + \boldsymbol{r}_n)^2 = \sum_{i=1}^{n} r_i^2 + 2 \sum_{i=1}^{n-1} \sum_{j=i+1}^{n} r_i r_j \cos \Theta_{i,i+j} \quad (4.62)$$

where $\Theta_{i,i+j}$ denotes the angle between the direction of the ith jump and the $(i + j)$th jump. Assuming the jump length is identical for all jumps ($|\boldsymbol{r}_i| = l =$ constant), we have

$$\begin{aligned}\langle R(t)^2 \rangle &= nl^2 + 2nl^2 \langle \cos \Theta_{i,i+1} \rangle + 2nl^2 \langle \cos \Theta_{i,i+2} \rangle + \cdots \\ &= nl^2 \{1 + 2\langle \cos \Theta_{i,i+1} \rangle + 2\langle \cos \Theta_{i,i+2} \rangle + \cdots\} \\ &= \frac{l^2}{\tau_{\text{eff}}} t \cdot f_{\text{t}}(c). \end{aligned} \quad (4.63)$$

Utilizing the relations $D_{\text{s}} = \langle R(t)^2 \rangle / 6t$ (Eq. 4.49) and $D_{\text{E}} = l^2/6\tau$ (Eq. 4.2), and replacing τ_{eff} by $\tau / V(c)$ (Eq. 4.60), we end up with Eq. (4.59). For an empty lattice all jump directions occur with the same probability; therefore $\langle \cos \Theta \rangle = 0$ and $f_{\text{t}}(0) = 1$; usually f_{t} is smaller than unity. We remember that the blocking factor in this case equals unity, $V(0) = 1$, and thus for the empty lattice the self-diffusion coefficient equals the Einstein diffusion coefficient. Compaan and Haven (1958) have shown that

$$\langle \cos \Theta_{i,i+k} \rangle = \langle \cos \Theta_{i,i+1} \rangle^k \quad (4.64)$$

and thus

$$\begin{aligned} f_{\text{t}} &= 1 + 2\langle \cos \Theta_{i,i+1} \rangle + 2\langle \cos \Theta_{i,i+1} \rangle^2 + 2\langle \cos \Theta_{i,i+1} \rangle^3 + \cdots \\ &= 1 + 2\langle \cos \Theta_{i,i+1} \rangle \{1 + \langle \cos \Theta_{i,i+1} \rangle + \langle \cos \Theta_{i,i+1} \rangle^2 + \cdots\} \\ &= 1 + \frac{2\langle \cos \Theta_{i,i+1} \rangle}{1 - \langle \cos \Theta_{i,i+1} \rangle} \end{aligned}$$

from which one finally obtains:

$$f_{\text{t}} = \frac{1 + \langle \cos \Theta_{i,i+1} \rangle}{1 - \langle \cos \Theta_{i,i+1} \rangle}. \quad (4.65)$$

For a concentrated lattice $\langle \cos \Theta_{i,i+1} \rangle$ deviates from zero and therefore the tracer correlation factor f_{t} deviates from unity because of correlation effects. A correlation between successive jumps appears since, after one jump of the self or tracer atom, the vacancy that promotes this jump is with certainty behind the self or tracer atom. It then effects more easily a backward jump of the self or tracer atom than a forward or sideways jump, resulting in a negative average $\langle \cos \Theta \rangle$ between two jumps so that $f_t < 1$. Note that in a non-cubic crystal the correlation factor depends on the direction of diffusion in the crystal. In the same way that generally $D_x \neq D_y \neq D_z$, one obtains in the calculations of $\langle x^2 \rangle$, $\langle y^2 \rangle$ or $\langle z^2 \rangle$ with the use of formulae of the type Eqs. (4.62) and (4.63) that $f_x \neq f_y \neq f_z$.

In the limit of vanishing vacancy concentration (fully occupied lattice) f_t can be calculated analytically. For the vacancy mechanism, Compaan and Haven (1958) obtained the following values:

	$-\langle \cos \Theta \rangle$	f_t
Diamond	0.3333	0.500
Simple cubic	0.20894	0.65311
Body-centred cubic	0.15793	0.72722
Face-centred cubic	0.12268	0.78146
Hexagonal honey comb	0.5	0.3333

The tracer correlation factor not only depends on the lattice type, but also on the atomistic diffusion mechanism. Therefore, from a measurement of f_t a hint of the actual diffusion mechanism is obtained; see below. For finite concentrations, $f_t(c)$ cannot be calculated exactly and various approximations are used. Monte Carlo simulations are particularly valuable because various blocking conditions can be taken into account. Fig. 4.7 displays the tracer correlation factor $f_t(x)$ for hydrogen in NbH_x for self-site blocking only and for multiple-site blocking up to the first, second, and third coordination sphere (Faux and Ross 1987).

The effects described until now are based, following Bardeen and Herring (1952), on the local configuration around a tagged atom migrating via point defects. When the mean square displacement is of interest, this leads to the introduction of a *spatial correlation* factor. Mean square displacements are essential for all kinds of macroscopic diffusion measurements and for QENS at small momentum transfer, see Chapter 5. If, on the other hand, effects connected to the jump frequency are measured, a *time-dependent correlation* has to be introduced, to connect the jump frequency with the effective frequency. This is the case for NMR T_1 experiments; it would also hold for QENS at large momentum transfer, but from those data diffusion coefficients are not usually derived.

In a gas, Bardeen–Herring type effects clearly do not exist, but there are so-called *dynamic correlations* between successive 'jumps' of a molecule because of the memory of its motion before the collision that produced the new 'jump'. In a crystalline solid, once the jump is effected, the energy of the jumping atom is dissipated into the lattice, i.e. the atom is 'thermalized' in its new site. Successive jumps of the particle are thus *dynamically uncorrelated.* But dynamic effects are possible at high temperatures. When the mean residence time between two jumps is of the same order of magnitude as the time it takes the atom to travel the interatomic distance at thermal velocity, the distribution of 'forward' and 'reverse' jumps, after a first jump has taken place, no longer correponds to a random distribution. The

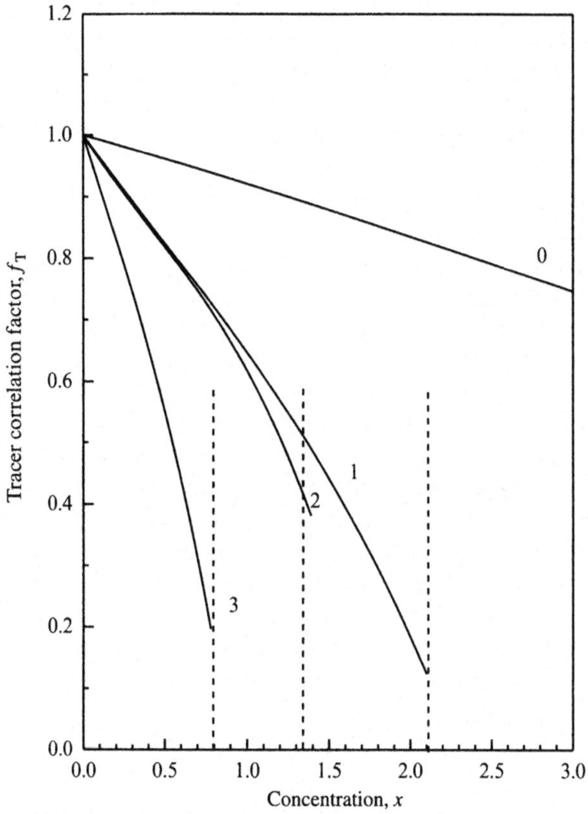

Fig. 4.7 Tracer correlation factor for hydrogen in NbH_x for self-site blocking only and for multiple-site blocking up to the first, second and third coordination sphere (Faux and Ross 1987, with permission from IOP Publishing Ltd) 0: self-site blocking only, 1, 2, 3: blocking of one/two/three coordination spheres.

forward jumps are far more frequent, as if the atom carried out double or multiple jumps at one time (see Section 8.2.1 on the diffusion of H in bcc metals at elevated temperatures).

Finally we comment on physical correlation (Sato and Kikuchi 1971; Murch 1980). This leads to the physical or intrinsic or, as we prefer to call it, to the *mobility correlation factor* f_m which appears in the expression of the chemical and conductivity diffusion coefficients, Eqs. (4.26) and (4.42), respectively (therefore sometimes also called the conductivity correlation factor). The chemical potential gradient (or the electrical field gradient, respectively) modifies the potential barriers around a site; this superimposes a drift velocity on the random walk of the diffusing particles due to the resulting net jump rate down the gradient. The *mobility correlation factor* f_m (Faux and Ross 1987) accounts for departures from a random walk with superimposed drift due to the correlated motion of indistinguishable particles. It should be stressed that the first jump is entirely random, because any external force is negligibly small compared to the height of the barrier between two sites. Thus f_m describes the change in the total particle flux when subsequent rearrangement jumps are included. Such a correlated motion of the indistinguishable particles is generated by the presence of short-range order between diffusing atoms. This may arise due to interactions between the diffusing atoms, or to the presence of different types of sites with different energies associated with them. Multiple-site blocking leads to short-range order at finite concentrations. In this case the atoms find particular modes of packing that are more efficient than others. Under these circumstances any disordering jump will then occur at a slower rate than the reverse jump. On the other hand, in the case of a gradient model with self-site blocking only and thus without short-range order, there is no mobility correlation, i.e. $f_m = 1$. Note that f_t will always deviate more from unity than will its counterpart f_m. This is because in the presence of a potential gradient an atom that has just jumped due to the gradient has its tendency of returning to its original site counteracted by the enhanced probability that atoms in front of it will move down the gradient and that atoms behind will diffuse into the space that has been left. For ionic conductors, on the one hand, the self-diffusion coefficient D_s (Eq. 4.59) can be measured using radioactive tracer isotopes (neglecting isotope effects); on the other hand, the conductivity diffusion D_σ (Eq. (4.42)) is obtained from conductivity measurements. The ratio of both diffusion coefficients yields the so-called Haven ratio

$$H_R = D_t/D_\sigma = f_t/f_m \tag{4.66}$$

or f_{cor} as some authors (Boureau 1985; Potzel *et al.* 1984) prefer to call it. Its value indicates a certain diffusion mechanism: a value of $H_R = 0.3$, for example, excludes a simple vacancy mechanism.

4.3 STOCHASTIC ASPECTS OF DIFFUSION

In this section the self-diffusion of a particle is described by stochastic methods. That is, probability concepts are used, and the information about the particle's dynamics is contained in a statistical quantity called a probability distribution on the lattice (Haus and Kehr 1987). The dynamics is formulated either in terms of enumerating the individual transitions of the particle from one site to another (a random walk description) or in terms of a set of rate equations for the probability distribution, the so-called master equation.

4.3.1 Some stochastic concepts

As an introduction to the stochastic treatment of solid state diffusion we first recall the definition of some stochastic quantities. For more details we refer to the books of van Kampen (1992), Haken (1982), and Röpke (1987), for instance.

The definition of a stochastic variable A consists in specifying the set of possible values (called 'set of states' or 'sample space' $\{a\}$) and the probability distribution over this set. In our case we essentially identify A with two kind of sets: (i) the set of lattice points $\boldsymbol{r}$ which is discrete and either one, two, or three dimensional; and (ii) the number of jumps of a tagged particle $n = \{0, 1, \ldots, n\}$ within a certain time; here the number is discrete, but the time is usually continuous. The probability distribution or, in short, the probability, is given by a function $P(a)$ that is normalized to unity. For continuous variables this would be a probability density, and the probability that A has a value between a and $a + \mathrm{d}a$ would be $P(a)\,\mathrm{d}a$; however, henceforth we shall deal mostly with discrete variables. The change of a state a with time is called a *process*. If we consider the time dependence of the value a, for a particular particle, then this is called a particular *realization* or complexion or sample function of this process and can alternatively be given by a sequence $\{a_i(t_i)\}, i = 1, \ldots, n$.

A *stochastic process* $A(t)$ means the ensemble of all possible realizations. To each possible sequence $\{a_i(t_i)\}$ we attribute a *joint probability* $P_n(a_1t_1; \ldots; a_nt_n)$. It can be reduced to lower joint probabilities by summing up:

$$\sum_{a_m} P_m(a_1t_1; \ldots; a_mt_m) = P_{m-1}(a_1t_1; \ldots; a_{m-1}t_{m-1}). \qquad (4.67)$$

The reduced joint probabilities $P_2(a_1t_1; a_2t_2), \ldots, P_n(a_1t_1; \ldots; a_nt_n)$ describe the probabilities, that a_1 occurs at t_1 and a_2 occurs at t_2, and, etc. The single time probability $P_1(a_1t_1) = P(a, t)$ is often called the a priori probability and is normalized.

The conditional probability

$$\begin{aligned} &p_{m|n}(a_1t_1; \ldots; a_nt_n|a_{n+1}t_{n+1}; \ldots; a_{n+m}t_{n+m}) \\ &\quad = P_{n+m}(a_1t_1; \ldots; a_{n+m}t_{n+m})/P_n(a_1t_1; \ldots; a_nt_n) \end{aligned} \qquad (4.68)$$

indicates the probability of the occurrence of a_{n+1} at $t_{n+1}, \ldots, a_{n+m}$ at t_{n+m} under the condition that a_1 occurred at $t_1, \ldots, a_n$ occurred at t_n. A special case of a conditional probability is the conditional transition probability $p_{n|1}(a_1t_1; \ldots; a_nt_n|a_{n+1}, t_{n+1})$ which indicates the transition probability to the state a_{n+1} at t_{n+1} if a_1 at $t_1, \ldots, a_n$ at t_n occurred. It can be shown that a stochastic process is determined by all conditional transition probabilities $p_{n|1}$ and by the a priori probability P_1.

For the classification of stochastic processes two cases are of special importance:

(a) Perfect random process

$$P_n(a_1t_1; \ldots; a_nt_n) = P_1(a_1t_1)P_1(a_2t_2)\cdots P_1(a_nt_n) \qquad (4.69)$$

$$p_{n|1}(a_1t_1; \ldots; a_nt_n|a_{n+1}t_{n+1}) = P_1(a_{n+1}t_{n+1}).$$

The perfect random process does not contain any correlations between events at different times. Thus there is no causality between different times.

(b) Markov process, $t_1 < t_2 < \cdots < t_{n+1}$:

$$p_{n|1}(a_1t_1; \ldots; a_nt_n|a_{n+1}t_{n+1}) = p_{1|1}(a_nt_n|a_{n+1}t_{n+1}). \tag{4.70}$$

The transition from the state at time t_n to the state at time t_{n+1} is independent of previous events. Just the a priori probability $P_1(a_0t_0)$ at a predetermined initial time t_0 and the conditional transition probability $p_{1|1}(a_1t_1|a_2t_2)$ are sufficient to completely determine this stochastic process. In particular we have

$$P_n(a_1t_1; \ldots; a_nt_n) = P_1(a_1t_1)\, p_{1|1}(a_1t_1|a_2t_2) \cdots p_{1|1}(a_{n-1}t_{n-1}|a_nt_n). \tag{4.71}$$

The Markov process is the simplest way to describe a causality relation between different times.

4.3.2 Master equation

Since for Markov processes transitions within two consecutive time intervals occur independently of each other, the time development is determined only by the actual state, not by previous ones. Mathematically, this can be expressed by a differential equation of first order, the so-called master equation. For its derivation we consider (with $t_1 < t_2 < t_3$) the reduction of the triple time joint probability

$$P_3(a_1t_1; a_2t_2; a_3t_3) = P_1(a_1t_1)\, p_{1|1}(a_1t_1|a_2t_2)\, p_{1|1}(a_2t_2|a_3t_3). \tag{4.72}$$

Summing up over of all states a_2 yields

$$P_2(a_1t_1; a_3t_3) = \sum_{a_2} P_1(a_1t_1)\, p_{1|1}(a_1t_1|a_2t_2)\, p_{1|1}(a_2t_2|a_3t_3) \tag{4.73}$$

and, after division by the a priori probability $P_1(a_1t_1)$, we obtain an expression for the conditional transition probability:

$$p_{1|1}(a_1t_1|a_3t_3) = \sum_{a_2} p_{1|1}(a_1t_1|a_2t_2)\, p_{1|1}(a_2t_2|a_3t_3). \tag{4.74}$$

This important equation, valid for Markov processes, is known as the *Chapman–Kolmogorov equation* (or sometimes the Smoluchovski equation) and means that the transition probability for two steps consecutive in time equals the transition probabilities from the initial state to an intermediate state multiplied by the transition probability from the intermediate to the final state, summed over all intermediate states, see Fig. 4.8.

Another form of the Chapman–Kolmogorov equation is obtained if Eq. (4.74) is multiplied by $P_1(a_1t_1)$ and summed over all a_1. Using

$$p_{1|1}(a_1t_1|a_3t_3)\, P_1(a_1t_1) = P_2(a_1t_1; a_3t_3). \tag{4.75}$$

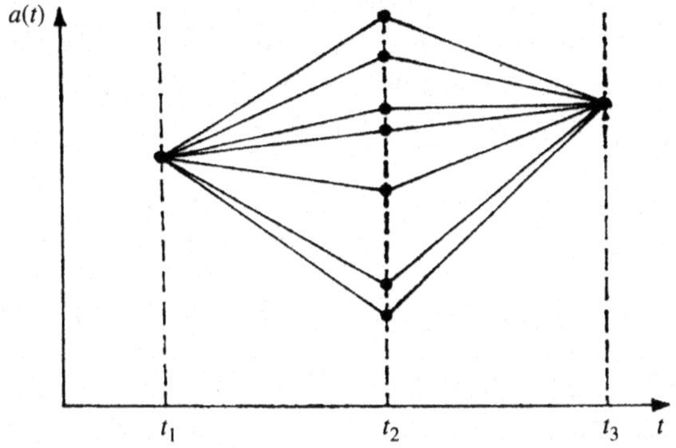

Fig. 4.8 Chapman–Kolmogorov equation: one has to sum up over all possible intermediate states.

we obtain

$$P_1(a_3t_3) = \sum_{a_2} p_{1|1}(a_2t_2|a_3t_3)\, P_1(a_2t_2). \tag{4.76}$$

We change the indices ($a_3 \to a, a_2 \to a'$) and henceforth omit the subscript 1 in the a priori probability:

$$P(a,t) = \sum_{a'} p_{1|1}(a't'|at)\, P(a',t'). \tag{4.77}$$

Now we consider infinitesimal time periods, i.e. we replace t by $t' + \tau$ and consider the expression

$$\frac{1}{\tau}[P(a,t+\tau) - P(a,t)]$$

in the limit $\tau \to 0$. In other words, we form the time derivative of Eq. (4.77) and obtain

$$\dot{P}(a,t) = \sum_{a'} \dot{p}_t(a',a)\, P(a',t) \tag{4.78}$$

with

$$\dot{p}_t(a',a) = \lim_{\tau\to 0} \tau^{-1}\big[p_{1|1}(a't'|a,t+\tau) - p_{1|1}(a't'|at)\big].$$

The discussion of $\dot{p}_t$ requires some care. If $a' \neq a$, then $\dot{p}_t$ is just the number of transitions from a' to a per second, which we express as

$$\dot{p}_t(a',a) = \Gamma_{a'a}; \tag{4.79}$$

thus Γ is the transition rate or jump rate. (In the following we always assume that Γ is time independent. Then the solution of the differential equation (4.78) is a simple exponential, from which by a recursion formalism the Poissson distribution is derived, see Eq. (4.96).) If, on the other hand, $a' = a$, i.e. if we consider $\dot{p}_t(a, a)$, then the discussion is more difficult. We have to go back to the definition of the conditional transition probability for a Markov process. Let us consider the difference

$$p_{1|1}(a, t'|a, t + \tau) - p_{1|1}(at'|at). \tag{4.80}$$

This difference means the change of the probability of finding a particle at the point a (or of finding the system in state a, respectively) at a later time $t + \tau$, if it has been at that point at time t. The corresponding change of the probability is caused by those events where the particle leaves its original site. If we divide Eq. (4.80) by τ and consider the limit $\tau \to 0$, then Eq. (4.78) must represent the sum over the rates for the transitions from a to all other states (the minus sign indicates a decrease of the probability):

$$\dot{p}_t(a, a) = -\sum_{a'} \Gamma_{aa'}. \tag{4.81}$$

Now we insert Eqs. (4.79) and (4.81) into Eq. (4.78) and obtain the so-called master equation

$$\frac{\partial P(a, t)}{\partial t} = \sum_{a' \neq a} \{\Gamma_{a'a} P(a', t) - \Gamma_{aa'} P(a, t)\}. \tag{4.82}$$

The first term on the right-hand side of Eq. (4.82) describes the increase of the probability of the system in the state a due to transitions out of other states a' into the state a. The second term correspondingly contains the decrease of the probability.

Markov processes are of such great importance because on the one hand they are sufficiently simple to allow a mathematical treatment and on the other hand they are sufficiently general to be valid for many applications, not only in physics. A precondition is that for small times Δt the transition probabilities are proportional to Δt. The proportionality constant is, in the case of diffusion, the jump rate Γ which is thus assumed to be time independent. Thus from the short-time behaviour of a system, characterized by the transition or jump rate, the long-time behaviour can be derived by means of the master equation, i.e. by stochastic means. This will be outlined in the remaining part of this section whereas the theoretical understanding of the jump rate and thus of the single diffusive step will be considered in the next section.

4.3.3 Application to jump diffusion

We now apply the above formalism to jump diffusion of a lattice gas. Here we consider the simplest case, a Bravais lattice, where all sites are energetically and crystallographically equivalent. We allow only jumps to the z nearest neighbour sites, which are separated from the site at $\boldsymbol{r}$ under consideration by the jump vector $\boldsymbol{s}_j$, $j = 1, \ldots, z$. The jump rate, Γ, is constant (precondition), which in this case is identical for all jumps, and the mean residence time on a site, τ, is related by $\Gamma = 1/z\tau$. The master equation, Eq. (4.82), is expressed as

$$\frac{\partial}{\partial t} P(\boldsymbol{r}, t) = \frac{-1}{z\tau} \sum_{j=1}^{z} \{P(\boldsymbol{r}, t) - P(\boldsymbol{r} + \boldsymbol{s}_j, t)\}. \tag{4.83}$$

Usually the boundary condition

$$P(\boldsymbol{r}, 0) = \delta(\boldsymbol{r}) \tag{4.84}$$

is valid. Equation (4.83) can easily be solved in Fourier space. With the Fourier transform

$$\int P(\boldsymbol{r}, t) \mathrm{e}^{\mathrm{i}\boldsymbol{Q}\boldsymbol{r}} \, \mathrm{d}^3 r = P(\boldsymbol{Q}, t) \tag{4.85}$$

and, replacing $\boldsymbol{r} + \boldsymbol{s}_j$ by $\boldsymbol{r}'$, with

$$\int P(\boldsymbol{r}', t) \mathrm{e}^{\mathrm{i}\boldsymbol{Q}\boldsymbol{r}'} \mathrm{e}^{-\mathrm{i}\boldsymbol{Q}\boldsymbol{s}_j} \, \mathrm{d}^3 r' = \mathrm{e}^{-\mathrm{i}\boldsymbol{Q}\boldsymbol{s}_j} P(\boldsymbol{Q}, t) \tag{4.86}$$

we obtain a differential equation for $P(\boldsymbol{Q}, t)$:

$$\frac{\partial}{\partial t} P(\boldsymbol{Q}, t) = \frac{-1}{z\tau} \sum_{j=1}^{z} (1 - \mathrm{e}^{-\mathrm{i}\boldsymbol{Q}\boldsymbol{s}_j}) P(\boldsymbol{Q}, t). \tag{4.87}$$

The boundary condition $P(\boldsymbol{r}, 0) = \delta(\boldsymbol{r})$ is Fourier transformed to $P(\boldsymbol{Q}, 0) = 1$; then the solution of Eq. (4.87) is obviously

$$P(\boldsymbol{Q}, t) = \exp\left\{ -\frac{t}{z\tau} \sum_{j=1}^{z} (1 - \mathrm{e}^{-\mathrm{i}\boldsymbol{Q}\boldsymbol{s}_j}) \right\}. \tag{4.88}$$

In the present connection Fourier transformation is just a mathematical procedure, but in the next chapter the Fourier variable Q will be identified with the scattering vector $\boldsymbol{Q} = \boldsymbol{k}_0 - \boldsymbol{k}_1$, and thus the above formalism will be used to derive the incoherent scattering function for jump diffusion; the choice of the scattering vector probes the function P. This application of the master equation is due to Chudley and Elliott (1961).

In order to show that the master equation for a jump diffusion process behaves properly in the macroscopic limit, we consider, for the sake of simplicity, one-dimensional diffusion in the x-direction, with the lattice constant or jump length

a. Since the coordination number is 2, the jump rate is given by $\Gamma = 1/2\tau$ where τ is the mean residence time on a site. Then the master equation reads

$$\frac{dP(x,t)}{dt} = \frac{1}{2\tau}[P(x-a,t) + P(x+a,t) - 2P(x,t)]. \tag{4.89}$$

The macroscopic limit means that the lattice constant a becomes very small; then Eq. (4.89) can be developed in a power series around $P(x,t)$:

$$\begin{aligned}\frac{dP(x,t)}{dt} &= \frac{1}{2\tau}\{P(x,t) + P(x,t) - 2P(x,t) \\ &\quad + [P'(x,t) - P'(x,t)]a \\ &\quad + \tfrac{1}{2}[P''(x,t) + P''(x,t)]a^2\}.\end{aligned} \tag{4.90}$$

In the limit $a \to 0$, the zero and first-order terms vanish. Then from the *microscopic* master equation we obtain the *macroscopic* diffusion equation, i.e. Fick's second law:

$$\frac{dP(x,t)}{dt} = \frac{a^2}{2\tau}\frac{d^2P(x,t)}{dx^2} = D_E\frac{d^2P(x,t)}{dx^2} \tag{4.91}$$

where, however, other than for collective diffusion, the Einstein diffusion coefficient or, at finite concentrations, the self-diffusion coefficient, respectively, enters. Generalization to three dimensions is straightforward.

For a random walk the Chudley–Elliott result Eq. (4.88) can alternatively be derived by recursion relations (Gissler and Rother 1970; for a comprehensive review see Haus and Kehr 1987). If a particle has moved from a lattice site $\boldsymbol{l}$ to a lattice site $\boldsymbol{n}$ within the time t, it can have done this in $1, 2, \ldots, \nu, \ldots$ jumps; therefore the conditional probability of finding the particle at lattice site $\boldsymbol{n}$ at time t when it started at site $\boldsymbol{l}$ at $t = 0$ is decomposed into the contributions of different numbers ν of transitions or jumps:

$$P(\boldsymbol{l}, 0|\boldsymbol{n}, t) = \sum_{\nu=0}^{\infty} P_\nu(\boldsymbol{l}, 0|\boldsymbol{n}, t). \tag{4.92}$$

$P\nu(\boldsymbol{l}, 0|\boldsymbol{n}, t)$ is the conditional probability of finding the particle at site $\boldsymbol{n}$ at time t when it performed exactly ν steps. For a Markov process the spatial transition probabilities and the time dependence are separable:

$$P_\nu(\boldsymbol{l}, 0|\boldsymbol{n}, t) = P_\nu(\boldsymbol{l}|\boldsymbol{n})p_\nu(t); \tag{4.93}$$

$p_\nu(t)$ is the probability that exactly ν steps have been performed until time t. In the following we will derive this probability by means of a recursion relation and thus obtain the well-known Poisson distribution. The probability $p_0(t)$ that the particle has not yet performed a transition until time t when it arrived at a site at $t = 0$

is $\exp(-t/\tau)$ which results from Eq. (4.78); thereby the sum over all jump rates $\Gamma_{aa'}$ (Eq. 4.81) is the negative reciprocal mean residence time on this site. Hence

$$p_0(t) = \exp(-t/\tau) \tag{4.94}$$

and

$$p_\nu(t) = \int_0^t \mathrm{d}t' \exp\left\{-\left(\frac{t-t'}{\tau}\right)\right\}\frac{1}{\tau}p_{\nu-1}(t'). \tag{4.95}$$

$1/\tau$ is the transition rate at an (arbitrary) time point t', and the first factor in the integral is the probability that no further transitions occur between t' and t. Mathematically, Eq. (4.95) is a ν-fold convolution integral which for more complex cases (see the treatment of the two-state model in Chapter 5) is solved by a Laplace transformation converting a convolution into a product. The infinite sum in Eq. (4.92) thus becomes a geometric series in Laplace space which can be expressed in a short formula. The present case is simple: Eq. (4.95) can be directly recursively integrated yielding

$$p_\nu(t) = \frac{1}{\nu!}\left(\frac{t}{\tau}\right)^\nu \exp\left(-\frac{t}{\tau}\right). \tag{4.96}$$

This is the Poisson distribution; it is valid for so-called Poisson processes, i.e. processes where for randomly distributed times the stochastic variable $\nu = 0, 1, \ldots$ grows only in steps of 1.

Now we consider the first factor on the right-hand side of Eq. (4.93). $P_\nu(\boldsymbol{l}|\boldsymbol{n})$ is the conditional spatial probability of finding the particle at lattice site $\boldsymbol{n}$ after ν steps when it started at site $\boldsymbol{l}$. The following recursion is obvious from the Chapman–Kolmogorov equation:

$$P_\nu(\boldsymbol{l}|\boldsymbol{n}) = \sum_{\boldsymbol{m}} P_{\nu-1}(\boldsymbol{l}|\boldsymbol{m})\, P_1(\boldsymbol{m}/\boldsymbol{n}). \tag{4.97}$$

Thus the probability $P_\nu(\boldsymbol{l}|\boldsymbol{n})$ can be expressed by iteration as a ν-fold product over the so-called spatial transition probabilities P_1 (the discrete analogue to convolution integral); in the simplest case (only nearest neighbour jumps, all sites energetically equivalent) the spatial transition probabilities are given by

$$P_1(\boldsymbol{m}|\boldsymbol{n}) = \begin{cases} 1/z & \text{if } \boldsymbol{m} \text{ is a nearest neighbour of } \boldsymbol{n} \\ 0 & \text{otherwise} \end{cases} \tag{4.98}$$

where z is the number of neighbouring sites (the coordination number). The Fourier transform of the spatial transition probability $P_1(\boldsymbol{m}|\boldsymbol{n})$ is defined as

$$\begin{aligned} p(\boldsymbol{Q}) &= \sum_{\langle n,m\rangle} \exp\{-\mathrm{i}\boldsymbol{Q}(\boldsymbol{R}_n - \boldsymbol{R}_m)\}P_1(\boldsymbol{m}|\boldsymbol{n}) \\ &= \sum_{i=1}^{z} \exp(-\mathrm{i}\boldsymbol{Q}\boldsymbol{s}_j)\frac{1}{z}; \end{aligned} \tag{4.99}$$

the notation $\langle \boldsymbol{n}, \boldsymbol{m} \rangle$ designates $\boldsymbol{m}$ as the nearest-neighbour site of $\boldsymbol{n}$; thus $\boldsymbol{R}_n - \boldsymbol{R}_m$ equals the jump vector $\boldsymbol{s}_j$. In Fourier space the iteration Eq. (4.97) leads to a ν-fold product

$$P_\nu(\boldsymbol{Q}) = (p(\boldsymbol{Q}))^\nu = p^\nu(\boldsymbol{Q}). \tag{4.100}$$

Now Eq. (4.92) can be written in Fourier space:

$$P(\boldsymbol{Q}, t) = \sum_{\nu=0}^{\infty} p^\nu(\boldsymbol{Q}) \frac{1}{\nu!} \left(\frac{t}{\tau}\right)^\nu \exp\left(-\frac{t}{\tau}\right). \tag{4.101}$$

If we consider this as the Taylor expansion of an exponential function we can rewrite it as

$$\begin{aligned} P(\boldsymbol{Q}, t) &= \exp\left(-\frac{t}{\tau}\right) \exp\left(p(\boldsymbol{Q})\frac{t}{\tau}\right) \\ &= \exp\left\{-\frac{t}{\tau} + \frac{t}{z\tau}\sum_{j=1}^{z} \mathrm{e}^{-\mathrm{i}\boldsymbol{Q}\boldsymbol{s}_j}\right\} \\ &= \exp\left\{-\frac{t}{z\tau}\sum_{j=1}^{z}(1 - \mathrm{e}^{-\mathrm{i}\boldsymbol{Q}\boldsymbol{s}_j})\right\}. \end{aligned} \tag{4.102}$$

This result is identical to the Chudley–Elliott result Eq. (4.88). Of course, the recursion relations and the master equation must yield identical results since the underlying assumptions of a Markov process are the same, as well as the basic transition probabilities.

For disordered systems, percolation is a standard theory (Bunde and Havlin 1996; Stauffer and Aharony 1992). Its applications range from transport in amorphous and porous media and composites to the properties of branched polymers, gels, and complex ionic conductors. Percolation clusters at the critical concentration are self-similar on all length scales and their structures as well as several substructures can be described with the concept of fractal dimensions (Bunde and Kantelhardt 1998). Because the clusters have loops and dangling ends on all length scales, diffusion processes on these structures are slowed down and become anomalous: the mean square displacement is sublinear in time. Such diffusional behaviour is also called subdiffusive.

Superdiffusive transport behaviour can be described in terms of Lévy flights or Lévy walks (Schlesinger *et al.* 1995; Klafter *et al.* 1996). A Lévy flight constituates a non-Brownian random motion which can be most easily visualized by considering a particle that at regularly spaced intervals performs jumps in random directions with jump length l drawn from a distribution having a slowly decaying algebraic tail. Thus the particle can perform rare but anomalously long flights which significantly contribute to the mean square displacement which turns out to be superlinear in time.

4.4 THE SINGLE DIFFUSIVE STEP

In the preceding section we have derived the long-time diffusional behaviour from the short-time behaviour, i.e. from the single diffusive step, which was treated, so to speak, as a black box characterized by one parameter, the jump rate Γ. The remaining task in this chapter about solid state diffusion is to investigate this single diffusive step. We start with a classical over-barrier jump process and show that this process, which is adequate for all but the lightest atoms, is closely related to lattice vibrations. Subsequently, we consider the single diffusive step for light interstitial atoms which exhibits, depending on the temperature, more or less pronounced quantum effects.

4.4.1 Classical over-barrier jump processes

Figure 4.9 displays interstitial diffusion in a schematic two-dimensional representation and the corresponding one-dimensional diffusion or reactive coordinate. We start with purely classical considerations. The interstitial particle in a minimum of the potential energy at x_0 has to surmount the energy barrier (which in three dimensions is a saddle point) midway between x_0 and x_1 in order to reach its neighbouring site at x_1. According to the conventional transition state theory (CTST) which goes back to Stearn and Eyring (1940) we consider, for an ensemble of solid state systems consisting of N particles, a quasichemical equilibrium between the initial state and the so-called activated state at the barrier during the short life time of the activated state; then we treat the motion through the saddle point as a very loose vibration. The quasichemical equilibrium between the particle in the initial state and in the activated state may be expressed by the equation

$$\frac{f^{\neq}}{f} = K^{\neq} \tag{4.103}$$

where $f^{\neq}$ is the fraction of systems with the particle under consideration in the activated state and $f \approx 1$ the fraction of systems in the ground state, and where $K^{\neq}$ means the chemical equilibrium constant. This constant may be expressed in terms of partition functions:

$$f^{\neq} = \frac{q^{\neq}}{q} \mathrm{e}^{-\Delta E/kT}, \tag{4.104}$$

where ΔE is the difference between the zero-point energy in the activated state and in the initial state. Here and in the following equations we omit the subscript B on the Boltzmann constant k. Both our initial state and our activated state systems consist of N particles and have $3N$ degrees of freedom which are all vibrational degrees of freedom. One of the activated system's degrees, vibration along the reactive coordinate, has a different character from the rest, since it

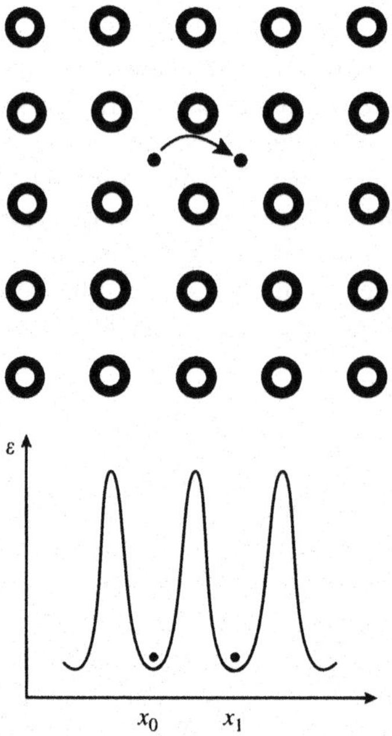

Fig. 4.9 Interstitial diffusion in a schematical two-dimensional representation and the corresponding one-dimensional diffusion path or reactive coordinate.

corresponds to such a loose vibration that there is no restoring force: the activated state decomposes without any restraint into the final state with the particle in the lattice plane at x_1, see Fig. 4.9. This particular vibrational degree of freedom is therefore replaced by a translational one; i.e. the transition state in the direction of the diffusion path is considered as a one-dimensional box of length δ. The one-dimensional translational partition function is given by

$$q_{\text{trans}} = \frac{(2\pi mkT)^{1/2}}{h}\delta. \tag{4.105}$$

We include this expression in the partition function $q^{\neq}$ and write the remaining product of factors as $q_{\neq}$ which now refers to only $3N - 1$ vibrational degrees of freedom. Equation (4.104) thus becomes

$$f^{\neq} = \frac{(2\pi mkT)^{1/2}}{h}\delta\frac{q_{\neq}}{q}\mathrm{e}^{-\Delta E/kT}. \tag{4.106}$$

The mean thermal velocity of the particle in the one-dimensional box is given by

$$\langle\dot{x}\rangle = \left(\frac{kT}{2\pi m}\right)^{1/2}. \tag{4.107}$$

Then the frequency $\langle \dot{x} \rangle / \delta$ indicates how often the activated particle passes through its box, i.e. how frequently the particle overcomes the potential barrier; this equals the rate of decomposition of the transition state. The product of 'fraction of particles in the activated state' times 'decomposition rate' represents the jump rate of diffusion:

$$\Gamma = \frac{\langle \dot{x} \rangle}{\delta} f^{\neq} = \frac{kT}{h} \frac{q_{\neq}}{q} \mathrm{e}^{-\Delta E/kT} \tag{4.108}$$

which is the CTST formula. The first factor, kT/h, has the dimension of a frequency and amounts to $6 \cdot 10^{12}\,\mathrm{s}^{-1}$ at room temperature. The second factor, $q_{\neq}/q \equiv \exp(S^{\mathrm{m}}/k)$, contains the migration entropy, and the third factor, $\exp(-\Delta E/kT) \equiv \exp(-H^{\mathrm{m}}/kT)$, contains the migration enthalpy. Since the migration entropy is a vibrational entropy, it is straightforward that solid state diffusion is connected with the phonons in the solid. Actually, a correlation between small Debye temperatures Θ_{D} (soft phonons) and high diffusivities is empirically well established (Tewary 1973; Klotsman 1983).

The connection between diffusion and lattice vibrations can be formulated in a comparatively simple way for high temperatures. There kT is much larger than $h\nu_i$ where ν_i is a typical phonon frequency. Therefore the partition functions for the different vibrational degrees of freedom are expanded in a power series, and higher-order terms are neglected:

$$\lim_{h\nu/kT \to 0} \left(\frac{1}{\mathrm{e}^{h\nu/kT} - 1} \right) = \frac{1}{(1 + h\nu/kT) - 1} = \frac{kT}{h\nu}. \tag{4.109}$$

Hence the diffusional jump rate yields (Vineyard 1957; LeClaire 1966)

$$\Gamma = \frac{kT}{h} \frac{\prod_{j=1}^{3N-1} (kT/h\nu_j^{\neq})}{\prod_{j=1}^{3N} (kT/h\nu_j)} \mathrm{e}^{-\Delta E/kT} = \frac{\prod_{j=1}^{3N} \nu_j}{\prod_{j=1}^{3N-1} \nu_j^{\neq}} \mathrm{e}^{-\Delta E/kT}. \tag{4.110}$$

From the $3N$ factors of the system in the initial state we separate the frequency ν_0 of the vibration along the reactive coordinate and obtain

$$\Gamma = \nu_0 \frac{\prod_{j=1}^{3N-1} \nu_j}{\prod_{j=1}^{3N-1} \nu_j^{\neq}} \mathrm{e}^{-\Delta E/kT} = \Gamma_0 \mathrm{e}^{-E_{\mathrm{a}}/kT}. \tag{4.111}$$

For a cubic crystal (sc, bcc, or fcc with jump lengths l of a_0, $a_0\sqrt{3}/2$, or $a_0/\sqrt{2}$, respectively, and coordination numbers z of 6, 8, or 12, respectively), the Einstein diffusion coefficient is given by

$$D_{\mathrm{E}} = \frac{l^2}{6\tau} = a_0^2 \Gamma = a_0^2 \Gamma_0 \mathrm{e}^{-E_{\mathrm{a}}/kT} = D_{\mathrm{E},0} \mathrm{e}^{-E_{\mathrm{a}}/kT}; \tag{4.112}$$

thus Eqs. (4.109) and (4.111) represent the classical derivation of the Arrhenius law for the temperature dependence of the diffusion coefficient, see Eq. (4.4). The jump rate Γ contains the pre-exponential factor Γ_0, which is—see Eq. (4.111)—composed of a vibrational frequency, the so-called attempt frequency, and an exponential function containing the migration entropy:

$$\Gamma_0 = \nu_0 e^{S^m/k} e^{-H^m/kT} = \nu_0 e^{G^m/kT} \tag{4.113}$$

where the high-temperature limit of the migration entropy is given by

$$S^m = k_B \ln \left(\frac{\prod_{j=1}^{3N-1} \nu_j}{\prod_{j=1}^{3N-1} \nu_j^{\neq}} \right). \tag{4.114}$$

For the vacancy mechanism of diffusion (see Eq. 4.56) the pre-exponential factor additionally contains an exponential with the vacancy formation entropy. For the high-temperature limit we obtain in a completely analogous way

$$S_v^f = k_B \ln \left(\frac{\prod_{j=1}^{3N} \nu_j}{\prod_{j=1}^{3N} \nu_j^v} \right) \tag{4.115}$$

by considering, for an ensemble of solid state systems consisting of N particles, the quasichemical equilibrium between an undistorted state and a state with a vacancy; both states have $3N$ regular vibrational degrees of freedom.

In general $S_v^f > 0$ since a softening of the lattice is expected when an empty lattice site is introduced. This is clear from the argument that cutting off a bond that connects an atom to its neighbours leads to a reduction of the vibrational frequency of the atom. The eigenmode frequencies ν_j can be calculated from the force constants ϕ_{ij} known from Born–von Karman fits to measured phonon dispersion curves, see Chapter 9 (Schober *et al.* 1992). The entropy of the defect crystal can be calculated by taking one atom out of a model crystal with periodic boundary conditions. The vacancy can then be mimicked by the missing force constants from the empty lattice site to all sites. An equivalent simulation for the calculation of the migration entropy S^m is difficult because the curvature of the potential at the saddle point is not known.

4.4.2 Quantum diffusion of light particles

Barrier limitation of kinetic processes is a phenomenon not only encountered in solid state diffusion, but also in chemical reactions, in nuclear reactions, and in biological transport. The Arrhenius law provides a very good description of the temperature dependence of these processes for the high and 'normal' temperatures usually encountered in natural environments. But the exponential decrease towards zero rate for the processes predicted by the Arrhenius law in the limit of high $1/T$ values is not always in agreement with experiment. Here, effects of quantum tunnelling through the barriers manifest themselves. They provide additional possibilities for reactions or transfers to occur at low temperatures, although most often at very low rates. However, for the translational motion of light interstitials, quantum effects are pronounced.

In classical diffusion theory the activation energy is taken as the difference in potential energy between the equilibrium and saddle point configurations. Due to the small mass, there is, in the case of interstitial hydrogen, also a considerable change in vibrational energy (zero-point motion) between the two configurations. The activation energy is the difference between the respective zero-point energy levels and, therefore, is isotope-dependent. The sign of this isotope effect will depend on the local structure and on the material. In the case of the Pd/H(D) system, for example, this effect suffices to explain the observed non-classical behaviour.

The basic theories for quantum diffusion of light particles in metals were derived some 20 years ago; for a recent review we refer to Schober (1995). The starting point is the Born–Oppenheimer approximation where it is assumed that the electrons adiabatically follow the nuclei and thus do not need to be considered. Therefore the diffusion process comprises a coupling of the light particle to the phonons of the host system. This defect–phonon problem is treated in the 'classical' papers on phonon-assisted tunnelling by Flynn and Stoneham (1970), Kagan and Klinger (1974), and Teichler and Seeger (1981). This work originates from the small polaron theory of Yamashita and Kurosawa (1958) and of Holstein (1959).

The elementary steps of hydrogen diffusion are treated as transitions between bound states at the potential minima of the interstitial sites. The particles are surrounded by lattice distortions and diffuse together with them; the quasiparticle 'interstitial atom + distortion cloud' is termed a *small polaron*. These small polarons are formed because hydrogen atoms push apart the surrounding metal atoms by 0.1–0.3 Å in order to diminish their potential energy, see Fig. 4.10. Since now the adjacent interstitial site exhibits a different site energy, direct tunnelling is impossible. If, however, certain lattice modes move the metal atoms adjacent to the interstitial atom apart by an amount q such that for a short period a so-called coincidence configuration is achieved, then tunnelling to the neighbouring site is possible, see Fig. 4.10. For temperatures above about half of the Debye temperature, many phonons contribute to the coincidence configuration; the probability of its occurrence is given by

$$P(q) \propto T^{-1/2} \exp\left(-\frac{E_q}{k_B T}\right) \tag{4.116}$$

where E_q is the potential energy of a classical harmonic oscillator $(E_q = m\omega^2 q^2/2)$ and the prefactor is due to normalization. Hence the diffusion coefficient is proportional to the product of the square of the tunnelling matrix

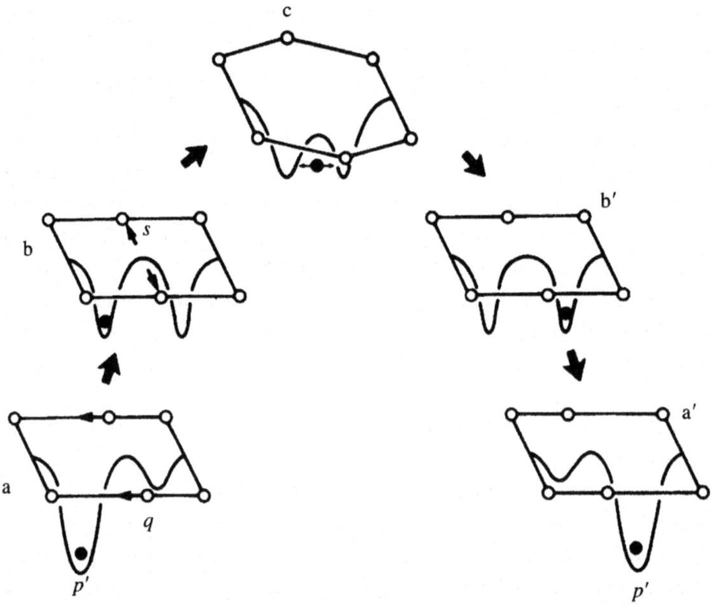

Fig. 4.10 Schematic representation of the elementary steps of hydrogen or muon diffusion for the phonon assisted tunnelling. Top: beyond the Condon-approximation; this mechanism is valid for hydrogen, deuterium, and tritium. Middle: tunnelling of bare muons out of a coincidence configuration; top and middle mechanisms are phonon-assisted processes, i.e. incoherent tunnelling. Bottom: coherent tunnelling of small polarons which is inhibited by phonons (from Hempelmann 1984, with permission from Elsevier Science).

element and the probability of occurrence of the coincidence configuration

$$D \propto J_0^2 T^{-1/2} \exp\left(-\frac{E_q}{k_B T}\right). \tag{4.117}$$

Here J_0 denotes the tunnelling matrix element for the bare interstitial atom in the coincidence configuration, i.e. according to Fermi's golden rule the transition matrix element $\langle 2|\boldsymbol{H}|1\rangle$ for the transition of the particle from site 1 to site 2 where $|1\rangle$ and $|2\rangle$ are the wavefunctions localized on sites 1 and 2, respectively, and $\boldsymbol{H}$ is the Hamiltonian of the system. The above result correctly describes the diffusion of positive muons in metals at elevated temperatures, but disagrees both quantitatively and qualitatively with the experimental results for the heavier hydrogen isotopes shown in Fig. 4.11. It turned out to be necessary to also consider those lattice modes which by pushing apart certain metal atoms 'open the door' between adjacent interstitial sites and thus lower the energy barrier. In other words, the naked tunnelling probability cannot be taken to be independent of the state of the phonons, i.e. the temperature. This so-called Condon approximation holds as long

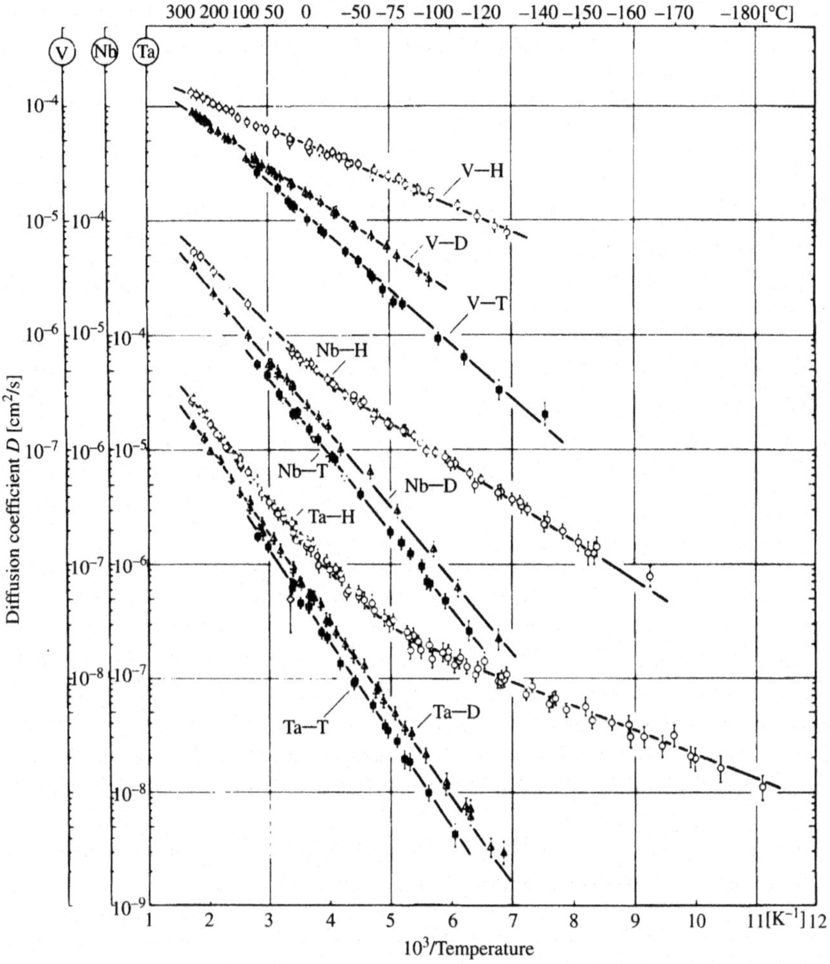

Fig. 4.11 Diffusion coefficients of the hydrogen isotopes H, D, and T in bcc metals as a function of inverse temperature. Note the large isotope effect and the low thermal activation energies (from Zh Qi *et al.* 1983, with permission from IOP Publishing Ltd).

as only long-wavelength phonons are populated, and breaks down at temperatures above about 50 K. Diffusion is only possible out of configurations where adjacent metal atoms have been displaced at least by an amount S. By integration of an expression analogous to Eq. (4.116) we obtain

$$P(s) \propto T^{1/2} \exp\left(-\frac{E_s}{k_B T}\right). \tag{4.118}$$

The diffusion coefficient is eventually proportional to the product of the probability of occurrence of the coincidence configuration and the critical distortion

$$D \propto J_0^2 \exp\left(-\frac{E_q + E_s}{k_B T}\right). \tag{4.119}$$

The activation energy is mainly a property of the metal lattice and is at most weakly dependent on the hydrogen isotope, whereas the tunnelling matrix element differs for different isotope masses, thus causing a strongly isotope-dependent prefactor.

At still higher temperatures the protons are excited so highly that the retardation by the barrier is negligible and the tunnelling matrix element does not limit the transition rate

$$D \propto \exp\left(-\frac{E_a}{k_B T}\right). \tag{4.120}$$

This is called the case of *adiabatic diffusion*. The metal lattice is more highly activated, $E_a > E_q + E_s$, than in the previous case of *non-adiabatic diffusion*. With increasing hydrogen isotope mass increasingly high energy coincidences are needed for diffusion; this explains the isotope dependence of the activation energy for the high-temperature hydrogen diffusion in bcc metals; in contrast the prefactor does not depend on the hydrogen isotope. The kink in the Arrhenius plot in Fig. 4.11 is, in the framework of this concept, related to the transition from non-adiabatic diffusion at somewhat lower temperatures to adiabatic diffusion at more elevated temperatures.

At temperatures immediately below the validity regime of phonon-assisted tunnelling the diffusion coefficient follows a power law

$$D \propto T^n \tag{4.121}$$

and diffusion is no longer a multiphonon process. For a two-phonon process $n = 7$ or $n = 3$ is obtained, depending on the details of the model and the relevant lattice properties (Flynn and Stoneham 1970; Fujii 1979). In an ideal crystal, for reasons of energy and momentum conservation, two phonons are necessary. In real crystals, however, due to impurities and lattice distortions, adjacent sites are no longer energetically equivalent, and one-phonon processes no longer violate conservation laws. If ΔE denotes the energy difference of two adjacent sites, then for $k_B T \gg \Delta E$, Teichler and Seeger (1981) obtained

$$D \propto (\Delta E)^2 k_B T \tag{4.122}$$

i.e. the diffusion coefficient is proportional to T and explicitly depends on ΔE.

The same temperature dependence and a slightly modified ΔE dependence also holds for jumps between crystallographically non-equivalent sites (for instance octahedral site → tetrahedral site). If one assumes that for doped but otherwise distortion-free materials the energy differences of adjacent sites depend on the long-range elastic interactions of the dopants ($V(r) \propto r^{-3}$), then

$$\Delta E \propto c_{\mathrm{d}}^{4/3} \tag{4.123}$$

with the dopant concentration c_{d}. All these phonon-assisted diffusion processes are called *incoherent* because there is no phase relationship of the particles' wave-function before and after the site change. The corresponding temperature range is sometimes called the *hopping regime. Phonon-assisted* means that phonons are conducive to the diffusion process: the more phonons exist (i.e. the higher the temperature), the faster diffusion proceeds.

Coherent tunnelling, on the other hand, was predicted to occur at low temperatures. If the light interstitial is in the small polaron state, it is possible that the whole small polaron tunnels to its neigbouring site as outlined in the bottom part of Fig. 4.10. Since the particle is now dressed with a distortion field comprising the surrounding host metal atoms, the matrix element is strongly reduced:

$$J = J_0 \mathrm{e}^{-S(T)}. \tag{4.124}$$

A simple estimate (Kehr 1984) yields

$$S(0) = \frac{5E_{\mathrm{a}}}{2\hbar\omega_{\mathrm{D}}} \tag{4.125}$$

where E_{a} is the activation energy mentioned above and ω_{D} the Debye frequency. For muons in aluminium, values for S(0) are 1.73 or 2.16 depending on whether one prefers the experimental value of E_{a} or a theoretical estimate. The tunnelling matrix element is decreased by one order of magnitude. The temperature dependence of $S(T)$ is given by the factor $(k_{\mathrm{B}}t/\hbar\omega_{\mathrm{D}})^4$ and is thus negligible at low temperatures (Stoneham 1972).

In a perfect crystal all equivalent sites are degenerated, and delocalized states (Bloch states) can be formed. By phonons this coherent propagation is inhibited, and in the framework of the polaron theory Kagan and Klinger (1974) predicted the temperature dependence

$$D \propto J^2 T^{-9}. \tag{4.126}$$

Contraproductive to coherent propagation are also electrons ($D \propto T^{-1}$) and impurities ($D \propto T$ or T^0). Furthermore, elastic lattice distortions inhibit the formation of band-like states because in this way the lattice sites are no longer degenerated (Kagan and Klinger 1974; Kehr *et al.* 1982). For muons in metals those coherent tunnelling states possibly exist in the superconducting state. These effects have been observed for muonium in KCl and NaCl, i.e. in systems without conduction electrons (Kiefl *et al.* 1989; Kadono *et al.* 1990). In this temperature regime phonons obstruct diffusion, i.e. with increasing temperature diffusion slows down due to the increasing number of phonons.

For muons in aluminium and copper at low temperatures a dependence of $D \propto T^{-0.7}$ and $T^{-0.6}$, respectively, had been measured; the first conjecture (Kehr *et al.* 1982) was that scattering on electrons inhibits coherent propagation. An estimate of this process, however, yielded a much too large value of the diffusion coefficient (Kehr 1984; Jäckle and Kehr 1983). Kondo (1984) and Yamada (1984) explained the temperature dependence: the electrons of the shielding cloud around the positive muon are not able to follow the muon adiabatically since the shielding clouds of two adjacent sites at $T = 0\,\mathrm{K}$ are mutually orthogonal (the orthogonality catastrophe, Anderson 1967). The very fundamental Born–Oppenheimer approximation, recognized as valid everywhere in condensed matter and molecular physics, is not valid here. Kondo calculated for the diffusion coefficient

$$D_\mu = \frac{4a^2 J_{\mathrm{eff}}^2}{\gamma_{\mathrm{eff}}} \tag{4.127}$$

with

$$J_{\mathrm{eff}} = J\left(\frac{\pi k_{\mathrm{B}} T}{E_{\mathrm{F}}}\right)^K = J_0 \mathrm{e}^{-S(T)}\left(\frac{\pi k_{\mathrm{B}} T}{E_{\mathrm{F}}}\right)^K$$

and with $\gamma_{\mathrm{eff}} = \pi K k_{\mathrm{B}} T$. E_{f} denotes the Fermi energy and K is an electron–particle coupling parameter. The tunnelling matrix element is reduced by the factor $(\pi k_{\mathrm{B}} T/E_{\mathrm{f}})^K$ which takes into account the non-adiabaticity of the shielding cloud. Since this reduction is appreciable, the formation of Bloch waves does not occur at measurable temperatures but the particle tunnels incoherently. The other aspect of the electron–particle interaction is dissipation which is taken into account by γ_{eff}; it denotes energetic level broadening. The mechanism of the broadening is essentially equal to that of the Korringa relaxation of solid state NMR spectroscopy. The particle can be scattered only on electrons near the Fermi energy, since at energies below E_{f} no unoccupied final states are available. The density of states of the electrons capable of interaction is proportional to the temperature. Hence in summary the temperature dependence of the diffusion coefficient is given by

$$D \propto T^{2K-1}. \tag{4.128}$$

Such a power law of the diffusion constant was observed for the long-range transport of μ^+ in Cu (Hartmann *et al.* 1980; 1981; Kadono *et al.* 1989) and Al (Hartmann *et al.* 1988) and, later, for the motion between two local potential minima for H in Nb doped with O or N (Wipf *et al.* 1987; Steinbinder *et al.* 1988; 1991) and in Sc (Anderson *et al.* 1990).

Experimentally, the dissipation to a conduction electron bath can be switched off by a transition to the superconducting state where the energy gap is an obstacle for all low-energy inelastic processes. For muon investigations the appropriate host is Al ($T_c = 1.2\,K$) and for hydrogen tunnelling studies Nb ($T_c = 9\,K$). In the superconducting state the normal conducting γ_{eff} in Eq. (4.127) has to be replaced by

$$\gamma_s = \frac{T}{1 + \exp(\Delta/k_B T)} \tag{4.129}$$

where Δ denotes the gap in the electronic energy levels due to superconductivity. Thus the expected strong exponential increase of the muon diffusion coefficient is in contrast to experimental observation.

Since narrow-band propagation is very sensitive to perturbations of the periodic potentials, Anderson localization (Anderson 1967) is expected to occur if the mean energy spread $\langle\xi\rangle$ of the particle levels exceeds the bandwith. Earlier treatments have considered the energy shifts ξ as randomly distributed over the lattice, but as pointed out by Kagan and Prokof'jev (1991) and recently more explicitly by Prokof'jev (1994), this is an inadequate approach since the inhomogeneities are in all practical cases concentrated around impurity centres in the host crystals, which cause elastic expansions ($dE/dr \propto r^{-3}$) for the metallic materials. Both phenomena affect the energy levels strongly near the centres of the impurities but allow regions of volume in between the centres where the perturbations are relatively weak, as long as the impurity concentration is low enough to avoid appreciable overlaps of impurity affected regions. For the distorted host lattice regions Eq. (4.127) has thus to be expanded:

$$D_\mu = \frac{4a^2 J_{eff}^2 \gamma_s}{\gamma_s^2 + \langle\xi\rangle^2}. \tag{4.130}$$

Thus at temperatures slightly below T_c the dissipative broadening γ_s could still exceed the mean energy spread $\langle\xi\rangle$ and Eq. (4.130) transforms back into Eq. (4.127), but at appreciably lower temperatures $\langle\xi\rangle \gg \gamma_s$, the muon diffusion coefficient decreases exponentially, and the muon becomes weakly localized. In the undisturbed host lattice region, however, coherent tunnelling might develop where the muon Bloch waves are elastically scattered at the defective regions without losing their phase relationship (Karlsson 1995).

5

Incoherent quasielastic neutron scattering

Quasielastic neutron scattering (QENS) is, strictly speaking, slightly inelastic scattering at Q values between Bragg reflections; it is centred at $\hbar\omega = 0$, but energetically broadened, see Fig. 5.1. As was pointed out in Chapter 3, QENS is connected to the van Hove correlation functions; because of their importance some definitions are given here again: $G(\boldsymbol{r}, t)$ is the conditional probability density that, given a particle at time $t = 0$ at the origin $\boldsymbol{r} = \boldsymbol{0}$, *any* particle is found at time t at the position $\boldsymbol{r}$ in a volume element $d\boldsymbol{r}$. 'Any particle' means the same or a different particle, i.e. the correlation function contains a self and a distinct part:

$$G(\boldsymbol{r}, t) = G_{\mathrm{s}}(\boldsymbol{r}, t) + G_{\mathrm{d}}(\boldsymbol{r}, t). \tag{5.1}$$

The correlation function $G(\boldsymbol{r}, t)$ is related to the intermediate scattering function, $I(\boldsymbol{Q}, t)$, by a Fourier transformation in space and to the coherent scattering function $S(\boldsymbol{Q}, \omega)$ by a further Fourier transformation in time. The solid state physics which can be extracted from *coherent* quasielastic neutron scattering will be treated in Chapter 6.

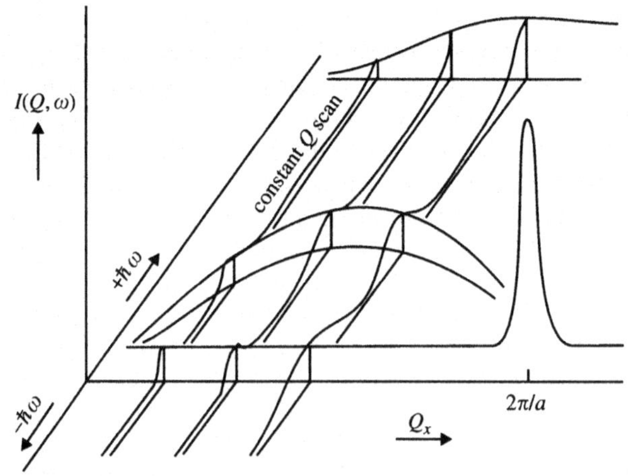

Fig. 5.1 Neutron scattering intensity in the $\hbar\omega - Q_x$ plane showing quasielastic scattering, inelastic scattering due to acoustical and optical phonons and Bragg scattering; also displayed are three constant Q scans.

Table 5.1 Elements (or isotopes) suited for incoherent QENS and their incoherent neutron scattering cross-sections, σ_i, in comparison with the coherent ones, σ_c; elements in brackets possess a coherent scattering cross-section larger than their incoherent one

Element	H	(D)	^{7}Li	(Na)	(Cl)	(Ar)	K	(Sc)	Ti	V	Cr
σ_i (barn)	79.9	2.0	0.7	1.6	5.2	0.2	0.3	4.7	2.7	5.2	1.8
σ_c (barn)	1.8	5.6	0.6	1.7	11.5	0.5	1.7	19.0	1.4	0.0	1.7

Element	Co	(Mn)	(Ni)	(Cu)	(Ag)	(In)	(La)	Nd	(Yb)	(W)
σ_i (barn)	4.8	0.4	5.2	0.5	0.6	0.5	1.1	11	3.0	2.0
σ_c (barn)	0.8	1.8	13.3	7.5	4.4	2.1	8.5	7.4	19.3	2.9

In the present chapter we are interested in the self-correlation function, $G_s(\boldsymbol{r}, t)$, which is the conditional probability density that, given a particle was at time $t = 0$ at the origin $\boldsymbol{r} = \boldsymbol{0}$, the *same* particle is found at time t at the position $\boldsymbol{r}$ in a volume element $d\boldsymbol{r}$. Analogously, the self-correlation function $G_s(\boldsymbol{r}, t)$ is related to the self-part of the intermediate scattering function, $I_s(\boldsymbol{Q}, t)$, and to the incoherent scattering function, $S_i(\boldsymbol{Q}, \omega)$, by Fourier transformations in space and time. Quasielastic neutron scattering is due to stochastic (non-periodic) motions of atoms, whereas periodic atomic motions give rise to inelastic neutron scattering.

In Table 5.1 those elements or isotopes, respectively, are listed which possess an incoherent neutron scattering cross-section sufficiently large for incoherent QENS.

Some elements consist entirely (Na, Co) or almost entirely (H, V) of a single isotope with $I \neq 0$ and exhibit spin incoherence. Hydrogen and vanadium are unique in having $\sigma_i \gg \sigma_c$. This results from the fact that b^+ and b^- are of opposite sign for these nuclides. On the other hand, isotope incoherence dominates for elements that consist entirely (Ar) or almost entirely (Ti, W, Ni) of isotopes with $I = 0$. Some elements or isotopes, like ^{6}Li, ^{10}B and many rare earth metals, exhibit appreciable incoherent neutron scattering cross-sections but are unsuited for neutron scattering experiments because of their huge absorption cross-sections.

The single-particle dynamics of solids, in addition to possibly occurring diffusional motions, always comprise vibrations. For solids we can presume that diffusional and vibrational motions are dynamically independent of each other because they occur on clearly different time scales. The self-part of the intermediate scattering function is obtained as the product of the intermediate scattering functions for vibrations and for diffusion:

$$I_s(\boldsymbol{Q}, t) = I_s^{\mathrm{v}}(\boldsymbol{Q}, t) \cdot I_s^{\mathrm{d}}(\boldsymbol{Q}, t). \tag{5.2}$$

By Fourier transformation into the frequency domain a convolution (symbolized by $\otimes$) of scattering functions is obtained:

$$S_i^{\mathrm{tot}}(\boldsymbol{Q}, \omega) = S_i^{\mathrm{v}}(\boldsymbol{Q}, \omega) \otimes S_i^{\mathrm{d}}(\boldsymbol{Q}, \omega). \tag{5.3}$$

Henceforth the superscript d on the diffusional scattering function is omitted.

The scattering function of the vibrations contains an elastic term and inelastic terms, see Section 3.4. The inelastic terms are outside the energy window of quasielastic neutron scattering and, therefore, are not considered here. The elastic term is called the Debye–Waller factor; it does not change the shape of the quasielastic scattering function, but only reduces its intensity. For details see Section 3.5.

The diffusional scattering function generally contains contributions originating from rotational diffusion and from translational diffusion, briefly denoted as rotation and diffusion.

For a general discussion of the incoherent quasielastic neutron scattering function with respect to rotation and diffusion we start from an expression derived in Chapter 3 for the self-part of the intermediate scattering function:

$$I_\mathrm{s}(\boldsymbol{Q}, t) := \frac{1}{N} \sum_{\substack{j \\ \text{particles}}} \left\langle \mathrm{e}^{-\mathrm{i}\boldsymbol{Q}\boldsymbol{R}_j(0)} \mathrm{e}^{\mathrm{i}\boldsymbol{Q}\boldsymbol{R}_j(t)} \right\rangle. \tag{5.4}$$

The brackets mean a thermal average, and the sum runs over all the N particles in the system. The correlations, which self-evidently exist for the positions of a tagged particle at short time intervals, are lost at infinite times such that the average in Eq. (5.4) can be formed separately for each factor:

$$I_\mathrm{s}(\boldsymbol{Q}, \infty) = \frac{1}{N} \sum_j \left\langle \mathrm{e}^{-\mathrm{i}\boldsymbol{Q}\boldsymbol{R}_j(0)} \mathrm{e}^{\mathrm{i}\boldsymbol{Q}\boldsymbol{R}_j(\infty)} \right\rangle = \frac{1}{N^2} \sum_i \sum_j \left\langle \mathrm{e}^{-\mathrm{i}\boldsymbol{Q}\boldsymbol{R}_j} \right\rangle \left\langle \mathrm{e}^{\mathrm{i}\boldsymbol{Q}\boldsymbol{R}_i} \right\rangle. \tag{5.5}$$

Because of translational and time invariance the average formations simply yield

$$I_\mathrm{s}(\boldsymbol{Q}, \infty) = \frac{1}{N^2} \left| \sum_j \mathrm{e}^{+\mathrm{i}\boldsymbol{Q}\boldsymbol{R}_j} \right|^2 \tag{5.6}$$

where N still means the number of particles in the system. But now we perform the ergodic transition from the ensemble to the time average, such that the sum now runs over all the sites which a tagged particle visits during its stochastic motion; N now means the number of these sites, and we obtain

$$I_\mathrm{s}(\boldsymbol{Q}, \infty) = \frac{1}{N^2} \left| \sum_{\substack{j \\ \text{sites}}} \mathrm{e}^{\mathrm{i}\boldsymbol{Q}\boldsymbol{R}_j} \right|^2 =: F(\boldsymbol{Q}); \tag{5.7}$$

this quantity is called the elastic incoherent structure factor. For a spatially restricted motion (like rotation) the number of sites is finite and so is $I_\mathrm{s}(\boldsymbol{Q}, \infty)$; for long-range diffusion with a huge number of sites involved $I_\mathrm{s}(\boldsymbol{Q}, \infty)$ vanishes. $I_\mathrm{s}(\boldsymbol{Q}, 0)$, on the other hand, equals unity in both cases. Therefore, as shown in

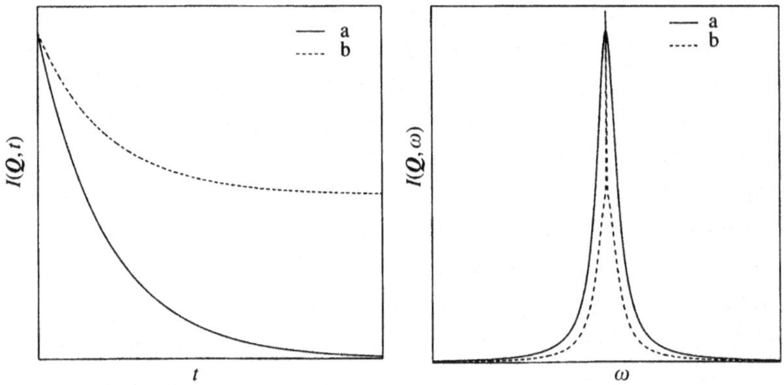

Fig. 5.2 Left: self-part of the intermediate scattering functions for a localized motion (a) and for long-range diffusion (b). Right: Fourier transforms, the incoherent quasielastic scattering function for a long-range diffusion (a) and for a localized motion (b).

Fig. 5.2, the self-part of the intermediate scattering function is a function which equals unity at $t = 0$ and decays to a finite value for a localized motion, but decays to zero for long-range diffusion as $t \to \infty$. Its Fourier transform consists of an elastic term and quasielastic terms in the case of localized motion and of only quasielastic contributions in the case of long-range diffusion. Therefore, an elastic scattering component observed in an incoherent neutron scattering spectrum indicates localized motion provided that the elastic scattering is not due to the immobile host lattice.

Rotations in solids are, strictly speaking, not the topic of the present book; a comprehensive presentation is given by Bée (1988). But in some complex solids clusters of equienergetic sites occur which are energetically well separated from other sites; an example is the hexagon of tetrahedral interstitial sites in $LaNi_5$, see Section 8.2.4. The diffusing particle spends some time τ_0 on such a cluster of sites and during this time performs a spatially restricted motion which formally equals a rotation. If the time τ_0 is large compared to the residence time on the other sites, then 'rotation' and diffusion can be considered as dynamically independent, i.e. the self-part of the intermediate scattering function is a product and the incoherent scattering function the convolution of the rotational and diffusional contributions. It should be emphasized, however, that a complete treatment of jump diffusion over energetically different sites yields a scattering function which 'automatically' comprises the 'rotational' motion.

The general strategy for the interpretation of incoherent QENS spectra as shown in Fig. 5.3 is very often applied; here we presume that necessary experimental corrections (see Chapter 7) have already been performed. From model-independent data evaluation, from knowledge of the system to be investigated, and from intuition one develops an idea about the diffusion mechanism. This idea is cast into a mathematical model, mostly a differential equation or a system of differential

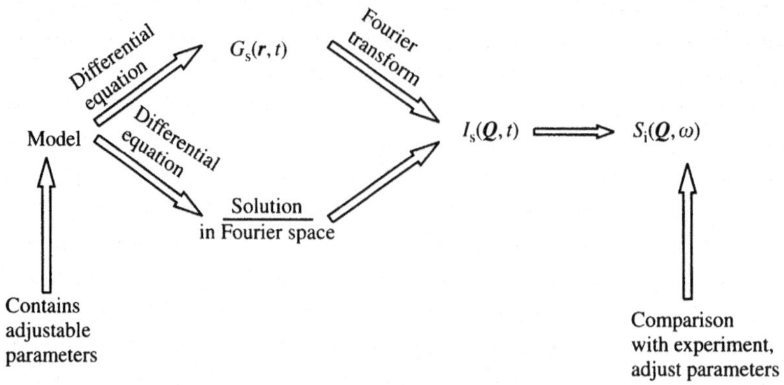

Fig. 5.3 General strategy for the evaluation of quasielastic neutron scattering data.

equations, for the self-correlation function; the differential equation can be solved in some cases immediately, see for instance Section 5.2, and yields the self-correlation function $G_S(\boldsymbol{r}, t)$; this function is analytically Fourier transformed into the incoherent scattering function. The free parameters in the model are adjusted in such a way that the best possible agreement with the experiment is achieved (a least squares fit). More numerous are cases where the set of differential equations cannot be solved immediately. Often, however, it can be solved after Fourier transformation with respect to space; see for instance Section 5.3. This property can be utilized because the Fourier transformation has to be performed anyway: the solution is the self-part of the intermediate scattering function which is subsequently analytically Fourier transformed with respect to time. Thus in these cases the self-correlation function is not known analytically. In other cases, see for instance Section 5.5, it can be favourable to primarily perform a Laplace transformation instead of a Fourier transformation and afterwards transcribe it into the Fourier transform.

In the following we first discuss spatially restricted jump diffusion processes (Section 5.1), because for complex translational jump diffusion processes we have to refer back to the former cases. In Section 5.2 we consider long-range translational diffusion, which in cubic solids is isotropic, but generally in solids it might be anisotropic. Section 5.3 deals with jump diffusion on primitive sub-lattices, whereas complex sublattices are the content of Section 5.4. Finally, in Section 5.5, we discuss diffusion in disordered systems. It has to be emphasized, however, that the microscopic mechanism of diffusion in amorphous systems is still an unsolved theoretical problem.

5.1 SPATIALLY LIMITED JUMP DIFFUSION, EISF

The overwhelming majority of solids are compounds and consist of two or more elements. In many of these systems the diffusing species (atom, ion) occupies crystallographically different sites. These different sites can also be energetically different. If several equivalent sites are adjacent and separated from other sites—by a larger distance and/or a different energy—then these sites can form a kind of trap: the diffusing particle jumps back and forth inside this trap, i.e. it performs a spatially limited jump diffusion. Of course the particle will eventually manage to escape because of thermal fluctuations. If the residence time in the extended 'trap' is much larger than the time for free diffusion then both types of motion can be considered as dynamically independent such that the combined incoherent neutron scattering function is the convolution of both separated scattering functions; otherwise both kinds of motion have to be cast into a complex jump diffusion model, see Section 5.5.

Here we treat spatially limited jump diffusion separately and consider several cases.

5.1.1 Jump model between two sites

The simplest case is the so-called dumb-bell, where the particle jumps back and forth back between two *isoenergetical* sites, at $\boldsymbol{r}_1$ and $\boldsymbol{r}_2$. We start with the master equation (4.83), which has to be applied to both sites; the coordination number z equals unity:

$$\frac{\partial P(\boldsymbol{r}_1, t)}{\partial t} = \frac{-1}{\tau}\{P(\boldsymbol{r}_1, t) - P(\boldsymbol{r}_2, t)\} \tag{5.8a}$$

$$\frac{\partial P(\boldsymbol{r}_2, t)}{\partial t} = \frac{-1}{\tau}\{P(\boldsymbol{r}_2, t) - P(\boldsymbol{r}_1, t)\} \tag{5.8b}$$

where $P(\boldsymbol{r}, t)$ means the probability of finding the particle at site $\boldsymbol{r}$ at time t. Since the particle with certainty is found on one of the two sites and the probabilities are normalized, we obtain the boundary conditions

$$P(\boldsymbol{r}_1, t) + P(\boldsymbol{r}_2, t) = 1 \tag{5.9}$$

and consequently

$$\frac{\partial}{\partial t}(P(\boldsymbol{r}_1, t) + P(\boldsymbol{r}_2, t)) = 0. \tag{5.10}$$

For the solution we make the following ansatz

$$\begin{aligned} P(\boldsymbol{r}_1, t) &= A + B\mathrm{e}^{-2t/\tau} \\ P(\boldsymbol{r}_2, t) &= A - B\mathrm{e}^{-2t/\tau} \end{aligned} \tag{5.11}$$

where the coefficients A and B have to be determined from the initial conditions. Assuming that the particle was initially at $\boldsymbol{r}_1$ at time $t = 0$

$$\begin{aligned} P(r_1, 0) &= A + B = 1 \\ P(r_2, 0) &= A - B = 0 \end{aligned} \tag{5.12}$$

we obtain the conditional probabilities (see Section 4.3)

$$\begin{aligned} p_{1|1}(r_1, 0|r_1, t) &= \tfrac{1}{2}(1 + e^{-2t/\tau}) \\ p_{1|2}(r_1, 0|r_2, t) &= \tfrac{1}{2}(1 - e^{-2t/\tau}). \end{aligned} \tag{5.13}$$

Vice versa, with the particle initially at $\boldsymbol{r}_2$, we get

$$\begin{aligned} p_{2|1}(r_2, 0|r_1, t) &= \tfrac{1}{2}(1 - e^{-2t/\tau}) \\ p_{2|2}(r_2, 0|r_2, t) &= \tfrac{1}{2}(1 + e^{-2t/\tau}). \end{aligned} \tag{5.14}$$

Irrespective of the initial site, at long times the equilibrium distribution is reached:

$$P(\boldsymbol{r}_1, \infty) = P(\boldsymbol{r}_2, \infty) = \tfrac{1}{2}. \tag{5.15}$$

Now the self-part of the intermediate scattering function can be calculated according to Eq. (5.4):

$$I_s(\boldsymbol{Q}, t) = \left\langle e^{-i\boldsymbol{Q}\boldsymbol{R}(0)} e^{i\boldsymbol{Q}\boldsymbol{R}(t)} \right\rangle. \tag{5.16}$$

The configurational average is calculated by means of joint probabilities of the kind

$$P_{1+1}(\boldsymbol{r}_1, 0|\boldsymbol{r}_1, t) = p_{1|1}(\boldsymbol{r}_1, 0|\boldsymbol{r}_1, t) \cdot P(\boldsymbol{r}_1, 0) \tag{5.17}$$

where in thermal equilibrium for two isoenergetic sites $P(\boldsymbol{r}_1, 0) = P(\boldsymbol{r}_2, 0) = \tfrac{1}{2}$; hence

$$\begin{aligned} I_s(Q, t) &= P(r_1, 0) \cdot \left[p_{1|1} e^0 + p_{1|2} e^{i\boldsymbol{Q}(r_2 - r_1)}\right] \\ &\quad + P(r_2, 0)\left[p_{2|1} e^{i\boldsymbol{Q}(r_1 - r_2)} + p_{2|2} e^0\right] \\ &= \tfrac{1}{2}\left[\tfrac{1}{2}\left(1 + e^{-2t/\tau}\right) + \tfrac{1}{2}\left(1 - e^{-2t/\tau}\right) e^{i\boldsymbol{Q}(r_2 - r_1)}\right. \\ &\quad \left. + \tfrac{1}{2}\left(1 + e^{-2t/\tau}\right) e^{-i\boldsymbol{Q}(r_2 - r_1)} + \tfrac{1}{2}\left(1 - e^{-2t/\tau}\right)\right] \\ &= \tfrac{1}{2}\left[\left(1 + e^{-2t/\tau}\right) + \left(1 - e^{-2t/\tau}\right) \cdot \cos\{\boldsymbol{Q}(\boldsymbol{r}_2 - \boldsymbol{r}_1)\}\right] \\ &= \tfrac{1}{2}\left[1 + \cos\{\boldsymbol{Q}(\boldsymbol{r}_2 - \boldsymbol{r}_1)\}\right] + \tfrac{1}{2}\left[1 - \cos\{\boldsymbol{Q}(\boldsymbol{r}_2 - \boldsymbol{r}_1)\}\right] e^{-2t/\tau} \\ &= A_0(\boldsymbol{Q}) + A_1(\boldsymbol{Q})\, e^{-2t/\tau}. \end{aligned} \tag{5.18}$$

Fourier transformation with respect to time gives

$$S_i(\boldsymbol{Q}, \omega) = A_0(\boldsymbol{Q})\, \delta(\hbar\omega) + A_1(\boldsymbol{Q}) \frac{1}{\pi} \frac{2\hbar/\tau}{(2\hbar/\tau)^2 + (\hbar\omega)^2}. \tag{5.19}$$

Thus $A_0(\boldsymbol{Q})$ is the elastic incoherent structure factor (EISF) for a jump motion over two equivalent sites; it can be calculated directly via Eq. (5.7) as will be shown in Section 5.1.4. We recall the trigonometric relation

$$A_0(\boldsymbol{Q}) = 1 + \cos\{\boldsymbol{Q}(\boldsymbol{r}_2 - \boldsymbol{r}_1)\} = \cos^2\{\boldsymbol{Q}(\boldsymbol{r}_2 - \boldsymbol{r}_1)/2\}. \tag{5.20}$$

For powder samples, an average has to be taken over all possible orientations of the vector $\boldsymbol{Q}$ (see Eq. (5.52)) yielding

$$S(Q, \omega) = A_0(Q)\,\delta(\hbar\omega) + A_1(Q)\frac{1}{\pi}\frac{2\hbar/\tau}{(2\hbar/\tau)^2 + (\hbar\omega)^2} \tag{5.21}$$

with

$$A_0(Q) = \tfrac{1}{2} + \tfrac{1}{2} j_0(Qd) \tag{5.22}$$

$$A_1(Q) = \tfrac{1}{2} - \tfrac{1}{2} j_0(Qd) \tag{5.23}$$

and with $j_0(x) = \sin(x)/x$ where $d = |\boldsymbol{r}_2 - \boldsymbol{r}_1|$ is the jump distance between two sites.

For jumps among two energetically different sites with the ratio of the residence times $\tau_1/\tau_2 = \rho$ again the scattering function Eq. (5.21) is valid but with the following replacements (Bée 1988):

$$\frac{1}{\tau} := \frac{1}{\tau_1}(1 + \rho) \tag{5.24}$$

$$A_0(Q) := \frac{1}{(1+\rho)^2}[1 + \rho^2 + 2\rho\, j_0(Qd)] \tag{5.25}$$

$$A_1(Q) := \frac{1}{(1+\rho)^2}[1 - j_0(Qd)]. \tag{5.26}$$

5.1.2 General jump model among N sites

After this introductory example we now treat the stochastic motion extending over a limited number of N sites in a very general matrix formulation. We denote by $P_n(t)$ the probability of finding a particle at site r_n at the time t; the temporal change of this probability results from the difference of the jump rates into and out of the site under consideration:

$$\frac{\partial}{\partial t} P_n(t) = \sum_{m=0}^{N-1} \Gamma_{mn} P_m(t), \quad n = 0, \ldots, N-1 \tag{5.27}$$

where Γ_{mn} is the jump rate from site m to site n.

Thus generally jumps to each of the other $N-1$ sites can be taken into account; mostly, however, the treatment is restricted to jumps into the z next-nearest neighbouring sites (z = coordination number). For a dumb-bell $z=1$, for a hexagon $z=2$, for an octahedron $z=4$.

In most cases all N sites involved in the localized motion are energetically equivalent, i.e. all residence times $\tau_n (n=0,\ldots,N-1)$ are identical. Then

$$\Gamma_{mn} = \begin{cases} +1/z\tau & \text{for neighbouring sites} \\ -(1/\tau)\delta_{mn} & \text{for the self-site} \\ 0 & \text{otherwise.} \end{cases} \tag{5.28}$$

The set of rate equations (5.27) can be cast into one matrix equation:

$$\dot{\boldsymbol{P}} = \boldsymbol{\Lambda}\boldsymbol{P} \tag{5.29}$$

where the matrix elements of the $N \times N$ matrix $\boldsymbol{\Lambda}$ are the jump rates

$$\Lambda_{mn} = \Gamma_{mn}. \tag{5.30}$$

The solution is immediately obvious:

$$\boldsymbol{P} = \exp(\boldsymbol{\Lambda}t)\boldsymbol{P}_0. \tag{5.31}$$

If all sites are energetically equivalent, $\Gamma_{mn} = \Gamma_{nm}$ and thus the matrix $\boldsymbol{\Lambda}$ is symmetric; by a similarity transformation it is diagonalized, i.e. transformed into a diagonal matrix $\mathbf{D}$

$$\boldsymbol{\Lambda} = \mathbf{TDT}^{-1}. \tag{5.32}$$

The transformation matrix $\mathbf{T}$ is orthogonal, i.e. $(T^{-1})_{ij} = T_{ji}$. In terms of $\mathbf{D}$ the vector $\boldsymbol{P}$, the components of which are the probabilities $P_n(t)$, is expressed as

$$\boldsymbol{P} = \mathbf{T}\exp(\mathbf{D}t)\mathbf{T}^{-1}\boldsymbol{P}_0 = \mathbf{T}\begin{pmatrix} e^{\lambda_0 t} & & \\ & \ddots & \\ & & e^{\lambda_{N-1} t} \end{pmatrix}\mathbf{T}^{-1}\boldsymbol{P}_0. \tag{5.33}$$

The respective components of $\boldsymbol{P}$ are hence

$$P_n(t) = \sum_{k=0}^{N-1}\sum_{m=0}^{N-1} T_{nk}\exp(\lambda_n t)(T^{-1})_{km}P_m(0). \tag{5.34}$$

The transformation matrix $\mathbf{T}$ is formed column-by-column out of the eigenvectors $\boldsymbol{E}_k$ of the matrix $\boldsymbol{\Lambda}$; hence

$$T_{nk} = E_k^n \tag{5.35}$$

is the nth component of the eigenvector belonging to the eigenvalue λ_k. Equation (5.34) can therefore be rewritten as

$$P_n(t) = \sum_{m=0}^{N-1} \sum_{k=0}^{N-1} E_k^n \exp(\lambda_k t) E_k^m P_m(0). \tag{5.36}$$

We now transfer to the conditional probability $p_{m|n}(t)$ to find a particle at site $\boldsymbol{r}_n$, if it has been at site $\boldsymbol{r}_m$ at time $t = 0$:

$$p_{m|n}(t) = P_n(t)\, \delta(\boldsymbol{r}_n - \boldsymbol{r}_m). \tag{5.37}$$

This equation can be made plausible by a consideration of the situation at time $t = 0$:

$$\begin{aligned} p_{m|n}(0) &= \sum_{m=0}^{N-1} \sum_{k=0}^{N-1} E_k^n E_k^m P_m(0)\, \delta(\boldsymbol{r}_n - \boldsymbol{r}_m) \\ &= \sum_{m=0}^{N-1} \delta_{nm} P_m(0)\, \delta(\boldsymbol{r}_n - \boldsymbol{r}_m) \\ &= P_n(0); \end{aligned} \tag{5.38}$$

the second line of this equation utilizes the property that eigenvectors are normalized and orthogonal, i.e. that a dot product between two different eigenvectors vanishes.

The self-correlation function is obtained by thermal averaging over all initial states; if all sites are energetically equivalent, then all $P_n(0) = 1/N$; hence

$$G_s(\boldsymbol{r}, t) = \sum_{n=0}^{N-1} \sum_{m=0}^{N-1} \sum_{k=0}^{N-1} \frac{1}{N} E_k^n E_k^m \exp(\lambda_k t)\, \delta(\boldsymbol{r} - (\boldsymbol{r}_n - \boldsymbol{r}_m)). \tag{5.39}$$

Fourier transformation with respect to space yields

$$\begin{aligned} I(\boldsymbol{Q}, t) &= \int_{-\infty}^{\infty} \sum_k \sum_m \sum_n \frac{1}{N} E_k^n E_k^m \exp(\lambda_k t)\, \delta(\boldsymbol{r} - (\boldsymbol{r}_n - \boldsymbol{r}_m))\, e^{i\boldsymbol{Q}\boldsymbol{r}}\, d^3r \\ &= \sum_k \sum_m \sum_n \frac{1}{N} E_k^n E_k^m \exp(i\boldsymbol{Q}(\boldsymbol{r}_n - \boldsymbol{r}_m)) \exp(\lambda_k t) \\ &= \sum_{k=0}^{N-1} A_k(\boldsymbol{Q}) \exp(\lambda_k |t|). \end{aligned} \tag{5.40}$$

Because of the spatial restriction we obtain $\lambda_0 = 0$, which constitutes a time-independent contribution, namely the EISF. By a Fourier transformation with respect to time the scattering function is obtained:

$$S_{\mathrm{i}}(\boldsymbol{Q},\omega) = A_0(\boldsymbol{Q})\,\delta(\omega) + \sum_{k=1}^{N-1} A_k(\boldsymbol{Q})\frac{1}{\pi}\frac{-\hbar\lambda}{(\hbar\lambda)^2 + (\hbar\omega)^2} \tag{5.41}$$

with

$$A_k(\boldsymbol{Q}) = \sum_m \sum_n \frac{1}{N} E_k^n E_k^m \exp(\mathrm{i}\boldsymbol{Q}(\boldsymbol{r}_n - \boldsymbol{r}_m)),$$
$$k = 0, 1, \ldots, N-1. \tag{5.42}$$

For demonstration purposes this formalism will now be applied to the problem of the dumb-bell-like motion with two sites located at $\boldsymbol{r}_0 = 0$ and $\boldsymbol{r}_1 = \boldsymbol{R}$. According to Eqs (5.30) and (5.28) with the coordination number $z = 1$ the jump matrix is

$$\boldsymbol{\Lambda} = \begin{pmatrix} -1/\tau & 1/\tau \\ 1/\tau & -1/\tau \end{pmatrix}. \tag{5.43}$$

The resulting eigenvectors and eigenvalues are

	0	1	← index k (subscript of E_k^n)
	0	$-2/\tau$	← eigenvalues λ_k
0	0.707	0.707	← E_k^n, i.e. components of the eigenvectors
1	0.707	−0.707	
↑			index n (superscript of E_k^n)

Thus we obtain:

$$\begin{aligned} A_0(\boldsymbol{Q}) &= \tfrac{1}{2}\{E_0^0 E_0^0 \exp(\mathrm{i}Q0) + E_0^0 E_0^1 \exp(\mathrm{i}\boldsymbol{Q}\boldsymbol{R}) \\ &\quad + E_0^1 E_0^0 \exp(-\mathrm{i}\boldsymbol{Q}\boldsymbol{R}) + E_0^1 E_0^1 \exp(\mathrm{i}Q0)\} \\ &= \tfrac{1}{2}\{(E_0^0)^2 + (E_0^1)^2 + 2E_0^0 E_0^1 \cos(\boldsymbol{Q}\boldsymbol{R})\} \\ &= \tfrac{1}{2}\{1 + \cos(\boldsymbol{Q}\boldsymbol{R})\} \end{aligned} \tag{5.44}$$

$$\begin{aligned} A_1(Q) &= \tfrac{1}{2}\{(E_1^0)^2 + (E_1^1)^2 + 2E_1^0 E_1^1 \cos(\boldsymbol{Q}\boldsymbol{R})\} \\ &= \tfrac{1}{2}\{1 - \cos(\boldsymbol{Q}\boldsymbol{R})\}. \end{aligned} \tag{5.45}$$

For two energetically non-equivalent sites the jump matrix is not symmetric:

$$\mathbf{\Lambda} = \begin{pmatrix} -1/\tau_1 & 1/\tau_2 \\ 1/\tau_1 & -1/\tau_2 \end{pmatrix}. \tag{5.46}$$

But the eigenvalues are nevertheless real:

$$\begin{aligned} -\lambda_0 &= 0 \\ -\lambda_1 &= \tau_1^{-1}\left(1 + \frac{\tau_1}{\tau_2}\right) = \tau_1^{-1}(1+\rho). \end{aligned} \tag{5.47}$$

For the eigenvectors one obtains

$$e_1 = \begin{pmatrix} \rho/\sqrt{1+\rho^2} \\ 1/\sqrt{1+\rho^2} \end{pmatrix}; \qquad e_2 = \begin{pmatrix} -1/\sqrt{2} \\ 1/\sqrt{2} \end{pmatrix}. \tag{5.48}$$

The results are identical to those of Section 5.1.1; the general formalism can easily be applied to any geometrical arrangement of sites. In hexagonal structures a hexagon of sites is possible, as has actually been observed for H in $LaNi_5$; the jump matrix can then be written down as

$$\mathbf{\Lambda} = \begin{pmatrix} -1/\tau & 1/2\tau & 0 & 0 & 0 & 1/2\tau \\ 1/2\tau & -1/\tau & 1/2\tau & 0 & 0 & 0 \\ 0 & 1/2\tau & -1/\tau & 1/2\tau & 0 & 0 \\ 0 & 0 & 1/2\tau & -1/\tau & 1/2\tau & 0 \\ 0 & 0 & 0 & 1/2\tau & -1/\tau & 1/2\tau \\ 1/2\tau & 0 & 0 & 0 & 1/2\tau & -1/\tau \end{pmatrix}$$

and can easily be solved numerically:

	0 0	1 $-1/2\tau$	2 $-1/2\tau$	3 $-3/2\tau$	4 $-3/2\tau$	5 $-4/2\tau$
0	0.408	0	0.577	−0.577	0	−0.408
1	0.408	0.500	0.289	0.289	0.500	0.408
2	0.408	0.500	−0.289	0.289	−0.500	−0.408
3	0.408	0	−0.577	−0.577	0	0.408
4	0.408	−0.500	−0.289	0.289	0.500	−0.408
5	0.408	−0.500	0.289	0.289	−0.500	0.408

For N sites located on a circle the eigenvalue problem can also be solved analytically:

$$\lambda_k = -\frac{2}{\tau}\left(1 - \cos\frac{k\pi}{N}\right) \tag{5.49}$$

$$E_0^n = \frac{1}{\sqrt{N}}, \quad n = 0, 1, \ldots, N-1, \tag{5.50}$$

$$E_k^n = \sqrt{\frac{2}{N}}\cos\left(\frac{2n+1}{2}\frac{k\pi}{N}\right), \quad \begin{array}{l} n = 0, 1, \ldots, N-1, \\ k = 1, \ldots, N-1. \end{array} \tag{5.51}$$

An octahedral arrangement of sites for localized motion can occur in cubic structures. The corresponding jump matrix can easily be written down:

$$\mathbf{\Lambda} = \begin{pmatrix} -1/\tau & 1/4\tau & 1/4\tau & 1/4\tau & 1/4\tau & 0 \\ 1/4\tau & -1/\tau & 1/4\tau & 0 & 1/4\tau & 1/4\tau \\ 1/4\tau & 1/4\tau & -1/\tau & 1/4\tau & 0 & 1/4\tau \\ 1/4\tau & 0 & 1/4\tau & -1/\tau & 1/4\tau & 1/4\tau \\ 1/4\tau & 1/4\tau & 0 & 1/4\tau & -1/\tau & 1/4\tau \\ 0 & 1/4\tau & 1/4\tau & 1/4\tau & 1/4\tau & -1/\tau \end{pmatrix}.$$

If the sample is polycrystalline, the structure factors $A_k(\boldsymbol{Q})$ have to be spatially averaged, which means a spatial average of the exponential function (with $|\boldsymbol{r}_n - \boldsymbol{r}_m| = \boldsymbol{r}_{nm}$):

$$\begin{aligned} \langle \exp(\mathrm{i}\boldsymbol{Q}(\boldsymbol{r}_n - \boldsymbol{r}_m))\rangle_{\text{sp. av.}} &= \frac{1}{4\pi}\int_0^{2\pi}\int_0^{\pi} \exp(\mathrm{i}Q|\boldsymbol{r}_n - \boldsymbol{r}_m|\cos\theta)\sin\theta\,\mathrm{d}\theta\,\mathrm{d}\phi \\ &= \frac{2\pi}{4\pi}\int_{-1}^{+1} \exp(\mathrm{i}Qr_{nm}\cos\theta)\,\mathrm{d}(\cos\theta) \\ &= \frac{1}{2}\frac{1}{\mathrm{i}Qr_{nm}}[\exp(\mathrm{i}Qr_{nm}) - \exp(-\mathrm{i}Qr_{nm})] \\ &= \frac{\sin(Qr_{nm})}{Qr_{nm}} = j_0(Qr_{nm}). \end{aligned} \tag{5.52}$$

Thus, with Eq. (5.42)

$$\langle A_k(Q)\rangle_{\text{sp. av.}} = \sum_{m=0}^{N-1}\sum_{n=0}^{N-1}\frac{1}{N}E_k^n E_k^m\, j_0(Qr_{nm}), \tag{5.53}$$

again with $j_0(x) = \sin(x)/x$. Specifically, for a dumb-bell, in this way the results Eqs. (5.22) and (5.23) are obtained.

For N sites located on the circumference of a circle in a powder sample (Barnes 1973; Dianoux *et al.* 1975; Leadbetter and Lechner 1979; Volino and Dianoux 1980) the resulting expression of the scattering function is

$$S_{\mathrm{i}}(Q, \omega) = A_0(Q)\,\delta(\hbar\omega) + \sum_{k=1}^{N} A_k(Q) \frac{1}{\pi} \frac{\hbar/\tau_k}{(\hbar/\tau_k) + (\hbar\omega)^2} \tag{5.54}$$

with

$$A_k(Q) = \frac{1}{N} \sum_{n=1}^{N} j_0(Qr_n) \cos\left(\frac{2kn\pi}{N}\right); \tag{5.55}$$

the r_n are the jump distances under the effect of $2n\pi/N$ rotations:

$$r_n = 2r \sin\left(\frac{n\pi}{N}\right). \tag{5.56}$$

The correlation rates are evaluated from (Barnes 1973)

$$\tau_k^{-1} = 2\tau^{-1} \sin^2\left(\frac{\pi k}{N}\right). \tag{5.57}$$

5.1.3 Continuous rotational diffusion

There are cases in the field of solid state diffusion where localized motion around an atom with a strong affinity to the diffusing atom or ion can approximately and conveniently be treated as rotational diffusion on the surface of a sphere. The position of the particle is conveniently described in spherical coordinates by means of the orientation vector $\boldsymbol{\Omega} = (\phi, \theta)$. We consider the occurrence probability $P(\boldsymbol{\Omega}, t)$ of the orientation $\boldsymbol{\Omega}$ at time t. It can be obtained as the solution of an appropriate diffusional differential equation, analogous to Fick's first law:

$$\frac{\partial}{\partial t} P(\boldsymbol{\Omega}, t) = D_{\mathrm{r}} \Delta_{\Omega} P(\boldsymbol{\Omega}, t) \tag{5.58}$$

where D_{r} means the rotational diffusion coefficient (dimension s^{-1}) and where Δ_{Ω} denotes the angular part of the Laplace operator in spherical coordinates:

$$\Delta_{\Omega} = \left(\frac{1}{\sin^2\theta} \frac{\partial^2}{\partial\phi^2} + \frac{\partial^2}{\partial\theta^2} + \cot\theta \frac{\partial}{\partial\theta}\right). \tag{5.59}$$

As is well known from the quantum mechanical textbook problem of the hydrogen atom, special functions called spherical harmonics are eigenfunctions of the above operator:

$$\Delta_\Omega Y_l^m(\theta,\phi) = -l(l+1)Y_l^m(\theta,\phi). \tag{5.60}$$

The spherical harmonics $Y_l^m(\theta,\phi)$ form a function system which is normalized, orthogonal, and complete:

$$\int_0^{2\pi} \mathrm{d}\phi \int_0^{\pi} \sin\theta \, \mathrm{d}\theta \; Y_l^{*m}(\theta,\phi)\, Y_{l'}^{m'}(\theta,\phi) = \delta_{ll'}\,\delta_{mm'} \tag{5.61}$$

$$\sum_{l=0}^{\infty}\sum_{m=-l}^{+l} Y_l^{*m}(\theta,\phi)\, Y_l^m(\theta_0,\phi_0) = \frac{\delta(\theta-\theta_0)\,\delta(\phi-\phi_0)}{\sin\theta}. \tag{5.62}$$

The diffusional equation (5.58) is solved with the ansatz $P_\mathrm{e}^m(\mathbf{\Omega},t) = Y_l^m \mathrm{e}^{\lambda_l t}$ such that

$$D_\mathrm{r} l(l+1)\, P(\mathbf{\Omega},t) = \lambda_l P(\mathbf{\Omega},t) \tag{5.63}$$

which requires that

$$\lambda_l = D_\mathrm{r} l(l+1).$$

The general solution consists of the superposition of all spherical harmonics Y_l^m and their complex conjugates with appropriate linear coefficients:

$$P(\mathbf{\Omega},t) = \sum_{l=0}^{\infty}\sum_{m=-l}^{+l} c_l^{*m}\, Y_l^m(\theta,\phi)\mathrm{e}^{-D_\mathrm{r}l(l+1)t} + \sum_{l=0}^{\infty}\sum_{m=-l}^{+l} c_l^m\, Y_l^{*m}(\theta,\phi)\mathrm{e}^{-D_\mathrm{r}l(l+1)t}. \tag{5.64}$$

We now proceed to the conditional probability density $P(\mathbf{\Omega}_0, \mathbf{\Omega}; t)$; it has to satisfy the initial condition

$$P(\mathbf{\Omega}_0, \mathbf{\Omega}; t=0) = \frac{\delta(\theta-\theta_0)\,\delta(\phi-\phi_0)}{\sin\theta} = \delta(\mathbf{\Omega}-\mathbf{\Omega}_0). \tag{5.65}$$

Comparison with Eq. (5.63) shows that this conditional probability can be written as

$$P(\mathbf{\Omega}_0, \mathbf{\Omega}; t) = \sum_{l=0}^{\infty}\sum_{m=-l}^{+l} Y_l^{*m}(\theta,\phi)\, Y_l^m(\theta_0,\phi_0)\mathrm{e}^{-D_\mathrm{r}l(l+1)t}. \tag{5.66}$$

The conditional probability density, Eq. (5.66), has to be considered as the rotational analogue of van Hove's self-correlation function; intergration over all initial orientations can be omitted because in this case all initial orientations are energetically equivalent. Spatial Fourier transformation yields the self-part of the intermediate scattering function:

$$I_{\mathrm{s}}^{\mathrm{rot}}(Q,t) = \frac{1}{4\pi}\int P(\boldsymbol{\Omega}_0,\boldsymbol{\Omega};t)\mathrm{e}^{\mathrm{i}\boldsymbol{Q}(\boldsymbol{R}(\boldsymbol{\Omega})-\boldsymbol{R}(\boldsymbol{\Omega}_0))}\,\mathrm{d}\boldsymbol{\Omega}. \tag{5.67}$$

We choose the coordinate system such that its z-axis coincides with the $\boldsymbol{Q}$-direction and expand the planar waves in terms of spherical harmonics:

$$\exp(\mathrm{i}Qz) = \sum_l (\mathrm{i})^l \sqrt{4\pi(2l+1)}\, j_l(\boldsymbol{QR})\, Y_l^0(\theta,\phi) \tag{5.68}$$

where $j_l(\boldsymbol{QR})$ is a spherical Bessel function of the first kind. Inserting this expansion into Eq. (5.67) we obtain

$$\begin{aligned} I_{\mathrm{s}}^{\mathrm{rot}}(Q,t) = &\frac{1}{4\pi}\int \mathrm{d}\boldsymbol{\Omega}\sum_{l'l''}(-\mathrm{i})^{l''}(\mathrm{i})^{l'}\sqrt{4\pi(2l'+1)}\sqrt{4\pi(2l''+1)} \\ &\times j_{l'}(\boldsymbol{QR})\, j_{l''}(\boldsymbol{QR})\, Y_{l'}^0(\theta,\phi)\, Y_{l''}^0(\theta_0,\phi_0) \\ &\times \sum_{l=0}^{\infty}\sum_{m=-l}^{+l} Y_l^{*m}(\theta,\phi)\, Y_l^m(\theta_0,\phi_0)\mathrm{e}^{-D_{\mathrm{r}}l(l+1)t}. \end{aligned} \tag{5.69}$$

The products of the spherical harmonics yield $\delta_{ll'}\delta_{m0}\delta_{ll''}\delta_{m0}$ because of the orthogonality property, Eq. (5.61); thus Eq. (5.69) simplifies to

$$I_{\mathrm{s}}^{\mathrm{rot}}(\boldsymbol{Q},t) = \sum_{l=0}^{\infty}(2l+1)j_l^2(\boldsymbol{QR})\mathrm{e}^{-D_{\mathrm{r}}l(l+1)}. \tag{5.70}$$

Temporal Fourier transformation eventually leads to the incoherent scattering function for rotational diffusion on the surface of a sphere:

$$\begin{aligned} S_{\mathrm{i}}^{\mathrm{rot}}(Q,\omega) = &\, j_0^2(\boldsymbol{QR})\,\delta(\omega) \\ &+\sum_{l=1}^{\infty}(2l+1)\, j_l^2(\boldsymbol{QR})\frac{1}{\pi}\frac{D_{\mathrm{r}}l(l+1)}{(D_{\mathrm{r}}l(l+1))^2+(\hbar\omega)^2}. \end{aligned} \tag{5.71}$$

One-dimensional rotational diffusion (on the circumference of a circle) can be treated in an analogous way resulting in

$$S_{\mathrm{i}}^{\mathrm{rot}}(\boldsymbol{Q},\omega) = J_0^2(QR\sin\theta)\,\delta(\hbar\omega) + \sum_{n=1}^{\infty} J_n^2(QR\sin\theta)\frac{1}{\pi}\frac{D_{\mathrm{r}}n^2}{(D_{\mathrm{r}}n^2)^2+(\hbar\omega)^2}. \tag{5.72}$$

$J_n(x)$ mean cylindrical Bessel functions, and θ denotes the angle between the vector $\boldsymbol{Q}$ and the plane of the circle; for $\theta = 90°$, i.e. $\boldsymbol{Q}$ perpendicular to the circle, the arguments of the Bessel functions vanish, $J_0(0) = 1$, $J_n(0) = 0$ for $n \geq 1$, and the scattering function equals a δ-function without $|\boldsymbol{Q}|$ dependence. For practical purposes, particularly in the case of powder samples, it is convenient to describe continuous rotational diffusion on the circumference of a circle by a large number, say 15, of discrete sites of the circle. The difference in the scattering function is not detectable, but the powder average can be taken easily; see Eqs (5.52)–(5.55).

5.1.4 Calculation of some EISFs

Experimentally, the elastic incoherent structure factor (EISF) is the fraction of the total 'quasielastic' intensity contained in the purely elastic peak. For an 'exact' experimental determination of the EISF the neutron spectra have to be fitted with the correct scattering functions; see the previous subsections. Approximately, however, the EISF can be determined in a kind of model-independent data evaluation by means of a fit of the neutron spectra with a single Lorentzian plus an elastic term. Although in this way the line shape is not correct except for a dumb-bell, the EISF thus obtained allows statements to be made about the geometry for the localized motion.

Theoretically, the EISF can be calculated as the square of the Fourier transform of the (normalized) probability density $\rho(\mathrm{r})$ of the particle, see Eq. (5.7), which can be discrete or continuous:

$$\mathrm{EISF} = \frac{1}{N^2}\left|\sum_j \mathrm{e}^{-\mathrm{i}\boldsymbol{Q}\boldsymbol{R}_j}\right|^2 = \left|\int \rho(r)\mathrm{e}^{\mathrm{i}Qr}\,\mathrm{d}^3r\right|^2. \tag{5.73}$$

This will be outlined for some simple examples.

Jump rotation

If the particle performs a jump rotation over only two sites (a dumb-bell is the simplest example) located at $\boldsymbol{r} = -\boldsymbol{s}/2$ and $\boldsymbol{r} = +\boldsymbol{s}/2$, its EISF is given by

$$\mathrm{EISF} = \tfrac{1}{4}\left|\mathrm{e}^{\mathrm{i}\boldsymbol{Q}(-\boldsymbol{s}/2)} + \mathrm{e}^{\mathrm{i}\boldsymbol{Q}(\boldsymbol{s}/2)}\right|^2 = \cos^2(\boldsymbol{Q}\boldsymbol{s}/2). \tag{5.74}$$

Equivalently, the particle could be located at the origin and at $\boldsymbol{s}$:

$$\begin{aligned}\mathrm{EISF} &= \tfrac{1}{4}\left|\mathrm{e}^{\mathrm{i}\boldsymbol{Q}0} + \mathrm{e}^{\mathrm{i}\boldsymbol{Q}\boldsymbol{s}}\right|^2 \\ &= \tfrac{1}{4}(1 + \mathrm{e}^{-\mathrm{i}\boldsymbol{Q}\boldsymbol{s}})(1 + \mathrm{e}^{\mathrm{i}\boldsymbol{Q}\boldsymbol{s}}) = \tfrac{1}{2}(1 + \cos \boldsymbol{Q}\boldsymbol{s}).\end{aligned} \tag{5.75}$$

Both results are of course identical, see Eq. (5.20). Analogously to Eq. (5.52), a spatial average yields

$$\begin{aligned}\langle \text{EISF} \rangle &= \frac{1}{4\pi} \int_0^{2\pi} \int_0^{\pi} \left\{ \frac{1}{2} + \frac{1}{2} \cos(Qs\cos\theta) \right\} \sin\theta \, \mathrm{d}\theta \, \mathrm{d}\phi \\ &= \frac{1}{2} + \frac{1}{4Qs} \int_{-Qs}^{+Qs} \cos x \, \mathrm{d}x \\ &= \frac{1}{2}\left(1 + \frac{\sin(Qs)}{Qs}\right) = \frac{1}{2}(1 + j_0(Qs)). \end{aligned} \tag{5.76}$$

Correspondingly, the elastic incoherent structure factors for jump rotation over any arrangement of discrete sites can be calculated.

Rotational diffusion on the surface of a sphere of radius R

The normalized probability density is given by

$$\rho(r) = \frac{1}{4\pi R^2} \delta(r - R). \tag{5.77}$$

Hence

$$\begin{aligned}\text{EISF} &= \left| \int \rho(r) \exp(\mathrm{i}\boldsymbol{Qr}) \, \mathrm{d}^3 r \right|^2 \\ \sqrt{\text{EISF}} &= \int_0^{2\pi} \mathrm{d}\phi \int_{-1}^{+1} \mathrm{d}(\cos\theta) \int_0^{\infty} r^2 \, \mathrm{d}r \left\{ \frac{1}{4\pi R^2} \delta(r - R) \exp(\mathrm{i}Qr\cos\theta) \right\} \\ &= \frac{2\pi}{4\pi R^2} \int_{-1}^{+1} \mathrm{d}(\cos\theta) R^2 \exp(\mathrm{i}QR\cos\theta) \\ &= \frac{1}{2} \frac{e^{\mathrm{i}QR} - \mathrm{e}^{-\mathrm{i}QR}}{\mathrm{i}QR} = \frac{\sin(QR)}{QR} \\ &\equiv j_0(QR), \\ \text{EISF} &= j_0^2(QR). \end{aligned} \tag{5.78}$$

Rotational diffusion on the circumference of a circle of radius R

Conveniently we use a cylindrical coordinate system, such that

$$\boldsymbol{r} = \begin{pmatrix} r\cos\phi \\ r\sin\phi \\ z \end{pmatrix}, \qquad \boldsymbol{Q} = \begin{pmatrix} Q\sin\theta \\ 0 \\ Q\cos\theta \end{pmatrix}.$$

The normalized probability density is given by

$$\rho(r) = \frac{1}{2\pi R}\delta(r - R)\,\delta(z). \tag{5.79}$$

Therefore we derive:

$$\begin{aligned}
\int \rho(r)\exp(\mathrm{i}\boldsymbol{QR})\,\mathrm{d}^3r &= \int_0^{2\pi}\mathrm{d}\phi \int_{-\infty}^{\infty}\mathrm{d}z \int_0^{\infty}\mathrm{d}r\, r\left\{\frac{1}{2\pi R}\delta(r - R)\,\delta(z)\exp(\mathrm{i}\boldsymbol{Qr})\right\} \\
&= \frac{1}{2\pi R}\int_0^{2\pi}\mathrm{d}\phi \int_0^{\infty}\mathrm{d}r\, r\exp(\mathrm{i}Qr\sin\theta\,\cos\phi)\,\delta(r - R) \\
&= \frac{1}{2\pi}\int_0^{2\pi}\mathrm{d}\phi\exp(\mathrm{i}QR\sin\theta\,\cos\phi) \\
&\equiv J_0(QR\sin\theta), \\
\mathrm{EISF} &= J_0^2(QR\sin\theta).
\end{aligned} \tag{5.80}$$

Diffusion within the volume of a sphere of radius R

This is an example of diffusion within a limited volume:

$$\rho(r) = \begin{cases} 1/z & \text{for } 0 < r \le R \text{ with } z = (4/3)\pi R^3 \\ 0 & \text{for } r > R \end{cases} \tag{5.81}$$

$$\begin{aligned}
\int \rho(r)\exp(\mathrm{i}\boldsymbol{Qr})\,d^3r &= \int_0^{2\pi}\mathrm{d}\phi \int_{-1}^{+1}\mathrm{d}(\cos\theta)\int_0^{R}\mathrm{d}r\, r^2\left\{\frac{1}{z}\exp(\mathrm{i}\boldsymbol{Qr})\right\} \\
&= \frac{4\pi}{zQ}\left\{\left[-\frac{r}{Q}\cos Qr\right]_0^R + \frac{1}{Q}\int_0^R \cos Qr\,\mathrm{d}r\right\} \\
&= \frac{4\pi}{(4/3)\pi}\left\{\frac{-1}{Q^2R^2}\cos QR + \frac{1}{Q^3R^3}\sin QR\right\} \\
&= \frac{3}{QR}\left\{\frac{\sin(QR)}{(QR)^2} - \frac{\cos(QR)}{QR}\right\} \\
&\equiv \frac{3}{QR}\,j_1(QR) \\
\mathrm{EISF} &= \left[\frac{3j_1(QR)}{QR}\right]^2 = \frac{9\{\sin(QR) - QR\cos(QR)\}^2}{(QR)^6}.
\end{aligned} \tag{5.82}$$

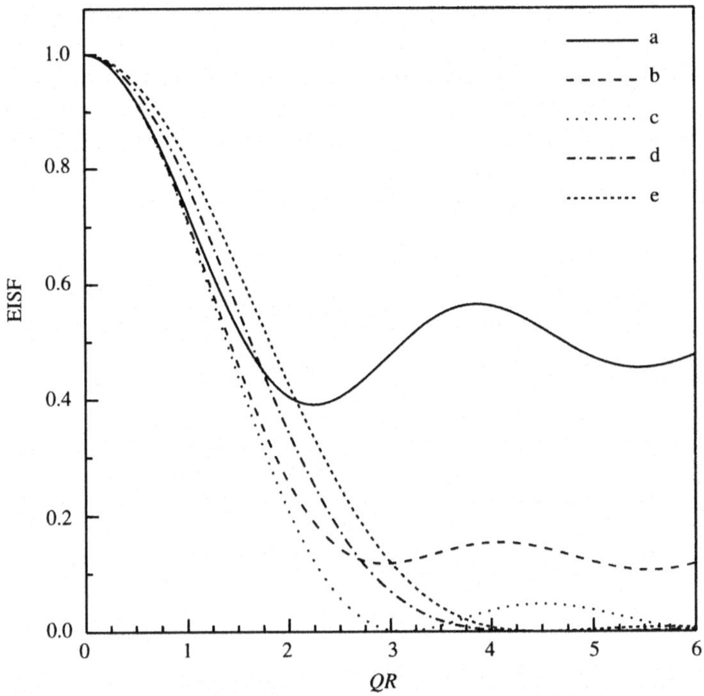

Fig. 5.4 Comparison of different EISFs: (a) dumb-bell, (b) hexagon, (c) circumference of a circle, (d) surface of a sphere and (e) volume of a sphere.

These formulas for the elastic *incoherent* structure factor for one atom distributed over several sites or over a certain space hold also for the *coherent* scattering of a correspondingly shaped molecule or particle. It is then called the particle form factor and plays an important role, e.g., in the field of small-angle scattering on mesoscopic particles. In particular, Eq.(5.82) represents the small-angle scattering pattern of monodisperse, non-interacting (dilute) spherical particles of radius R. Figure 5.4 displays a comparison of different orientationally averaged EISFs.

5.2 'LONG RANGE' TRANSLATIONAL DIFFUSION

Single particle diffusion extending over 'large' distances r (several nm) can be studied by means of incoherent QENS at small momentum transfers Q. The small Q value range of a backscattering spectrometer comprises $0.1\,\text{Å}^{-1} \leq Q \leq 0.5\,\text{Å}^{-1}$, i.e. the investigated distances are between 60 Å and 10 Å. Since the ratio between the mean square displacement and the square of the jump length equals the number of jumps involved, a mean square displacement of 10 Å means 25 jumps assuming a jump length of 2 Å; after 25 or more jumps details of the jump mechanism are lost by averaging, and only a global quantity characterizing the single-particle diffusion process is obtained, namely the macroscopic self-diffusion coefficient measured over microscopic distances. As has been derived in Section 4.3.4 (Eq. 4.91) the master equation, which describes jump diffusion processes microscopically, in the macroscopic limit transforms into Fick's second law, and the self-correlation function obeys this differential equation:

$$\frac{\partial}{\partial t} G_s(r,t) = D_s \Delta G_s(r,t) \tag{5.83}$$

with the initial condition $G_s(r, t = 0) = \delta(r)$. Thus we treat the diffusion process as if all diffusing particles start at time zero at the origin; this leads to the correct results because of temporal and spatial invariance. The solution is

$$G_s(r,t) = (4\pi D_s |t|)^{-3/2} \exp(-r^2/4D_s|t|). \tag{5.84}$$

Spatial Fourier transformations yields the self-part of the intermediate scattering function:

$$I_s(Q,t) = \int_{-\infty}^{\infty} (4\pi D_s|t|)^{-3/2} \exp(-r^2/4D_s|t|) \mathrm{e}^{\mathrm{i}\boldsymbol{Qr}} \, \mathrm{d}\boldsymbol{r}. \tag{5.85}$$

This three-dimensional integral is factorized into three one-dimensional ones by taking into account that $r^2 = x^2 + y^2 + z^2$ and that $\boldsymbol{Qr} = Q_x x + Q_y y + Q_z z$; eventually we obtain

$$I_s(Q,t) = \exp(-Q^2 D_s |t|). \tag{5.86}$$

Subsequently a Fourier transformation with respect to time has to be performed:

$$\begin{aligned} S_i(Q,\omega) &= \frac{1}{2\pi\hbar} \int_{-\infty}^{\infty} \mathrm{e}^{-Q^2 D_s |t|} \mathrm{e}^{-\mathrm{i}\omega t} \, \mathrm{d}t \\ &= \frac{1}{2\pi\hbar} \left[\int_{-\infty}^{0} \mathrm{e}^{D_s Q^2 t} \mathrm{e}^{-\mathrm{i}\omega t} \, \mathrm{d}t + \int_{0}^{\infty} \mathrm{e}^{-D_s Q^2 t} \mathrm{e}^{-\mathrm{i}\omega t} \, \mathrm{d}t \right] \\ &= \frac{1}{\pi} \frac{\hbar D_s Q^2}{(\hbar D_s Q^2)^2 + (\hbar\omega)^2}. \end{aligned} \tag{5.87}$$

The incoherent scattering function is thus represented by a single Lorentzian function with the linewidth (HWHM)

$$\Lambda = \hbar D_s Q^2. \tag{5.88}$$

At sufficiently small Q this so-called Q^2 law is generally valid irrespective of the details of the diffusion mechanism, if the diffusion is a Poisson process. If, on the other hand, the time correlation function is a stretched exponential or if the network of diffusive paths has a fractal character, then the Q^2 law does not apply, and the line shape is not exactly Lorentzian-like. Deviations from a Lorentzian line shape of the scattering function at small Q can also be due to multiple scattering, see Chapter 7.

In non-cubic solids diffusion is generally anisotropic; the self-diffusion coefficient D_s has to be replaced by the diffusion tensor $\mathbf{D}$ with the elements D_{ij}; the diffusion equation (Fick's second law) is expressed as

$$\nabla[\mathbf{D}\nabla G_s(r,t)] = \frac{\partial}{\partial t} G_s(\boldsymbol{r},t) \tag{5.89a}$$

what is equivalent to

$$\begin{pmatrix} \partial/\partial x_1 \\ \partial/\partial x_2 \\ \partial/\partial x_3 \end{pmatrix} \begin{pmatrix} \sum_{j=1}^{3} D_{1j}\partial/\partial x_j \\ \sum_{j=1}^{3} D_{2j}\partial/\partial x_j \\ \sum_{j=1}^{3} D_{3j}\partial/\partial x_j \end{pmatrix} G_s(\boldsymbol{r},t) = \frac{\partial}{\partial t} G_s(\boldsymbol{r},t). \tag{5.89b}$$

In components this can be written as

$$\sum_{i=1}^{3}\sum_{j=1}^{3} D_{ij}\frac{\partial^2}{\partial x_i\,\partial x_j} G_s(\boldsymbol{r},t) = \frac{\partial}{\partial t} G_s(\boldsymbol{r},t). \tag{5.89c}$$

Because of translational invariance the initial condition can be formulated as

$$G_s(\boldsymbol{r},0) = \delta(\boldsymbol{r}). \tag{5.90}$$

Equation (5.89) can be Fourier transformed into

$$\frac{\partial}{\partial t} I_s(\boldsymbol{Q},t) = -\sum_{i=1}^{3}\sum_{j=1}^{3} D_{ij} Q_i Q_j\, I_s(\boldsymbol{Q},t) \tag{5.91}$$

with the Fourier transformed initial condition, Eq. (5.90), of $I_s\ (\boldsymbol{Q},0) = 1$. The solution is straightforward:

$$I_s(\boldsymbol{Q},t) = \exp\left(-\sum_{i=1}^{3}\sum_{j=1}^{3} D_{ij} Q_i Q_j |t|\right) = \exp(-\Lambda^{\#}|t|). \tag{5.92}$$

A further Fourier transformation yields the incoherent scattering function

$$S_i(\boldsymbol{Q},\omega) = \frac{1}{\pi}\left(\frac{\hbar\Lambda^{\#}}{(\hbar\Lambda^{\#})^2 + (\hbar\omega)^2}\right). \tag{5.93}$$

The trigonal, tetragonal, or hexagonal crystal systems exhibit uniaxial crystal symmetry. The diffusion tensor then contains only two components: the diffusion coefficient $D_\perp$ perpendicular to the axis and the diffusion coefficient $D_\parallel$ parallel to the axis:

$$\mathsf{D} = \begin{pmatrix} D_\perp & 0 & 0 \\ 0 & D_\perp & 0 \\ 0 & 0 & D_\parallel \end{pmatrix}. \tag{5.94}$$

If the angle between the Q vector and the z-axis of the system is denoted by θ, then the components of $\boldsymbol{Q}$ are

$$Q = \begin{pmatrix} Q \sin\theta/\sqrt{2} \\ Q \sin\theta/\sqrt{2} \\ Q \cos\theta \end{pmatrix}. \tag{5.95}$$

The linewidth in Eqs. (5.92), (5.93) is therefore given by

$$\Lambda^{\#} = \left(D_\parallel \cos^2\theta + D_\perp \sin^2\theta\right) Q^2. \tag{5.96}$$

Using a single crystalline sample both diffusion coefficients can be determined by proper alignment of the sample.

5.3 TRANSLATIONAL JUMP DIFFUSION IN BRAVAIS LATTICES

At larger Q values ($Q \geq 1\,\text{Å}^{-1}$) incoherent QENS data contain information on the diffusion mechanism. As was pointed out in Chapter 4, solid state diffusion at not too elevated temperatures can be considered as jump diffusion. This idea was cast into a model by Chudley and Elliott (1961); although originally developed for diffusion in liquids, this model found more applications in solid state diffusion. The Chudley–Elliott model comprises the simplest case of jump diffusion and involves the following postulates:

1. The sublattice on which the diffusion takes place is a Bravais lattice, i.e. all sites involved are crystallographically equivalent; this includes energetic equivalence.
2. The diffusion is dynamically independent of other kinds of motion, in particular of the vibrations.
3. Only jumps to nearest neighbour sites are allowed; these jumps are geometrically characterized by so-called jump vectors $\boldsymbol{s}_i$, $i = 1, \ldots, z$, where z is the coordination number; all jumps have the same jump length, i.e. $|\boldsymbol{s}_i| = l$ for all i.
4. The particle stays at a site for a certain mean residence time τ; then instantaneously it jumps to a neighbouring site; i.e. the jump time is negligibly small compared to the residence time.
5. Successive jumps are uncorrelated, i.e. the jump direction of each jump is completely random.

Thus the self-correlation function obeys the master equation, Eq. (4.83). According to the scheme, in Fig. 5.3, this equation has to be solved either directly or after Fourier transformation, because Fourier transformations have to be performed anyway. In Eqs. (4.83) ff the solution in Fourier space was derived, which we now identify with the self-part of the intermediate scattering function:

$$I_s(\boldsymbol{Q}, t) = \exp\left\{-\frac{t}{z\tau}\sum_{j=1}^{z}(1 - e^{-i\boldsymbol{Q}\boldsymbol{s}_i})\right\}. \tag{5.97}$$

This exponential function can be Fourier transformed with respect to time yielding

$$S_i(\boldsymbol{Q}, \omega) = \frac{1}{\pi}\frac{\Lambda(\boldsymbol{Q})}{\Lambda^2(\boldsymbol{Q}) + (\hbar\omega)^2}, \tag{5.98}$$

i.e. a single Lorentzian function with linewidth (HWHM)

$$\Lambda(\boldsymbol{Q}) = \frac{\hbar}{z\tau}\sum_{j=1}^{z}\left(1 - e^{-i\boldsymbol{Q}\boldsymbol{s}_j}\right). \tag{5.99}$$

Bravais lattices exhibit inversion symmetry, and each site is an inversion centre; therefore, for each jump vector $\boldsymbol{s}_j$ there exists a jump vector $-\boldsymbol{s}_j$, and the sum over z exponential functions in Eq. (5.79) is transformed into a sum over $z/2$ cosine functions.

Before we discuss some examples, we consider the limiting cases of small and large Q values. For small Q the exponential in Eq. (5.99) is expanded up to the square term:

$$\text{small } Q: \qquad \Lambda(\boldsymbol{Q}) = \frac{\hbar}{z\tau}\sum_{j=1}^{z}\left(1 - 1 + i\boldsymbol{Q}\boldsymbol{s}_j + \frac{1}{2}(\boldsymbol{Q}\boldsymbol{s}_j)^2\right). \tag{5.100}$$

Since each Bravais lattice exhibits inversion symmetry, the back and forth jumps in the linear term mutually cancle; we denote by $l = |\boldsymbol{s}_j|$ the mean jump length and take into account that the spatial average of $|\boldsymbol{Q}\boldsymbol{s}|^2$ yields $\frac{1}{3}Q^2l^2$:

$$\text{small } Q: \qquad \Lambda(\boldsymbol{Q}) = \hbar\frac{l^2}{6\tau}Q^2 = \hbar D_s Q^2. \tag{5.101}$$

Thus the Chudley–Elliott model at small Q transforms into the Q^2 law, which, as was pointed out previously, is generally valid at small Q. For large Q values, on the other hand, the linewidth oscillates with Q; if we average over a certain range of large Q values, then

$$\text{large } Q: \qquad \Lambda(\boldsymbol{Q}) = \frac{\hbar}{z\tau}\sum_{j=1}^{z}(1 - 0) = \frac{\hbar}{\tau} \tag{5.102}$$

i.e. the linewidth mirrors the jump rate.

5.3.1 Isotropic Chudley–Elliott model

Chudley and Elliott originally derived their model for jump diffusion in liquids. Thereby they assume that an atom or molecule in a liquid is enclosed in a cage formed from other atoms or molecules and from time to time performs a jump into a neighbouring cage. The jump length l is identical for all sites whereas the jump direction is random. Therefore, Eq. (5.99) has to be spatially averaged:

$$\langle \Lambda(Q) \rangle = \frac{\hbar}{\tau}\left(1 - \left\langle e^{iQl\cos\theta}\right\rangle\right) = \frac{\hbar}{\tau}\left(1 - \frac{1}{2}\int_{-1}^{1} e^{iQl\cos\theta}\, d\cos\theta\right)$$

$$= \frac{\hbar}{\tau}\left(1 - \frac{\sin(Ql)}{Ql}\right) = \frac{6\hbar D_s}{l^2}\left(1 - \frac{\sin(Ql)}{Ql}\right). \qquad (5.103)$$

This function is plotted in Fig. 5.5. For small Q values the sine function can be expanded, resulting in the Q^2 law. For large Q values, Eq. (5.102) is reproduced, i.e. the linewidth corresponds to the jump rate.

In this model also a distribution of jump lengths, $g(l)$, can be taken into account (Egelstaff 1992). The result, Eq. (5.103), is sometimes applied to incoherent QENS data from polycrystalline solid samples. From a pragmatic point of view, this turned out to be successful. The jumplength l is then an effective quantity which somehow also implies the periodicity of the lattice. Theoretically correct would be a spatial

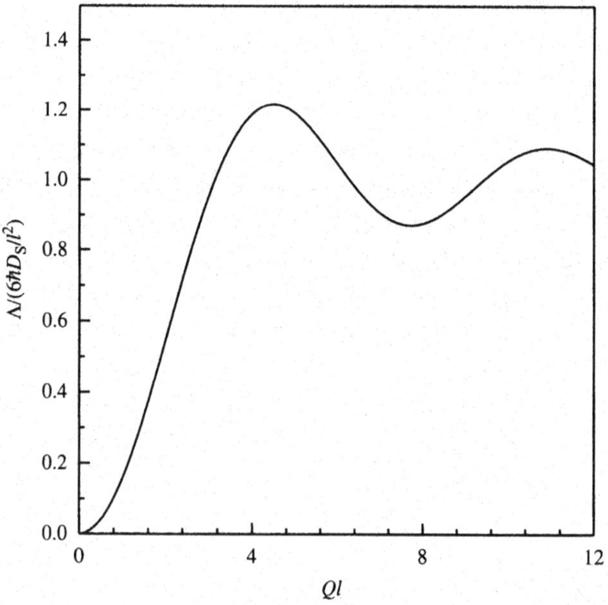

Fig. 5.5 Q-dependence of the linewidth of the isotropic Chudley–Elliott model.

average of the whole scattering function, i.e. of the appropriate form of Eq. (5.98) which, however, is only numerically feasible.

5.3.2 Jump diffusion on cubic lattices

For a primitive cubic lattice the lattice constant is denoted by a, and the coordination number is $z = 6$. The six jump vectors are $\pm(a, 0, 0)$, $\pm(0, a, 0)$ and $\pm(0, 0, a)$. Equation (5.99) then reads

$$\begin{aligned}\Lambda(\boldsymbol{Q}) &= \frac{\hbar}{6\tau}(6 - \mathrm{e}^{\mathrm{i}Q_x a} - \mathrm{e}^{-\mathrm{i}Q_x a} - \mathrm{e}^{\mathrm{i}Q_y a} - \mathrm{e}^{-\mathrm{i}Q_y a} - \mathrm{e}^{\mathrm{i}Q_z a} - \mathrm{e}^{-\mathrm{i}Q_z a}) \\ &= \frac{\hbar}{3\tau}(3 - \cos Q_x a - \cos Q_y a - \cos Q_z a). \qquad (5.104)\end{aligned}$$

If the scattering vector $\boldsymbol{Q}$ is directed parallel to the x-axis, this simplifies to

$$\Lambda(Q_x) = \frac{\hbar}{3\tau}(1 - \cos Q_x a). \qquad (5.105)$$

This function is displayed in Fig. 5.6. In the limit of small Q with $\cos x = 1 - x^2/2$ we obtain

$$\begin{aligned}&\text{small } Q_x: \quad \Lambda(Q_x) = \frac{\hbar}{3\tau}\frac{Q_x^2 a^2}{2} = \hbar D_{\mathrm{E}} Q_x^2; \\ &\text{small } Q: \quad \Lambda(Q) = \frac{\hbar}{3\tau}\left(\frac{Q_x^2 a^2}{2} + \frac{Q_y^2 a^2}{2} + \frac{Q_z^2 a^2}{2}\right) = \hbar\frac{a^2}{6\tau}Q^2. \qquad (5.106)\end{aligned}$$

Generally, the linewidth oscillates with maxima at the Brillouin zone boundaries and minima (zeros) at the Q values of the reciprocal lattice points. Thus the linewidth of the *incoherent* QENS mirrors the periodicity of the lattice. This might appear curious, since incoherent scattering means single-particle scattering, i.e. no mutual interference. However, the diffusing particle visits several lattice sites during the time of the passage of the neutron wave packet, and it is the self-interference from the distributed single particle which gives rise to the periodicity of the linewidths.

An fcc lattice is found for H diffusion in palladium and thus represents the standard example of the Chudley–Elliott model. Figure 5.7 shows one unit cell; the hydrogen sublattice is denoted by crosses and consists of the site in the centre of the cube and of the sites on the middle of the edges. The coordination number is $z = 12$, and the twelve jump vectors are:

$$\begin{aligned}&s_{1,7} = \pm(\tfrac{1}{2}, \tfrac{1}{2}, 0), \quad s_{2,8} = \pm(\tfrac{1}{2}, -\tfrac{1}{2}, 0), \quad s_{3,9} = \pm(\tfrac{1}{2}, 0, \tfrac{1}{2}), \\ &s_{4,10} = \pm(-\tfrac{1}{2}, 0, \tfrac{1}{2}), \quad s_{5,11} = \pm(0, \tfrac{1}{2}, \tfrac{1}{2}), \quad s_{6,12} = \pm(0, -\tfrac{1}{2}, \tfrac{1}{2}).\end{aligned}$$

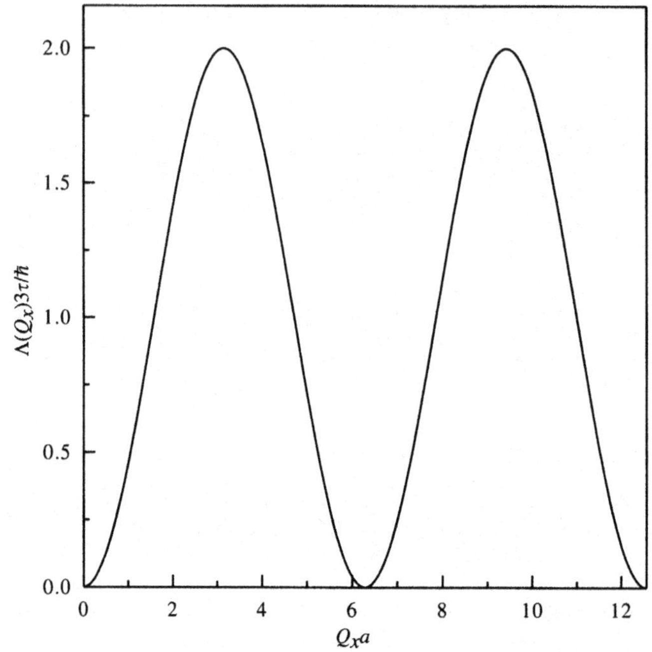

Fig. 5.6 Q-dependence of the linewidth for jump diffusion on a simple cubic lattice.

For the linewidth we thus obtain:

$$\Lambda(\boldsymbol{Q}) = \frac{\hbar}{6\tau}\left\{6 + \cos\left(\frac{Q_x a}{2} + \frac{Q_y a}{2}\right) + \cos\left(\frac{Q_x a}{2} - \frac{Q_y a}{2}\right)\right.$$
$$+ \cos\left(\frac{Q_x a}{2} + \frac{Q_z a}{2}\right) + \cos\left(\frac{Q_y a}{2} + \frac{Q_z a}{2}\right)$$
$$\left. + \cos\left(-\frac{Q_x a}{2} + \frac{Q_z a}{2}\right) + \cos\left(-\frac{Q_y a}{2} + \frac{Q_z a}{2}\right)\right\}. \qquad (5.107)$$

In this case a comparison with experimental results is possible and will be done in Section 8.2.

5.3.3 Jump diffusion on two-dimensional lattices

We proceed to two-dimensional jump diffusion and consider a hexagonal layer. The jump vectors are $\pm\boldsymbol{a}$, $\pm\boldsymbol{b}$, and $\pm(\boldsymbol{a}+\boldsymbol{b})$, and the resulting linewidth is

$$\Lambda(\boldsymbol{Q}) = \frac{\hbar}{6\tau}\{6 - 2\cos \boldsymbol{Q}\boldsymbol{a} - 2\cos \boldsymbol{Q}\boldsymbol{b} - 2\cos \boldsymbol{Q}(\boldsymbol{a}+\boldsymbol{b})\}. \qquad (5.108)$$

Thus an incoherent QENS spectrum with the scattering vector $\boldsymbol{Q}$ perpendicular to the layer exhibits zero line broadening, whereas a Lorentzian is obtained if

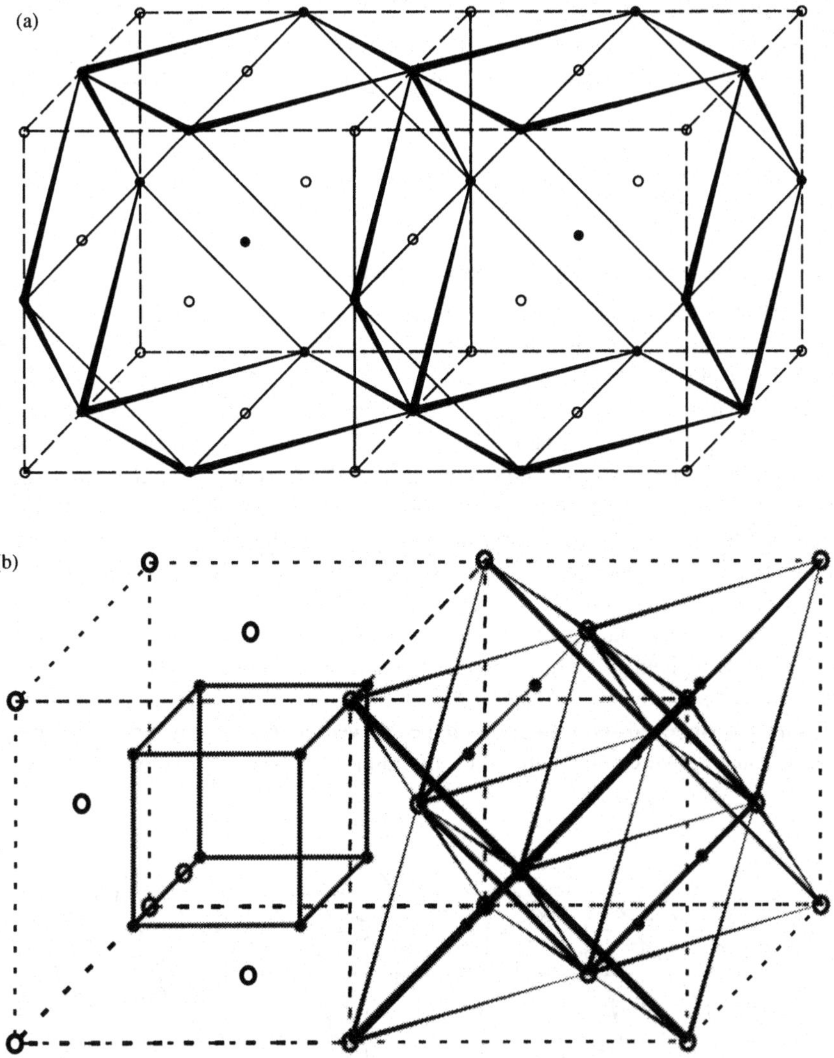

Fig. 5.7 Structure of fcc-Pd (host metal atoms: open circles) and two possibly hydrogen sites: (a) octahedral interstitial sites (full circles) and (b) tetrahedral interstitial sites (full circles).

the Q vector lies in the plane of the layer. For arbitrary directions a superposition of the Lorentzian and the δ-function represents the scattering function. Correspondingly, for a polycrystal of a two-dimensional diffusion system a peaked line shape is obtained (the spatial average of the (Lorentzian) scattering function which is different from a Lorentzian with a spatially averaged linewidth);

generally a peaked line shape indicates low-dimensional diffusion. However, as was pointed out by Lechner (1994) such a peaked line shape is very difficult to prove experimentally because of the finite instrumental linewidth of a neutron spectrometer.

5.4 TRANSLATIONAL JUMP DIFFUSION IN COMPLEX LATTICES

5.4.1 Lattices with all sites energetically equivalent

The Chudley–Elliott model can be generalized in order to deal with crystallographically different and later (Section 5.4.2) also with energetically different sites. Crystallographically different sites occur, e.g., for H diffusion over the tetrahedral interstitial sites in bcc metals like V, Nb, Ta, and for this example the extended Chudley–Elliott model will be explained in detail. Figure 5.8 displays two unit cells where the large spheres represent the metal atoms and the small ones the system of H sites. This system consists of a superposition of six bcc sublattices, which are denoted by the subscripts m, n; $Y = 6$ is the number of sublattices. We denote by $\boldsymbol{s}_{mn}^{\boldsymbol{k}}$ the jump vectors from sublattice m to sublattice n with a distance vector of the elementary cells $\boldsymbol{k}$; in the case of hydrogen occupying tetrahedral sites there are only four nearest neighbour jumps, i.e. $z = 4$; these jumps occur partly within a given cell, but also to neighbouring elementary cells such that the vector $\boldsymbol{k}$ takes values out of $\{000, 100, \bar{1}00, 010, 0\bar{1}0, 001, 00\bar{1}\}$. Also the jump rates must be indexed additionally with the cell vectors $\boldsymbol{k}$ as superscript in order to make clear into which elementary cell the jump takes place. If we only consider

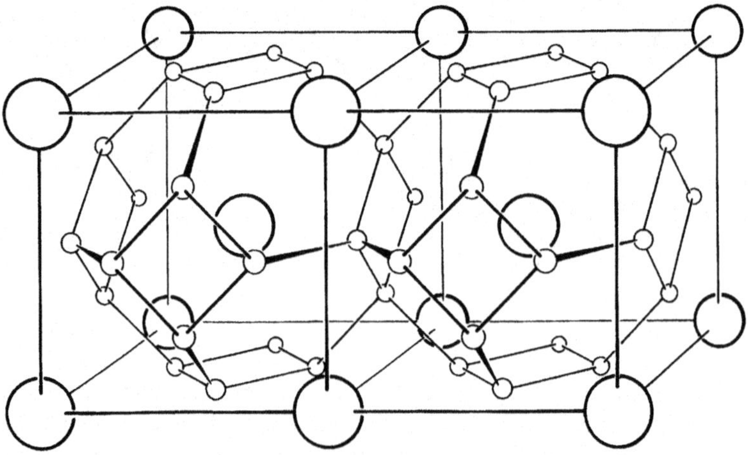

Fig. 5.8 Two unit cells of a bcc lattice; the large spheres represent the metal atoms, the small ones the system of tetrahedral H sites.

jumps to adjacent sites, then we have

$$\Gamma^{\boldsymbol{k}}_{mn} = \begin{cases} 1/z\tau & \text{if } m \text{ and } n \text{ are neighbouring sites} \\ 0 & \text{otherwise.} \end{cases} \tag{5.109}$$

For each of the $Y = 6$ sublattices a master equation is valid:

$$\frac{\partial}{\partial t} P_n(\boldsymbol{r}, t) = \sum_{\substack{m,\boldsymbol{k} \\ m \neq n}}^{4} \{\Gamma^{\boldsymbol{k}}_{mn} P_m(\boldsymbol{r} + \boldsymbol{s}^{\boldsymbol{k}}_{mn}, t) - \Gamma^{\boldsymbol{k}}_{nm} P_n(\boldsymbol{r}, t)\}, \quad n = 1, \ldots, 6 \tag{5.110}$$

which is simplified to

$$\frac{\partial}{\partial t} P(\boldsymbol{r}, t)_n = \frac{1}{z\tau} \sum_{\substack{m,\boldsymbol{k} \\ m \neq n}}^{4} \{P_m(\boldsymbol{r} + \boldsymbol{s}^{\boldsymbol{k}}_{mn}, t) - P_n(\boldsymbol{r}, t)\}, \quad n = 1, \ldots, 6 \tag{5.111}$$

because of Eq. (5.109). In spite of the assumption that Y $= 6$ the sum contains only $z = 4$ terms, because each interstitial site has only four nearest neighbour sites. The prefactor multiplied by the first term in the curly bracket means the jumps from four neighbouring sites to be specified to the site under consideration, the prefactor multiplied by the second term in the curly bracket means the jumps out of the site under considerations into four unspecified other sites. $P_n(\boldsymbol{r}, t)$ means the conditional probability density of finding an H atom at time t at position $\boldsymbol{r}$ on sublattice n, if it was at the origin at time $t = 0$, no matter which sublattice.

The conditonal probability densities are related to the self-correlation function $G_{\mathrm{s}}(\boldsymbol{r}, t)$ by

$$G_{\mathrm{s}}(\boldsymbol{r}, t) = \sum_{n=1}^{Y} P_n(\boldsymbol{r}, t) \tag{5.112}$$

which denotes the conditional probability density of finding a hydrogen atom at time t at position $\boldsymbol{r}$ (whatever sublattice) if it was at the origin at time $t = 0$ (whatever sublattice). The initial conditions are

$$G_{\mathrm{s}}(\boldsymbol{r}, 0) = \delta(\boldsymbol{r}) \tag{5.113}$$

and therefore

$$P_n(\boldsymbol{r}, 0) = \rho_n = \frac{1}{Y}\delta(\boldsymbol{r}), \quad n = 1, \ldots, 6. \tag{5.114}$$

All sites are energetically equivalent and ρ_n is their thermal occupation number. The system of equations (5.112) can be solved by Fourier transformation into momentum space; for each sublattice we obtain

$$\frac{\partial}{\partial t} P_n(\boldsymbol{Q}, t) = \frac{1}{z\tau} \sum_{m,\boldsymbol{k}}^{z} \left\{ \int P_m(\boldsymbol{r}' + \boldsymbol{s}_{nm}^{\boldsymbol{k}}, t) \mathrm{e}^{\mathrm{i}\boldsymbol{Q}\boldsymbol{r}'} \mathrm{d}\boldsymbol{r}' - P_n(\boldsymbol{Q}, t) \right\}$$

$$= \frac{1}{z\tau} \sum_{m}^{Y} \left\{ P_m(\boldsymbol{Q}, t) \sum_{\boldsymbol{k}} \exp(-\mathrm{i}\boldsymbol{Q}\boldsymbol{s}_{nm}^{\boldsymbol{k}}) - P_n(\boldsymbol{Q}, t) \right\} \tag{5.115}$$

where in the second row the substitution $\boldsymbol{r}' = \boldsymbol{r} + \boldsymbol{s}_{nm}^{\boldsymbol{k}}$ was used.

In order to transfer into matrix notation the second term in the sum is expressed by means of the Kronecker symbol

$$P_n(\boldsymbol{Q}, t) = \sum_{m} P_m(\boldsymbol{Q}, t) \delta_{mn}. \tag{5.116}$$

Hence

$$\frac{\partial}{\partial t} P_n(\boldsymbol{Q}, t) = \frac{1}{z\tau} \sum_{m}^{Y} \left\{ P_m(\boldsymbol{Q}, t) \sum_{\boldsymbol{k}} \left\{ \exp\left(-\mathrm{i}\boldsymbol{Q}\boldsymbol{s}_{nm}^{\boldsymbol{k}} \right) - \delta_{mn} \right\} \right\}; \tag{5.117}$$

the right-hand side of this equation represents the multiplication of a matrix $\boldsymbol{\Lambda}$ by a vector $\boldsymbol{P}$, hence Eq. (5.117) can be translated into

$$\dot{\boldsymbol{P}} = \boldsymbol{\Lambda} \boldsymbol{P}. \tag{5.118}$$

The matrix $\boldsymbol{\Lambda}$ is called the jump matrix, with matrix elements

$$\Lambda_{nm} = \frac{1}{z\tau} \sum_{\boldsymbol{k}} \exp\left(-\mathrm{i}\boldsymbol{Q}\boldsymbol{s}_{nm}^{\boldsymbol{k}} \right) - \frac{1}{\tau} \delta_{mn} \tag{5.119}$$

where the first term represents the non-diagonal elements and the second term the diagonal ones. On permutation of the subscripts n and m the jump vector $\boldsymbol{s}_{nm}^{\boldsymbol{k}}$ changes its sign, i.e.

$$\Lambda_{nm} = \Lambda_{mn}^{*}. \tag{5.120}$$

Thus the jump matrix $\boldsymbol{\Lambda}$ is hermitean and exhibits real eigenvalues. The solution of Eq. (5.118) is obvious:

$$\boldsymbol{P} = \exp(\boldsymbol{\Lambda} t) \boldsymbol{P}_0. \tag{5.121}$$

A similarity transformation transforms $\boldsymbol{\Lambda}$ into the diagonal matrix $\mathbf{D}$

$$\boldsymbol{\Lambda} = \mathbf{T}\mathbf{D}\mathbf{T}^{-1}. \tag{5.122}$$

Correspondingly we have

$$P = \mathbf{T}\exp(\mathbf{D}t)\mathbf{T}^{-1}P_0 = \mathbf{T}\begin{pmatrix} e^{\lambda_1 t} & & \\ & \ddots & \\ & & e^{\lambda_Y t}\end{pmatrix}\mathbf{T}^{-1}P_0. \tag{5.123}$$

This can be written component-by-component:

$$P_n(Q,t) = \sum_{\delta,l} T_{n\delta}\exp\left(\lambda_\delta t\right)\left(T^{-1}\right)_{\delta l} P_l(Q,0). \tag{5.124}$$

The transformation matrix $\mathbf{T}$ is formed column-by-column out of the eigenvectors ν_δ of the matrix $\mathbf{\Lambda}$; therefore

$$T_{n\delta} = \nu_\delta^n \tag{5.125}$$

is the nth component of the eigenvector belonging to the eigenvalue λ_δ. Since $\mathbf{T}$ is a unitary matrix, we obtain

$$\left(T^{-1}\right)_{\delta l} = \nu_\delta^{*l}. \tag{5.126}$$

Equation (5.124) can thus be rewritten as

$$P_n(Q,t) = \sum_{\delta l} \nu_\delta^n \exp(\lambda_\delta t)\nu_\delta^{*l} P_l(Q,0). \tag{5.127}$$

with the initial condition (5.114), which after Fourier transformation corresponds to

$$P_l(Q,0) = \frac{1}{Y}. \tag{5.128}$$

Because of Eq. (5.112) the self-part of the intermediate scattering function is given by

$$I_s(Q,t) = \sum_{n=1}^{Y} P_n(Q,t) = \sum_{\delta=1}^{Y}\left(\frac{1}{Y}\sum_{n,l}\nu_\delta^n \nu_\delta^{*l}\right)\exp(\lambda_\delta t). \tag{5.129}$$

Fourier transformation with respect to time yields the incoherent scattering function for jump diffusion on the tetrahedral interstitial sites of a bcc lattice:

$$S_i(Q,\omega) = \frac{1}{\pi}\sum_{\delta=1}^{Y} w_\delta \frac{-\hbar\lambda_\delta}{(\hbar\lambda_\delta)^2 + (\hbar\omega)^2}. \tag{5.130}$$

Thus a diffusional system consisting of a superposition of six Bravais sublattices, like hydrogen on the tetrahedral interstitial sites in Vb metals, leads to a 6×6 jump matrix and to a scattering function consisting of a superposition of six Lorentzians.

The linewidths are the negative eigenvalues of the jump matrix, whereas the respective weights are given by the eigenvectors according to

$$w_\delta = \frac{1}{6}\left|\sum_{l=1}^{6} v_\delta^l\right|^2. \tag{5.131}$$

The jump matrix is explicitly given by (Bée 1988)

$$\mathbf{\Lambda} = \frac{1}{\tau}\begin{pmatrix} -1 & 0 & T_{yz} & T_{\bar{y}z} & T^*_{xz} & T^*_{\bar{x}z} \\ 0 & -1 & T_{\bar{y}z} & T^*_{yz} & T_{\bar{x}z} & T_{xz} \\ T^*_{yz} & T_{\bar{y}z} & -1 & 0 & T_{xy} & T_{\bar{x}z} \\ T^*_{\bar{y}z} & T_{yz} & 0 & -1 & T^*_{\bar{x}y} & T^*_{xz} \\ T_{xz} & T_{\bar{x}z} & T^*_{xy} & T_{\bar{x}y} & -1 & 0 \\ T_{\bar{x}z} & T^*_{xz} & T^*_{\bar{x}y} & T_{xy} & 0 & -1 \end{pmatrix}$$

For instance, $\Lambda_{13} = T_{yz} = \frac{1}{4}\exp(\mathrm{i}a(Q_y + Q_z)/4)$. This matrix is hermitean, but not symmetric because there is no inversion symmetry. For the main crystallographic directions the linewidths (eigenvalues) and relative intensities (weight) are displayed in Fig. 5.9.

5.4.2 Lattices with energetically different sites

Now we proceed to a further generalization of the Chudley–Elliott model and allow the sites to be not only crystallographically but also energetically different. An example is hydrogen diffusion in $LaNi_5H_\alpha$, and we will here develop the model for this example. The starting point is Eq. (5.110). Each of the jump vectors $\boldsymbol{s}^{\boldsymbol{k}}_{mn}$ is associated with a jump rate $\Gamma^{\boldsymbol{k}}_{mn}$, but now, in contrast to the previous example, back and forth jump rates are not equal; they are, however, related by the detailed balance condition to the different site energies:

$$\Gamma_{nm}\exp(-\beta E_n) = \Gamma_{mn}\exp(-\beta E_m) \tag{5.132}$$

with $\beta = (k_B T)^{-1}$; E_n and E_m are the potential energies of sites n and m, respectively. Fourier transformation of Eq. (5.110) into momentum space yields

$$\frac{\partial}{\partial t}P_n(\boldsymbol{Q},t) = \sum_m P_m(\boldsymbol{Q},t)\left[\sum_{\boldsymbol{k}}\Gamma^{\boldsymbol{k}}_{mn}\exp(\mathrm{i}\boldsymbol{Q}\boldsymbol{s}^{\boldsymbol{k}}_{nm}) - \delta_{mn}\sum_{l,\boldsymbol{k}}\Gamma^{\boldsymbol{k}}_{nm}\right],$$

$$n = 1,\ldots,Y \tag{5.133}$$

(with $\boldsymbol{s}^{\boldsymbol{k}}_{nm} = -\boldsymbol{s}^{\boldsymbol{k}}_{\mathrm{mn}}$). In matrix notation this expression can be written as $\dot{\boldsymbol{p}} = \mathbf{\Lambda}\boldsymbol{p}$. The elements of the jump matrix $\mathbf{\Lambda}$ are given by

$$\Lambda_{mn} = \sum_{\boldsymbol{k}}\Gamma^{\boldsymbol{k}}_{mn}\exp(\mathrm{i}\boldsymbol{Q}\boldsymbol{s}^{\boldsymbol{k}}_{nm}) - \delta_{mn}\sum_{m,\boldsymbol{k}}\Gamma^{\boldsymbol{k}}_{nm}. \tag{5.134}$$

Because of Eq. (5.132) the jump rates corresponding to jumps in opposite directions are not necessarily equal, so that $\Lambda_{mn} \neq \Lambda^*_{nm}$. Thus the matrix $\mathbf{\Lambda}$ is not hermitean.

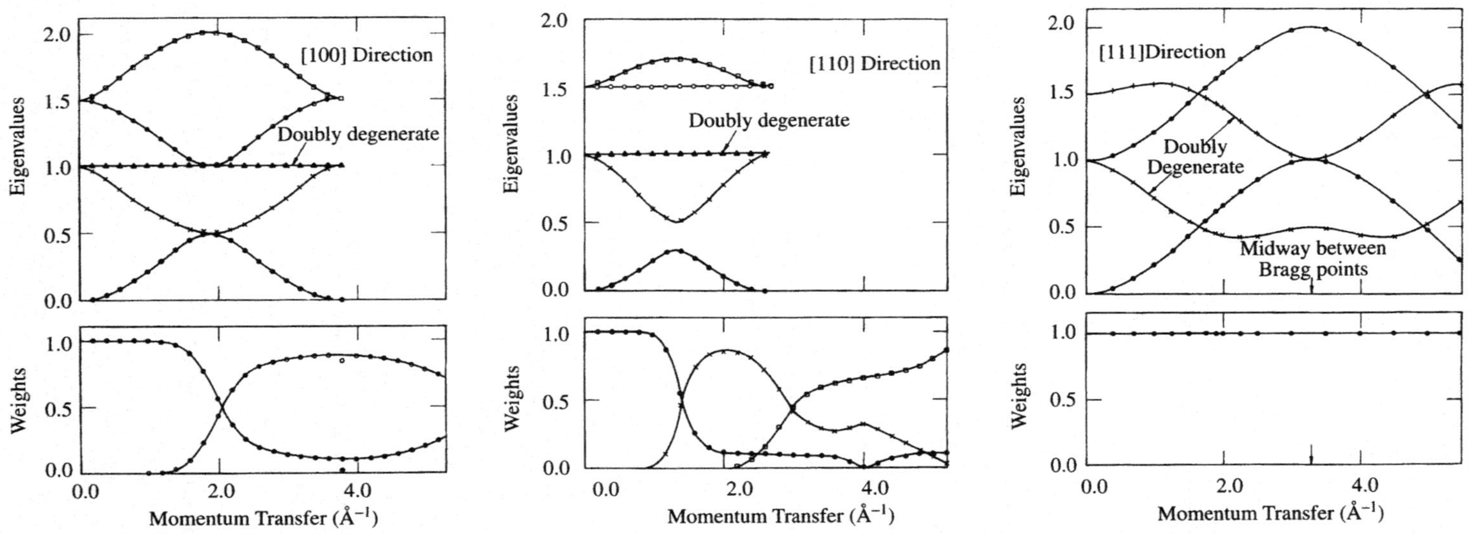

Fig. 5.9 Eigenvalues and weights of the jump matrix in the main crystallographic directions for diffusion on the tetrahedral sites of a bcc lattice (from Kehr 1985, unpublished).

To convert $\mathbf{\Lambda}$ into a hermitean form a similarity transformation with the matrix $\boldsymbol{\rho}$ of the thermal occupation numbers is performed:

$$\boldsymbol{\rho} = \rho_n \delta_{nm} \quad \text{with } \rho_n = \mathrm{e}^{-\beta E_n} \Big/ \sum_m \mathrm{e}^{-\beta E_m} \tag{5.135}$$

such that

$$\tilde{\mathbf{\Lambda}} = \boldsymbol{\rho}^{-1/2} \mathbf{\Lambda} \, \boldsymbol{\rho}^{1/2}. \tag{5.136}$$

With the matrix $\tilde{\mathbf{\Lambda}}$ we write the system of differential equations (5.133) as

$$\dot{\tilde{\boldsymbol{p}}} = \tilde{\mathbf{\Lambda}} \tilde{\boldsymbol{p}}; \tag{5.137}$$

the solution is obvious:

$$\tilde{\boldsymbol{p}} = \exp\left(\tilde{\mathbf{\Lambda}} |t|\right) \cdot \tilde{\boldsymbol{p}}_0. \tag{5.138}$$

As a hermitean matrix $\tilde{\mathbf{\Lambda}}$ can be diagonalized by means of a transformation matrix $\mathbf{T}$:

$$\tilde{\mathbf{\Lambda}} = \mathbf{T} \mathbf{D} \mathbf{T}^{-1}. \tag{5.139}$$

$\mathbf{D}$ is a diagonal matrix and contains the eigenvalues λ_δ of the matrix $\tilde{\mathbf{\Lambda}}$ which, because of Eq. (5.136), are also eigenvalues of the jump matrix $\mathbf{\Lambda}$. Equation (5.138) thus transforms into

$$\tilde{\boldsymbol{p}} = \mathbf{T} \exp(\mathbf{D} t) \mathbf{T}^{-1} \cdot \tilde{\boldsymbol{p}}_0 = \mathbf{T} \begin{pmatrix} \mathrm{e}^{\lambda_1 t} & & \\ & \ddots & \\ & & \mathrm{e}^{\lambda_Y t} \end{pmatrix} \mathbf{T}^{-1} \cdot \tilde{\boldsymbol{p}}_0. \tag{5.140}$$

With Eq. (5.136) we transform $\tilde{\boldsymbol{p}}$ back into $\boldsymbol{p}$, hence

$$\boldsymbol{p} = \boldsymbol{\rho}^{1/2} \mathbf{T} \begin{pmatrix} \mathrm{e}^{\lambda_1 t} & & \\ & \ddots & \\ & & \mathrm{e}^{\lambda_Y t} \end{pmatrix} \mathbf{T}^{-1} \, \boldsymbol{\rho}^{-1/2} \boldsymbol{p}_0. \tag{5.141}$$

Written in terms of components this expression yields

$$P_n(\boldsymbol{Q}, t) = \sum_{\delta, l} \rho_n^{1/2} T_{n\delta} \exp(\lambda_\delta t) \left(T^{-1}\right)_{\delta l} \rho_l^{-1/2} P_l(\boldsymbol{Q}, 0). \tag{5.142}$$

The transformation matrix is formed column-by-column out of the eigenvectors $\tilde{v}_\delta$ of the matrix $\tilde{\mathbf{\Lambda}}$. Therefore

$$T_{n\delta} = \tilde{v}_\delta^n \tag{5.143}$$

is the nth component of the eigenvector belonging to the eigenvalue λ_δ. $\mathbf{T}$ is a unitary matrix, hence

$$(T^{-1})_{\delta l} = \tilde{v}_\delta^{*l}. \tag{5.144}$$

Thus Eq. (5.142) transforms into

$$P_n(\boldsymbol{Q}, t) = \sum_{\delta,l} \rho_n^{1/2} \tilde{v}_\delta^n \exp(\lambda_\delta t) \tilde{v}_\delta^{*l} \rho_l^{1/2} P_l(\boldsymbol{Q}, 0). \tag{5.145}$$

With the initial condition that the diffusing particles at time $t = 0$ are in thermodynamic equilibrium,

$$P_l(\boldsymbol{r}, 0) = \rho_l \delta(\boldsymbol{r}), \quad \text{thus } P_l(\boldsymbol{Q}, 0) = \rho_l. \tag{5.146}$$

We obtain the self-part of the intermediate scattering function by summing over all sites n:

$$I_s(\boldsymbol{Q}, t) = \sum_{\delta=1}^{Y} \sum_{l=1}^{Y} \sum_{n=1}^{Y} \rho_l^{1/2} \tilde{v}_\delta^n \tilde{v}_\delta^{*l} \rho_n^{1/2} \exp(\lambda_\delta t). \tag{5.147}$$

Fourier transformation with respect to time yields the incoherent scattering function (5.130), i.e. a superposition of Y (number of sites) Lorentzians with linewidths (HWHM) $-\hbar\lambda_\delta$ (negative eigenvalues of the jump matrix) and weights w_δ which result from the corresponding eigenvectors:

$$w_\delta = \sum_{l=1}^{Y} \sum_{n=1}^{Y} \rho_l^{1/2} \tilde{v}_\delta^n \tilde{v}_\delta^{*l} \rho_n^{1/2} = \left| \sum_{l=1}^{Y} \rho_l^{1/2} \tilde{v}_\delta^l \right|^2. \tag{5.148}$$

5.5 DISORDERED SYSTEMS; MULTIPLE TRAPPING

From the previous subsections it became clear that QENS is a particularly powerful experimental technique for the investigation of highly ordered systems if samples in the form of large single crystals are available. But QENS is also very useful for the study of diffusion in disordered systems; the approach is different, and so are the possible results.

A very general and widely used approach is the so-called two-state model; this model was originally developed by Singwi and Sjölander (1960) in order to describe the diffusion of water molecules in liquid water and later adapted by Richter and Springer (1978) to the diffusion of hydrogen in metals in the presence of impurities.

We follow this latter derivation and consider an interstitial lattice, in which a small fraction of sites, distributed at random, is energetically lowered compared to the 'regular' sites. The particle resides for some time at a trap site; this is called the trapped state. Due to thermal fluctuations, after an average time τ_0 it manages to escape (escape rate τ_0^{-1}) and performs a random walk over the regular lattice site; this is called the free state. After an average time τ_1, i.e. with the trapping rate τ_1^{-1}, it hits a trap site and is trapped again. Thus the diffusion process comprises a sequence of changes, random in time and space, between the free and trapped states.

For the derivation of the respective incoherent scattering function we again start with the self-correlation function. We note that at time $t = 0$ the particle can be found either at a trap (with the probability $\tau_0/(\tau_1 + \tau_0)$) or at a regular site (with the probability $\tau_1/(\tau_1 + \tau_0)$). If the particle is in the trapped state at $t = 0$, then

$$p(t) = \exp\left(-\frac{t}{\tau_0}\right) \tag{5.149}$$

denotes the probability that it can still be found there after the time t; thus τ_0^{-1} is the escape rate. Correspondingly, if the particle is in the free state at $t = 0$, then

$$q(t) = \exp\left(-\frac{t}{\tau_1}\right) \tag{5.150}$$

denotes the probability that it can still be found in the free state after the time t; thus τ_1^{-1} is the trapping rate or τ_1 is the average time between two trapping events. Equations (5.149) and (5.150) correspond to Eq. (4.94) which is valid for the case that all sites are energetically equivalent. The self-correlation function for the trapped state is simply

$$G_{\mathrm{T0}}(\boldsymbol{r}, t) = \delta(\boldsymbol{r}) \tag{5.151}$$

whereas the self-correlation function for the free state, $G_{\mathrm{F0}}(\boldsymbol{r}, t)$, is represented by the appropriate van Hove self-correlation function, for instance by Eq. (5.84). However, we do not specify it here, but only consider its Fourier transform

$$G_{\mathrm{F0}}(\boldsymbol{Q}, t) = \exp\left(-\frac{t}{\hbar}\Lambda(Q)\right) \tag{5.152}$$

where the result of the appropriate diffusion model has to be inserted, e.g. Eq. (5.88) or Eq. (5.103).

$$p(t) - p(t + \mathrm{d}t) = -p'(t)\,\mathrm{d}t = \frac{1}{\tau_0}\,\mathrm{d}t \tag{5.153}$$

denotes the transition probability that the particle changes from the trapped to the free state between the time t and the time $t + \mathrm{d}t$, and

$$q(t) - q(t + \mathrm{d}t) = -q'(t)\,\mathrm{d}t = \frac{1}{\tau_1}\,\mathrm{d}t \tag{5.154}$$

correspondingly denotes the transition probability that the particle changes from the free to the trapped state between t and $t + \mathrm{d}t$. Let us assume that the particle starts in the trapped state at $t = 0$ at the origin of our coordinate system, $\boldsymbol{r} = \mathbf{0}$. Within the time period t it can reach the position $\boldsymbol{r}$ by $0, 1, 2, \ldots, \infty$ changes of state. Starting in the trapped state, we denote by $F_{\mathrm{T},i}\,(\boldsymbol{r}, t)$ the probability of reaching the position $\boldsymbol{r}$ within time t after i transitions of state. Hence

$$G_{\mathrm{T}}(\boldsymbol{r}, t) = \sum_{i=0}^{\infty} F_{\mathrm{T},i}(\boldsymbol{r}, t); \tag{5.155}$$

as examples the first three $F_{\mathrm{T},i}$'s are given explicitly:

(i) zero transitions within the time period t:

$$F_{\mathrm{T},0}(\boldsymbol{r}, t) = G_{\mathrm{T0}}(\boldsymbol{r}, t)\, p(t);$$

(ii) one transition at position r' at time t', integrated over all r' and t':

$$F_{\mathrm{T},1}(r, t) = -\int_0^t \mathrm{d}t' \int_{-\infty}^{\infty} \mathrm{d}^3 r' q(t - t')\, G_{\mathrm{F0}}(r - r', t - t')\, p'(t')\, G_{\mathrm{T0}}(r', t');$$

(iii) two transitions, the first one at r', t', the second one at r'', t'':

$$\begin{aligned} F_{\mathrm{T},2}(r, t) = (-1)^2 \int_0^t \mathrm{d}t'' \int_0^{t''} \mathrm{d}t' \int_{-\infty}^{\infty} & \mathrm{d}^3 r''\, \mathrm{d}^3 r'\, p(t - t'') \\ & \cdot G_{\mathrm{T0}}(r - r'', t - t'')\, q'(t'' - t') \\ & \cdot G_{\mathrm{F0}}(r'' - r', t'' - t')\, p'(t')\, G_{\mathrm{T0}}(r', t') \end{aligned}$$

and so on.

The separation into a spatial transition probability and a time dependence is admissible, as was pointed out in connection with Eq. (4.93). Thus Eq. (5.155) consists of a very complex sum of integral equations but, analogously to Feyman's graphs, it can be written as

$$G_{\mathrm{T}} = \text{——} + \text{——}\otimes\!\sim\!\sim\!\sim + \text{——}\otimes\!\sim\!\sim\!\sim\!\otimes\text{——} + \cdots \tag{5.156}$$

Here symbolically a straight line represents the trapped state, a wavy line the free state, and $\otimes$ the transition; mathematically $\otimes$ means a convolution integral (in space and time) and a multiplication by τ_0^{-1} or τ_1^{-1}, respectively. It is appropriate to specify not only the initial state of the particle by a subscript, but also the final state by a superscript, such that we define the following self-correlation functions:

$$
\begin{aligned}
G_{\mathrm{F}}^{\mathrm{F}} &= \sim\sim\sim + \sim\sim\sim\otimes\text{——}\otimes\sim\sim\sim + \sim\sim\sim\otimes\text{——}\otimes\sim\sim\sim\otimes\text{——}\otimes\sim\sim\sim + \cdots \\
G_{\mathrm{F}}^{\mathrm{T}} &= \sim\sim\sim\otimes\text{——} + \sim\sim\sim\otimes\text{——}\otimes\sim\sim\sim\otimes\text{——} \\
&\quad + \sim\sim\sim\otimes\text{——}\otimes\sim\sim\sim\otimes\text{——}\otimes\sim\sim\sim\otimes\text{——} + \cdots \\
G_{\mathrm{T}}^{\mathrm{T}} &= \text{——} + \text{——}\otimes\sim\sim\sim\otimes\text{——} + \text{——}\otimes\sim\sim\sim\otimes\text{——}\otimes\sim\sim\sim\otimes\text{——} + \cdots \\
G_{\mathrm{T}}^{\mathrm{F}} &= \text{——}\otimes\sim\sim\sim + \text{——}\otimes\sim\sim\sim\otimes\text{——}\otimes\sim\sim\sim + \cdots
\end{aligned}
\tag{5.157}
$$

These equations look very complicated, but they can be simplified; in symbols we obtain

$$
\begin{aligned}
G_{\mathrm{F}}^{\mathrm{F}} &= \sim\sim\sim + G_{\mathrm{T}}^{\mathrm{F}}\otimes\sim\sim\sim \\
G_{\mathrm{F}}^{\mathrm{T}} &= G_{\mathrm{F}}^{\mathrm{F}}\otimes\text{——} \\
G_{\mathrm{T}}^{\mathrm{T}} &= \text{——} + G_{\mathrm{T}}^{\mathrm{F}}\otimes\text{——} \\
G_{\mathrm{T}}^{\mathrm{F}} &= G_{\mathrm{T}}^{\mathrm{T}}\otimes\sim\sim\sim
\end{aligned}
\tag{5.158}
$$

which corresponds to the recursion formula (4.95); mathematically it is expressed as:

$$
\begin{aligned}
G_{\mathrm{F}}^{\mathrm{F}}(\boldsymbol{r},t) &= \exp\left(-\frac{t}{\tau_1}\right)G_{\mathrm{F}0}(\boldsymbol{r},t) \\
&\quad + \int_0^t \mathrm{d}t' \int_{-\infty}^{\infty} \mathrm{d}^3r' \exp\left(-\frac{t-t'}{\tau_1}\right)G_{\mathrm{F}0}(r-r',t-t')\frac{1}{\tau_0}G_{\mathrm{F}}^{\mathrm{T}}(r',t').
\end{aligned}
\tag{5.159}
$$

$1/\tau_0$ is the transition rate into the free state at an (arbitrary) time point t', and the factor $\exp(-(t-t')/\tau_1)$ is the probability that no further transitions occur between t' and t. Correspondingly we obtain

$$
G_{\mathrm{F}}^{\mathrm{T}}(\boldsymbol{r},t) = \int_0^t \mathrm{d}t' \int_{-\infty}^{\infty} \mathrm{d}^3r' \exp\left(-\frac{t-t'}{\tau_0}\right)G_{\mathrm{T}0}(r-r',t-t')\frac{1}{\tau_1}G_{\mathrm{F}}^{\mathrm{F}}(r',t')
\tag{5.160}
$$

$$G_{\mathrm{T}}^{\mathrm{T}}(\boldsymbol{r},t)=\exp\left(-\frac{t}{\tau_0}\right)G_{\mathrm{T0}}(\boldsymbol{r},t)$$
$$+\int_0^t \mathrm{d}t'\int_{-\infty}^{\infty}\mathrm{d}^3r'\exp\left(-\frac{t-t'}{\tau_0}\right)G_{\mathrm{T0}}(r-r',t-t')\frac{1}{\tau_1}G_{\mathrm{T}}^{\mathrm{F}}(r',t') \tag{5.161}$$

$$G_{\mathrm{T}}^{\mathrm{F}}(\boldsymbol{r},t)=\int_0^t \mathrm{d}t'\int_{-\infty}^{\infty}\mathrm{d}^3r'\exp\left(-\frac{t-t'}{\tau_1}\right)G_{\mathrm{F0}}(r-r',t-t')\frac{1}{\tau_0}G_{\mathrm{T}}^{\mathrm{T}}(r',t'). \tag{5.162}$$

The total self-correlation function is composed of the above functions by thermal averaging over the initial states:

$$G_{\mathrm{s}}(\boldsymbol{r},t)=\frac{\tau_0}{\tau_0+1}\left[G_{\mathrm{T}}^{\mathrm{T}}+G_{\mathrm{T}}^{\mathrm{F}}\right]+\frac{\tau_1}{\tau_1+\tau_0}\left[G_{\mathrm{F}}^{\mathrm{T}}+G_{\mathrm{F}}^{\mathrm{F}}\right]. \tag{5.163}$$

Other than the ν-fold convolution integral, Eq. (4.95), this complex integral equation cannot be integrated directly recursively but it can be solved by Fourier transformation in space:

$$F^{\mathrm{r}}\{f(r)\}=\int_{-\infty}^{\infty}\mathrm{e}^{-\mathrm{i}\boldsymbol{Q}\boldsymbol{r}}f(r)\,\mathrm{d}^3r=f(\boldsymbol{Q}),$$

together with Laplace transformation in time,

$$L^{\mathrm{t}}\{f(t)\}=\int_0^{\infty}\mathrm{e}^{-st}f(t)\,\mathrm{d}t=\tilde{f}(s).$$

We thereby make use of the convolution laws

$$L^{\mathrm{t}}\left\{\int_0^t f_1(t-t')f_2(t')\,\mathrm{d}t'\right\}=L^{\mathrm{t}}\{f_1(t)\}\cdot L^{\mathrm{t}}\{f_2(t)\}$$

$$F^{\mathrm{r}}\left\{\int_{-\infty}^{\infty} f_1(r-r')f_2(r')\,\mathrm{d}^3r'\right\}=F^{\mathrm{r}}\{f_1(r)\}\cdot F\{f_2(r)\}$$

and of the relation

$$L^{\mathrm{t}}\{\mathrm{e}^{-\alpha t}\}=\frac{1}{s+\alpha}.$$

The Fourier and Laplace transforms of the partial self-correlation function $G_{\mathrm{F}}^{\mathrm{F}}(\boldsymbol{r}, t)$ can thus be evaluated in the following way:

$$
\begin{aligned}
\tilde{G}_{\mathrm{F}}^{\mathrm{F}}(\boldsymbol{Q}, s) &= L^{\mathrm{t}}\{F^{\mathrm{r}}\{G_F^F(\boldsymbol{r}, t)\}\} \\
&= L^{\mathrm{t}}\left\{F^{\mathrm{r}}\left\{\exp\left(-\frac{t}{\tau_1}\right)\right\}G_{\mathrm{F0}}(\boldsymbol{r}, t)\right\} \\
&\quad + L^{\mathrm{t}}\left\{F^{\mathrm{r}}\left\{\frac{1}{\tau_0}\int_0^t \mathrm{d}t' \int_{-\infty}^{\infty} \mathrm{d}^3r' \exp\left(\frac{t-t'}{\tau_1}\right)G_{\mathrm{F0}}(\boldsymbol{r}-\boldsymbol{r}', t-t')\right.\right. \\
&\quad \left.\left. \cdot G_{\mathrm{F}}^{\mathrm{T}}(\boldsymbol{r}', t')\right\}\right\} \\
&= L^{\mathrm{t}}\left\{\exp\left(-\frac{t}{\tau_1}\right)G_{\mathrm{F0}}(\boldsymbol{Q}, t)\right\} \\
&\quad + L^{\mathrm{t}}\left\{\frac{1}{\tau_0}\int_0^t \mathrm{d}t' \exp\left(-\frac{t-t'}{\tau_1}\right)F^{\mathrm{r}}\{G_{\mathrm{F0}}(\boldsymbol{r}, t-t')\}F^{\mathrm{r}}\{G_{\mathrm{F}}^{\mathrm{T}}(\boldsymbol{r}, t')\}\right\} \\
&= L^{\mathrm{t}}\left\{\exp\left(-\frac{t}{\tau_1}\right)\exp(-t\Lambda(\boldsymbol{Q}))\right\} \\
&\quad + L^{\mathrm{t}}\left\{\frac{1}{\tau_0}\int_0^t \mathrm{d}t' \exp\left(-\frac{t-t'}{\tau_1}\right)G_{\mathrm{F0}}(\boldsymbol{Q}, t-t')\,\tilde{G}_{\mathrm{F}}^{\mathrm{T}}(\boldsymbol{Q}, t')\right\} \\
&= L^{\mathrm{t}}\left\{\exp\left[-t\left(\frac{1}{\tau_1}+\Lambda(\boldsymbol{Q})\right)\right]\right\} \\
&\quad + \frac{1}{\tau_0}L^{\mathrm{t}}\left\{\exp\left(-\frac{t}{\tau_1}\right)G_{\mathrm{F0}}(\boldsymbol{Q}, t)\right\}\cdot L^{\mathrm{t}}\left\{\tilde{G}_{\mathrm{T}}^{\mathrm{T}}(\boldsymbol{Q}, t)\right\} \\
&= L^{\mathrm{t}}\left\{\exp\left[-t\left(\frac{1}{\tau_1}+\Lambda(\boldsymbol{Q})\right)\right]\right\} \\
&\quad + \frac{1}{\tau_0}L^{\mathrm{t}}\left\{\exp\left(-t\left(\frac{1}{\tau_1}+\Lambda(\boldsymbol{Q})\right)\right)\right\}\tilde{G}_{\mathrm{F}}^{\mathrm{T}}(\boldsymbol{Q}, s) \\
&= \frac{1}{s+1/\tau_1+\Lambda(\boldsymbol{Q})} + \frac{1}{\tau_0}\frac{1}{s+1/\tau_1+\Lambda(\boldsymbol{Q})}\tilde{G}_{\mathrm{F}}^{\mathrm{T}}(\boldsymbol{Q}, s) \\
&= B(\boldsymbol{Q}, s) + \frac{1}{\tau_0}B(\boldsymbol{Q}, s)\,\tilde{G}_{\mathrm{F}}^{\mathrm{T}}(\boldsymbol{Q}, s)
\end{aligned}
\tag{5.164}
$$

with the abbreviation

$$
B(\boldsymbol{Q}, s) = \frac{1}{s+1/\tau_1+\Lambda(\boldsymbol{Q})} = \frac{\tau_1}{1+[s+\Lambda(\boldsymbol{Q})]\tau_1}.
$$

Correspondingly we obtain

$$\tilde{G}_{\mathrm{F}}^{\mathrm{T}}(\boldsymbol{Q},s) = \frac{1}{\tau_1} A(\boldsymbol{Q},s)\, \tilde{G}_{\mathrm{F}}^{\mathrm{F}}(\boldsymbol{Q},s) \tag{5.165}$$

with

$$A(\boldsymbol{Q},s) = \frac{1}{s+1/\tau_0} = \frac{\tau_0}{1+s\tau_0}$$

$$\tilde{G}_{\mathrm{T}}^{\mathrm{T}}(\boldsymbol{Q},s) = A(\boldsymbol{Q},s) + \frac{1}{\tau_1} A(\boldsymbol{Q},s)\, \tilde{G}_{\mathrm{T}}^{\mathrm{F}}(\boldsymbol{Q},s) \tag{5.166}$$

$$\tilde{G}_{\mathrm{T}}^{\mathrm{F}}(\boldsymbol{Q},s) = \frac{1}{\tau_0} B(\boldsymbol{Q},s)\, \tilde{G}_{\mathrm{T}}^{\mathrm{T}}(\boldsymbol{Q},s). \tag{5.167}$$

Now in a simple way the determination of the different $\tilde{G}(\boldsymbol{Q},s)$ functions is possible. We omit the variables and insert the expression $\tilde{G}_{\mathrm{T}}^{\mathrm{F}} = \tau_1^{-1} A \tilde{G}_{\mathrm{F}}^{\mathrm{F}}$ into Eq. (5.163) which eventually leads to

$$\tilde{G}_{\mathrm{F}}^{\mathrm{F}} = \frac{B}{1 - AB/(\tau_0\tau_1)} = \frac{B}{1-CD} \tag{5.168}$$

with $C = A/\tau_0$ and $D = B/\tau_1$; this result is used to obtain

$$\tilde{G}_{\mathrm{F}}^{\mathrm{T}} = \frac{1}{\tau_1} AB \frac{1}{1-CD} = \frac{AD}{1-CD}. \tag{5.169}$$

Correspondingly Eqs. (5.165) and (5.166) also form a set of two equations with two unknown quantities, yielding

$$\tilde{G}_{\mathrm{T}}^{\mathrm{T}} = \frac{A}{1-CD} \tag{5.170}$$

$$\tilde{G}_{\mathrm{T}}^{\mathrm{F}} = \frac{BC}{1-CD}. \tag{5.171}$$

The Fourier and Laplace transformed equation (5.163) is thus given by

$$\tilde{G}(\boldsymbol{Q},s) = \frac{\tau_0}{\tau_0+\tau_1}\frac{A+BC}{1-CD} + \frac{\tau_1}{\tau_0+\tau_1}\frac{B+AD}{1-CD}. \tag{5.172}$$

The temporal Laplace transformation has now to be translated into a Fourier transformation in order to derive the incoherent scattering function. This is done by

$$\pi S_{\mathrm{i}}(\boldsymbol{Q},\omega) = \mathrm{Re}\{\tilde{G}(\boldsymbol{Q}, s=\mathrm{i}\omega)\}.$$

We rescind the abbreviations A, B, C, D and obtain

$$\tilde{G}(\boldsymbol{Q},\mathrm{i}\omega) = \frac{(\tau_0^2 + 2\tau_0\tau_1 + \tau_1^2 + \Lambda(\boldsymbol{Q})\tau_0^2\tau_1) + \mathrm{i}\omega(\tau_0^2\tau + \tau_0\tau_1^2)}{(\Lambda(\boldsymbol{Q})\tau_1 - \omega^2\tau_0\tau_1) + \mathrm{i}\omega(\tau_0+\tau_1+\Lambda(\boldsymbol{Q})\tau_0\tau_1)} \frac{1}{(\tau_0+\tau_1)}. \tag{5.173}$$

This can be separated into a real part and an imaginary part multiplying numerator and denominator by the complex conjugate of the denominator and thus yielding

$$\mathrm{Re}\{\tilde{G}(\boldsymbol{Q}, \mathrm{i}\omega)\} = \frac{[\Lambda(\boldsymbol{Q})\tau_1(\tau_0+\tau_1)^2 + \Lambda^2(\boldsymbol{Q})\tau_0^2\tau_1^2] + \omega^2\Lambda(\boldsymbol{Q})\tau_0^2\tau_1^3}{\omega^4\tau_0^2\tau_1^2 + \omega^2[(\tau_0+\tau_1)^2 + 2\Lambda(\boldsymbol{Q})\tau_0\tau_1^2 + \Lambda^2(\boldsymbol{Q})\tau_0^2\tau_1^2] + \Lambda^2(\boldsymbol{Q})\tau_1^2} \cdot \frac{1}{\tau_0+\tau_1}. \tag{5.174}$$

After enlargement by $(\tau_0^2\tau_1^2)^{-1}$ this equation is of the shape

$$\begin{aligned}\mathrm{Re}\{\tilde{G}(\boldsymbol{Q}, \mathrm{i}\omega)\} &= \frac{\omega^2(R_1\Lambda_1 + R_2\Lambda_2) + (R_1\Lambda_2\Lambda_2^2 + R_2\Lambda_2\Lambda_2^2 + R_2\Lambda_2\Lambda_1^2)}{\omega^4 + (\Lambda_1^2+\Lambda_2^2)\omega^2 + \Lambda_1\Lambda_2} \\ &= \frac{R_1\Lambda_1(\Lambda_2^2+\omega^2) + R_2\Lambda_2(\Lambda_1^2+\omega^2)}{(\Lambda_1^2+\omega^2)(\Lambda_2^2+\omega^2)} \\ &= \frac{R_1\Lambda_1}{\Lambda_1^2+\omega^2} + \frac{R_2\Lambda_2}{\Lambda_2^2+\omega^2}.\end{aligned} \tag{5.175}$$

In this way we obtain the scattering function for the two-state model:

$$S_i(\boldsymbol{Q}, \omega) = \frac{1}{\pi}R_1\frac{\Lambda_1}{\Lambda_1^2+\omega^2} + \frac{1}{\pi}R_2\frac{\Lambda_2}{\Lambda_2^2\omega^2} \tag{5.176a}$$

with

$$\Lambda_{1,2} = \frac{1}{2}\left[\frac{1}{\tau_0} + \frac{1}{\tau_1} + \Lambda(\boldsymbol{Q})\right] \pm \frac{1}{2}\sqrt{\left[\frac{1}{\tau_0} + \frac{1}{\tau_1} + \Lambda(\boldsymbol{Q})\right]^2 - 4\Lambda(\boldsymbol{Q})/\tau_0} \tag{5.176b}$$

$$R_{1,2} = \frac{1}{2} \pm \frac{1}{2}\left[\Lambda(\boldsymbol{Q})\frac{\tau_1-\tau_0}{\tau_1+\tau_0} - \frac{1}{\tau_0} - \frac{1}{\tau_1}\right] \Bigg/ \sqrt{\left[\frac{1}{\tau_0} + \frac{1}{\tau_1} + \Lambda(\boldsymbol{Q})\right]^2 - 4\Lambda(\boldsymbol{Q})/\tau_0}. \tag{5.176c}$$

Both Singwi and Sjölander (1960) and Richter and Springer (1978) use the Q^2 law for $\Lambda(\boldsymbol{Q})$, such that $\Lambda(\boldsymbol{Q}) = D_s Q^2$. This restricts the validity of the model to the small Q range. A very convenient approximation for an extension to larger Q is the isotropic Chudley–Elliott model, Eq. (5.103), which was used in later applications of the two-state model. Fig. 5.10 gives a graphical representation of the scattering function; the thickness of the stroke indicates the intensity. At very

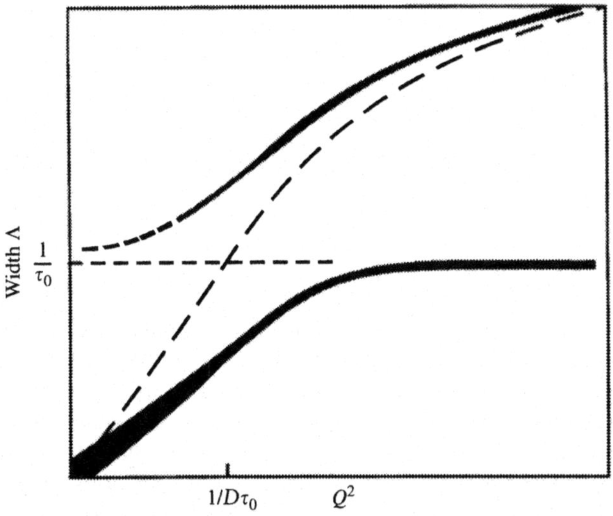

Fig. 5.10 Graphical representation of the scattering function of the two-state-model; the thicknesses of the strokes indicate the intensity.

low Q values, Taylor expansion of the square root according to $(1-x)^{1/2} = 1-\frac{1}{2}x$ leads to $R_1 \to 0$, $R_2 \to 1$, hence

$$\Lambda_2 = \frac{\tau_1}{\tau_0+\tau_1}\Lambda(Q) = \frac{\tau_1}{\tau_0+\tau_1}D_{\text{free}}Q^2 \tag{5.176}$$

where D_{free} is the self-diffusion coefficient in the free state and

$$D_{\text{eff}} = \frac{\tau_1}{\tau_0+\tau_1}D_{\text{free}} \tag{5.177}$$

is the effective self-diffusion coefficient. This form is highly plausible: on the one hand, if the long-time behaviour is considered, it describes the reduction of the free diffusion coefficient by the mean time spent in the immobile state; on the other hand, if the short-time behaviour is considered, it indicates the fraction of particles in the mobile state. Generally, however, there are two components. Their interpretation becomes especially simple at large Q:

$$\Lambda_1 \approx \tau^{-1}; \qquad \Lambda_2 \approx \tau_0^{-1},$$
$$R_1 \approx \frac{\tau_1}{\tau_0+\tau_1}; \qquad R_2 \approx \frac{\tau_0}{\tau_0+\tau_1}.$$

Evidently for large Q values one component exhibits the effect of release from traps with the appropriate weight of trap occupation. The other component represents essentially the free transition to neighbouring sites, again with the correct weight.

6

Coherent quasielastic neutron scattering

In the preceding chapters we dealt with the motion of *individual* particles investigated by *incoherent* scattering. The particles can be interstitial atoms diffusing over empty interstitial sites, or substitutional atoms migrating via encounters which leads to vacancy-induced diffusion. In this chapter, a more complicated situation will be described, namely diffusing atoms with a sizable coherent scattering cross-section. The corresponding scattering, as described by $S(\boldsymbol{Q}, \omega)$, then yields information on interference effects and on the correlated motion of the particles. The problem is difficult to formulate theoretically, and molecular dynamics may be a useful way to get quantitative results. As concerns the materials, we have to deal with the following situation: (i) coherent scattering on *deuterium* atoms diffusing over the interstitial lattice of metals or alloys (see Section 8.1), (ii) coherent scattering on *moving ions* in solid state ionic conductors, and (iii) coherent scattering on *vacancies* or *substitutional atoms* diffusing over the host lattice. In these cases, there is a coherent superposition of the scattering contributions of the diffusing atoms or vacancies themselves, and, eventually, coherent scattering caused by the *time-dependent displacement field* of the host atoms surrounding the diffusing particle. This contribution is supposed to be small except near Bragg reflections where it is called quasielastic Huang scattering. This is a new and still developing field.

In the following pages we introduce the theory of *coherent quasielastic neutron scattering*. This needs a short introduction to *elastic* coherent scattering due to short-range order, first neglecting distortion effects for simplicity, and, afterwards, also an introduction to elastic distortion scattering. As concerns the scattering on static defects there are many monographs and textbooks, and we quote in particular Krivoglaz (1969), Dederichs (1973), Schmatz (1973), Schwartz and Cohen (1977), Bauer (1979), and the short review of Schweika (1998) and references therein.

6.1 COHERENT ELASTIC SCATTERING DUE TO SHORT-RANGE ORDER

We formulate the intensity of coherent elastic scattering of interacting particles distributed over the sites of an interstitial lattice. In order to describe the mutual correlation caused by the interaction potential we introduce a short-range order parameter. As a simple example we consider an interstitial lattice occupied by deuterium atoms and vacancies and write for the Cowley short-range order parameters

(Cowley 1950) with regard to the mth coordination shell

$$\alpha_m = \frac{\langle \sigma_i \sigma_{i+m} \rangle - c^2}{c(1-c)} \tag{6.1}$$

where

$$\sigma = \begin{cases} 1 & \text{for an occupied site} \\ 0 & \text{for an empty site.} \end{cases}$$

$c = \langle \sigma \rangle$ is the average occupancy of the sites. Obviously, in a crystal the correlations of defects depend only on the mutual distance $\boldsymbol{R}_m = \boldsymbol{R}_i - \boldsymbol{R}_j$ of sites. $\langle \sigma_i \sigma_{i+m} \rangle$ is the probability of finding a particle pair on two sites having the distance R_m. For the self-term ($m = 0$) one always obtains $\langle \sigma_i \sigma_i \rangle = c$ and $\alpha_0 = 1$. Otherwise, for $m = 1, 2, \ldots$, three cases have to be distinguished:

- random occupation: $\langle \sigma_i \sigma_{i+m} \rangle = c^2$ and $\alpha_m = 0$;
- attractive interaction: $\langle \sigma_i \sigma_{i+m} \rangle > c^2$ and $\alpha_m > 0$;
  in the limit of cluster formation: $\langle \sigma_i \sigma_{i+m} \rangle = c$ and $\alpha_m = 1$;
- repulsive interaction: $\langle \sigma_i \sigma_{i+m} \rangle = c^2$ and $\alpha_m < 0$;
  in the limit of complete site blocking: $\langle \sigma_i \sigma_{i+m} \rangle = 0$ and $\alpha_m = -c/(1-c)$.

Thus for a preference of particle–particle pairs at distance $\boldsymbol{R}_m$ the short-range order parameters α_m have positive values, whereas they have negative ones for the preference of particle–vacancy pairs.

The structure factor for a lattice with particles with purely real scattering lengths b_i is given by

$$\frac{d\sigma}{d\Omega} = \left\langle \sum_i \sum_j b_i b_j \sigma_i \sigma_j e^{iQ(\boldsymbol{R}_i - \boldsymbol{R}_j)} \right\rangle \tag{6.2}$$

(see Section 3.2, Eq. (3.45)). Introduction of the coefficients α_m and assuming that all b_i on the occupied sites are equal, we obtain, with Eq. (6.1):

$$\frac{d\sigma}{d\Omega} = b^2 c(1-c) \sum_i \sum_j \alpha_{i-j} e^{i\boldsymbol{Q}(\boldsymbol{R}_i - \boldsymbol{R}_j)} + b^2 c^2 \sum_i \sum_j e^{i\boldsymbol{Q}(\boldsymbol{R}_i - \boldsymbol{R}_j)}. \tag{6.3}$$

The second term yields the Bragg reflections, the intensities of which are

$$\frac{d\sigma}{d\Omega} = b^2 c^2 N^2 V_c^{-1} (2\pi)^3 \sum_{\boldsymbol{G}} \delta(\boldsymbol{Q} - \boldsymbol{G}); \tag{6.4}$$

V_c is the volume of the unit cell and $\boldsymbol{G}$ are the reciprocal lattice vectors; extinction is not taken into account so that this term goes with $(cN)^2$. We are interested in the first term of Eq. (6.3) describing diffuse scattering. Since the correlation coefficients depend only on the mutual distance $\boldsymbol{R}_m = \boldsymbol{R}_i - \boldsymbol{R}_j$ we write for the differential neutron scattering cross-section due to short-range order

$$\left(\frac{\mathrm{d}\sigma}{\mathrm{d}\Omega}\right)_{\mathrm{SRO}} = b^2 c(1-c) N \sum_{m=0}^{\infty} \alpha_m \cos(\boldsymbol{Q}\boldsymbol{R}_m) \tag{6.5}$$

provided the scatteres are at sites that exhibit inversion symmetry. For a polycrystalline sample orientational averaging yields

$$\left(\frac{\mathrm{d}\sigma}{\mathrm{d}\Omega}\right)_{\mathrm{SRO}} = b^2 c(1-c) N \left\{1 + \sum_{i=1}^{\infty} z_i \alpha_i \frac{\sin(QR_i)}{QR_i}\right\} \tag{6.6}$$

where the sum runs over the different coordination spheres with the respective coordination number z_i and radius R_i. The self-term $i = 0$ has been taken out of the sum. The quantity N in Eqs. (6.5) and (6.6) refers to the number of sites, which in the case of interstitial systems is not identical with the number of particles, cN. Thus the structure factor per interstitial atom is given by

$$S_{\mathrm{SRO}}(Q) = (1-c)\left\{1 + \sum_{i=1}^{\infty} z_i \alpha_i \frac{\sin(QR_i)}{QR_i}\right\}. \tag{6.7}$$

As can be seen from the examples in Fig. 6.1 this leads to a liquid-like structure factor if, due to repulsive interactions, a given particle blocks the sites of the first and second coordination spheres. For a random distribution the sum in Eq. (6.6) vanishes and one obtains the well-known Laue term as already discussed under 'incoherent scattering' in Section 3.2, namely

$$\left(\frac{\mathrm{d}\sigma}{\mathrm{d}\Omega}\right)_{\mathrm{Laue}} = b^2 c(1-c) N. \tag{6.8}$$

Complete disorder in a perfect gas or in a liquid at large distance from the reference atom yields $S(Q) = 1$; so in a lattice gas the degree of disorder is less and thus $S(Q) < 1$.

For two atomic species (a) and (b) on interstitial sites (where we can identify one of them as a 'defect') formula (6.1) can be maintained, but the squared scattering length b^2 has to be replaced by $|b_{\mathrm{a}} - b_{\mathrm{b}}|^2$ and σ_i, σ_{i+m} by $\sigma_i^{\mathrm{a}}, \sigma_{i+m}^{\mathrm{b}}$ where $\sigma^{\mathrm{a}}, \sigma^{\mathrm{b}}$ are the occupancies of a site by atom (a) or (b), respectively, such that $\langle\sigma_{\mathrm{a}}\rangle = c^{\mathrm{a}}$, $\langle\sigma^{\mathrm{b}}\rangle = c^{\mathrm{b}}$, and $\langle\sigma_i^{\mathrm{a}}\sigma_{i+m}^{\mathrm{b}}\rangle = c^{\mathrm{a}}c^{\mathrm{b}}$ if the correlations disappear for large distances.

6.2 THE DIFFUSING LATTICE-GAS OR LATTICE-LIQUID

Since coherent scattering on atoms is determined by the interference of the scattered waves of diffusing atoms or particles, the corresponding scattering function

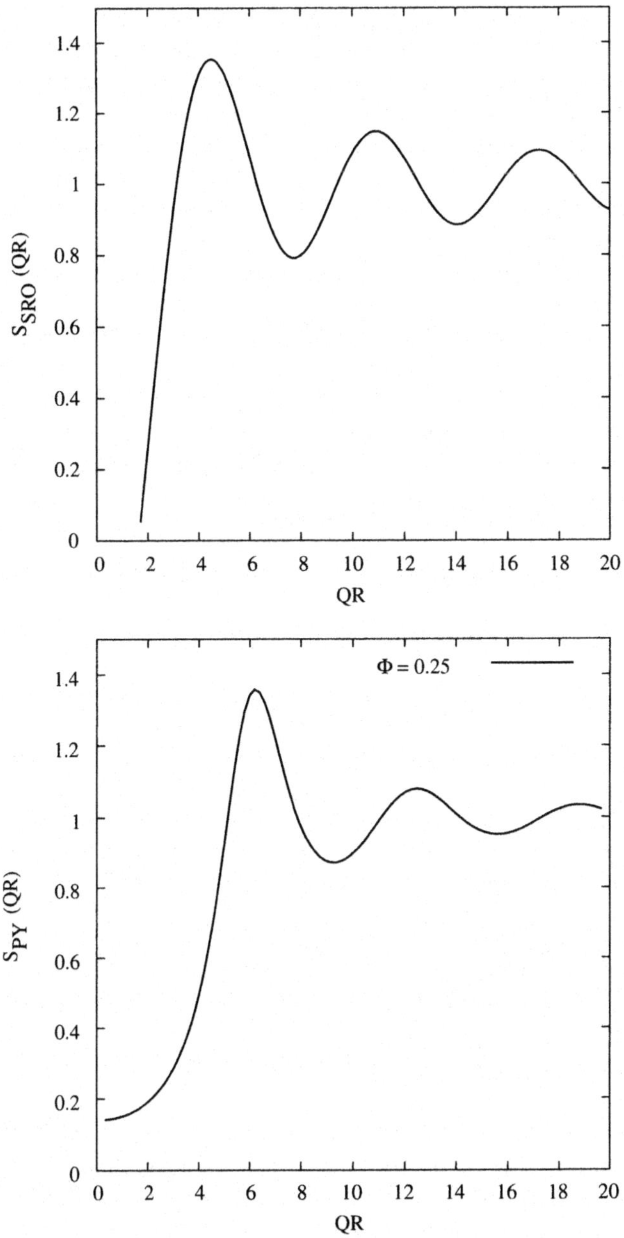

Fig. 6.1 Liquid-like structure factor from short-range order in a lattice fluid (upper part); for comparison: structure factor for hard-sphere particles according to the Percus-Yevick theory (lower part).

depends on the mutual interaction and on the correlations between the atoms. In the simplest case there is no interaction except 'self-site blocking' (excluded volume effects) which means that a site of the interstitial lattice over which diffusion occurs can be empty or occupied by one atom only, whereas double occupancy is forbidden. This model was worked out by Ross and Wilson (1978), by Kutner (1981) and by Kehr *et al.* (1981) on the basis of rate equations. Later, a general formulation of the problem will be presented which introduces a mutual potential between the particles.

In order to obtain the correlation function and the coherent scattering function for the self-site blocking model, a rate equation is formulated for the conditional probabilities. We denote by $P(\boldsymbol{r}, \boldsymbol{r}', t)$ the conditional probability of finding a diffusing particle at a site $\boldsymbol{r}$ at time t if, at the same time, there is another particle at an adjacent site $\boldsymbol{r}'$. Furthermore, $P(\boldsymbol{r}, \bar{\boldsymbol{r}}', t)$ is the conditional probability of finding a particle at site $\boldsymbol{r}$ at time t if at the same time there is *no* particle at $\boldsymbol{r}'$. In the following we use an interstitial Bravais lattice where the sites are connected by jump vectors $\boldsymbol{s}_k$, namely

$$\boldsymbol{r}' = \boldsymbol{r} + \boldsymbol{s}_k. \tag{6.9}$$

Obviously, the *unconditional* probability of finding a particle for a time t at sites $\boldsymbol{r}'$ or $\boldsymbol{r}$ is

$$P(\boldsymbol{r}', t) = P(\boldsymbol{r}', \bar{\boldsymbol{r}}, t) + P(\boldsymbol{r}', \boldsymbol{r}, t)$$

and

$$P(\boldsymbol{r}, t) = P(\boldsymbol{r}, \bar{\boldsymbol{r}}', t) + P(\boldsymbol{r}, \boldsymbol{r}', t). \tag{6.10}$$

Furthermore, we have

$$P(\boldsymbol{r}', \boldsymbol{r}, t) = P(\boldsymbol{r}', t)\, P(\boldsymbol{r}, t) \tag{6.11}$$

where $P(\boldsymbol{r}, t)$ is the unconditional probability of having the particle at $\boldsymbol{r}$. Consequently, one gets

$$P(\boldsymbol{r}', \boldsymbol{r}, t) = P(\boldsymbol{r}, \boldsymbol{r}', t). \tag{6.12}$$

Since a particle only jumps by a vector $\boldsymbol{s}_k$ between an occupied and an empty site, the rate equation then reads

$$\frac{\mathrm{d}P(\boldsymbol{r}, t)}{\mathrm{d}t} = \frac{1}{z\tau} \sum_{\boldsymbol{r}'} [P(\boldsymbol{r}', \boldsymbol{r}, t) - P(\boldsymbol{r}, \boldsymbol{r}', t)] \tag{6.13}$$

where the sum goes over all z neighbouring sites $\boldsymbol{r}' = \boldsymbol{r} + \boldsymbol{s}_k$. The quantity τ is the mean rest time for a particle on a site, assuming that all z neighbouring sites are unoccupied. $1/z\tau$ is the corresponding (empty-lattice) jump rate. Equations (6.10)–(6.13) finally lead to (Kutner 1981)

$$\frac{\mathrm{d}P(\boldsymbol{r}, t)}{\mathrm{d}t} = \frac{1}{z\tau} \sum_{\boldsymbol{r}'} [P(\boldsymbol{r}', t) - P(\boldsymbol{r}, t)]. \tag{6.14}$$

Obviously, this is just the Chudley–Elliott rate equation as obtained for a single particle in a lattice (i.e. the highly dilute case). However, the boundary condition

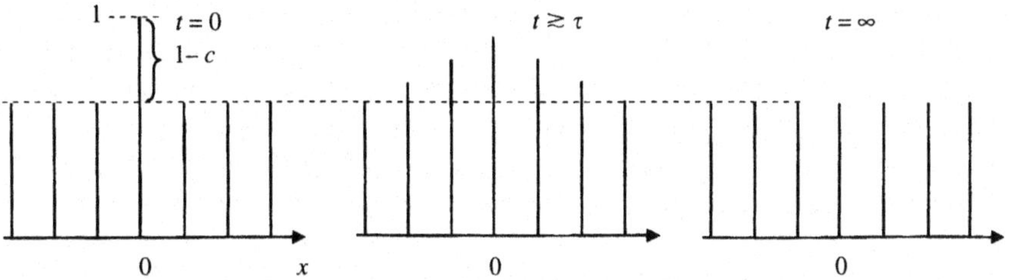

Fig. 6.2 Correlation function $G(\boldsymbol{r}, t)$ for particle diffusion on regular lattice, schematically for one-dimensional lattice. The particle is certainly at $x = 0$ for $t = 0$ whereas the probability to find another particle at any other site, provided that there is no interaction, is an equal distribution corresponding to the concentration c. With increasing time, the probability of occupancy at $x = 0$ decays.

for $t = 0$ is different from that for the single-particle case: for $t = 0$ the definition of van Hove's correlation function requires (compare Eq. (3.65))

$$G(\boldsymbol{r}, 0) = \delta(\boldsymbol{r}) + c \sum_{r_i \neq 0} \delta(\boldsymbol{r} - \boldsymbol{r}_i) = (1 - c)\, \delta(\boldsymbol{r}) + c \sum_{\text{all } r_i} \delta(\boldsymbol{r} - \boldsymbol{r}_i) \qquad (6.15)$$

where c is the average number of particles per available site. This equation means that for $t = 0$ there is *certainly* a particle at $\boldsymbol{r} = 0$ or, in other words, $G(0, 0) = 1$, whereas for $\boldsymbol{r} \neq \boldsymbol{0}$ the probability of finding another particle at the lattice sites is c. This situation is shown in Fig. 6.2.

Consequently, the resulting coherent scattering function (per site) is the sum of a term which is the single-particle (Chudley–Elliott) solution $S_{\mathrm{i}}^{\mathrm{CE}}(\boldsymbol{Q}, \omega)$ and of a Bragg term; for a *diffusing lattice-gas* we therefore obtain:

$$S(\boldsymbol{Q}, \omega) = c(1 - c) S_{\mathrm{i}}^{\mathrm{CE}}(\boldsymbol{Q}, \omega) + c^2 \frac{(2\pi)^3}{V_{\mathrm{c}}} \sum_{\boldsymbol{G}} \delta(\boldsymbol{Q} - \boldsymbol{G}); \qquad (6.16)$$

in this equation the sum goes over all reciprocal lattice vectors $\boldsymbol{G}$. In order to formulate the scattering function *per site* of the interstitial lattice we have introduced the factor c into Eq. (6.16). Integrating over all energy transfers yields the resulting diffuse Laue scattering proportional to $c(1 - c)$, the well-known Laue factor for coherent scattering of a lattice with atoms randomly distributed, see Eq. (6.8). The Chudley–Elliott scattering function is a single Lorentzian with linewidth

$$\Gamma_0(\boldsymbol{Q}) = \Gamma^{\mathrm{CE}}(\boldsymbol{Q}) = \frac{\hbar}{z\tau} \sum_n (1 - \mathrm{e}^{-\mathrm{i}\boldsymbol{Q}\boldsymbol{s}_n}) \qquad (6.17)$$

which does not depend on the concentration. It is interesting to note that Γ^{CE} is the result as obtained for incoherent scattering in the dilute case, see Eq. (5.99). This shows that at large concentration the line for coherent quasielastic scattering is *broader* than it would be for incoherent scattering because, in the latter case, the blocking factor reduces the incoherent width Γ^{CE} (see Eqs. (4.14) and (4.60)).

The formalism of rate equations is elegant and straightforward but only in the case of interaction-free systems. Therefore we now turn to a quite different formalism, the linear response theory (Lovesey 1987), which allows a more general treatment of the diffusion process. We start again with the case of no internal particle–particle interactions. The particles, however, interact with an 'external force', i.e. an external time- and space-dependent potential $V_{ext}(\boldsymbol{r}_i, t)$ which in the present context is due to the neutron waves passing through the sample. The response of the system consists in a change of the density or, for a lattice gas/fluid, in a change of the occupancy, δc. Of course, a site is either full or empty; however, in the spirit of the mean field approximation, δc is considered as an average quantity. The potential is characterized by periodic fluctuations in space and time with wavevector $\boldsymbol{Q}$ and frequency ω by

$$V_{ext}(\boldsymbol{r}_i, t) = V_{ext}(\boldsymbol{Q}, \omega)e^{i\boldsymbol{Q}\boldsymbol{r}_i - i\omega t} \tag{6.18}$$

where the interstitial sites are characterized by lattice vectors $\boldsymbol{r}_i$. For a linear response the average occupancy follows the potential, so that

$$\delta c(\boldsymbol{r}_i, t) = \delta c(\boldsymbol{Q}, \omega)e^{i\boldsymbol{Q}\boldsymbol{r}_i - i\omega t}. \tag{6.19}$$

The susceptibility is then defined by the relation

$$\delta c(\boldsymbol{Q}, \omega) = -\chi(\boldsymbol{Q}, \omega)\, V_{ext}(\boldsymbol{Q}, \omega). \tag{6.20}$$

Normally, the time dependence of δc is phase-shifted against V_{ext} so that $\chi(\boldsymbol{Q}, \omega)$ is a complex quantity.

Due to the interaction of the neutron wave with the diffusing particles the scattering process probes this system with a wavevector $\boldsymbol{Q}$, as determined by the experimental choice of the neutron wavevectors $\boldsymbol{k}_0$ and $\boldsymbol{k}_1$, and with a frequency ω, as determined by $\hbar\omega = \varepsilon_0 - \varepsilon_1$. Then one derives a simple relation between the imaginary or dissipative part of $\chi(\boldsymbol{Q}, \omega)$ and the scattering function which is the Fourier transform of the particle fluctuation (Kubo 1966; Lovesey 1987). This relation can be written as

$$S(\boldsymbol{Q}, \omega) = \frac{-1}{\pi}\frac{\mathrm{Im}\chi(\boldsymbol{Q}, \omega)}{e^{\hbar\omega/k_B T} - 1}. \tag{6.21}$$

We follow the derivation as given by Sinha and Ross (1988), but without the effects of phonon coupling, and start by considering the scattering particles hopping over the interstitial sites r_i of the lattice *without* mutual interaction (index 0). Under these conditions the coherent scattering function can be written in the form of Eq. (6.16):

$$S(\boldsymbol{Q}, \omega) = \frac{c(1-c)\Gamma_0(\boldsymbol{Q})/\pi}{\Gamma_0^2(\boldsymbol{Q}) + \hbar^2\omega^2}. \tag{6.22}$$

The width of the quasielastic line is given in Eq. (6.17). In the classical limit ($\hbar\omega/k_B T \ll 1$), Eq. (6.21) yields the corresponding interaction-free density response function, namely

$$\chi_0(\boldsymbol{Q},\omega) = \frac{c(1-c)\Gamma_0(\boldsymbol{Q})}{(k_B T/\hbar)[\Gamma_0(\boldsymbol{Q}) - \mathrm{i}\hbar\omega]}. \tag{6.23}$$

Now the external potential, as characterized by $V_{\mathrm{ext}}(\boldsymbol{Q},\omega)$ (Eq. 6.18), creates a density response δc. This response, in the sense of feedback, generates an additional pair potential. In the mean field approximation this potential is the product of $\delta c(\boldsymbol{Q},\omega)$ and a *particle–particle pair potential* $V_p(\boldsymbol{Q})$ in Fourier space,

$$V_p(\boldsymbol{Q}) = \sum_n V(\boldsymbol{r}_n)\exp(\mathrm{i}\boldsymbol{Q}\boldsymbol{r}_n). \tag{6.24}$$

This additional pair potential has to be added to the external potential which, as an expansion of Eq. (6.20), leads to a *self-consistent equation* of the form

$$\delta c(\boldsymbol{Q},\omega) = -\chi_0(\boldsymbol{Q},\omega)[V_{\mathrm{ext}}(\boldsymbol{Q},\omega) + V_p(\boldsymbol{Q})\,\delta c(\boldsymbol{Q},\omega)]. \tag{6.25}$$

Obviously, the concentration fluctuations caused by the external potential depend on the concentration fluctuations themselves. Solving for the response of the system *with* mutual interaction, we are led to the susceptibility and the corresponding scattering function

$$\chi(\boldsymbol{Q},\omega) = \frac{\chi_0(\boldsymbol{Q},\omega)}{1 + V_p(\boldsymbol{Q})\,\chi_0(\boldsymbol{Q},\omega)}. \tag{6.26}$$

From Eq. (6.23) we recognize the analogy with the Curie law, $\chi \propto 1/T$; with this in mind, the last equation is formally analogous to the Curie–Weiss law.

Equation (6.26) finally leads to

$$S(\boldsymbol{Q},\omega) = \frac{c(1-c)\Gamma_0(\boldsymbol{Q})/\pi}{\Gamma_{\mathrm{coh}}^2(\boldsymbol{Q}) + \hbar^2\omega^2} \tag{6.27}$$

where the width is given by

$$\Gamma_{\mathrm{coh}}(\boldsymbol{Q}) = \Gamma_0(\boldsymbol{Q})[1 + c(1-c)V_p(\boldsymbol{Q})/k_B T]. \tag{6.28}$$

The interaction potential $V(\boldsymbol{Q})$ can be related to the experimentally known structure factor of the interacting particles. Integrating $S(\boldsymbol{Q},\omega)$ over ω leads to the high-temperature and mean field approximation of Clapp and Moss (1968), namely

$$S_{\mathrm{CM}}(\boldsymbol{Q}) = \frac{c(1-c)}{1 + c(1-c)V_p(\boldsymbol{Q})/k_B T}. \tag{6.29}$$

In practice, as an alternative to the Clapp–Moss structure factor S_{CM}, the short-range order structure factor S_{SRO} of Eq. (6.7) is also employed. This allows the elimination of the potential in the expression (6.28) for Γ_{coh}, and we finally get

$$\Gamma_{\text{coh}}(\boldsymbol{Q}) = c(1-c)\Gamma_0(\boldsymbol{Q})/S_{\text{SRO}}(\boldsymbol{Q}) \tag{6.30}$$

which for Γ_0 in the numerator of Eq. (6.27) means

$$\Gamma_0 = S_{\text{SRO}}(\boldsymbol{Q})\,\Gamma_{\text{coh}}(\boldsymbol{Q})/\{c(1-c)\}. \tag{6.31}$$

So we can finally write for the coherent quasielastic scattering function of a *diffusing lattice-fluid*:

$$S(\boldsymbol{Q},\omega) = S_{\text{SRO}}(\boldsymbol{Q})\frac{\Gamma_{\text{coh}}(\boldsymbol{Q})/\pi}{\Gamma^2_{\text{coh}}(\boldsymbol{Q}) + \hbar^2\omega^2}. \tag{6.32}$$

For the interaction-free case ($V_p/k_B T \ll 1$ in Eq. (6.29) or $\alpha_0 = 1$ and all other $\alpha_i = 0$ in Eq. (6.7) and in Eq. (6.30)) we have $S_{\text{SRO}}(Q) = c(1-c)$ and $\Gamma_{\text{coh}} = \Gamma_0$ as derived in Eq. (6.16). Because of the structure factor in Eq. (6.30) in the denominator the quasielastic width oscillates and has minima for those Q values where $S_{\text{SRO}}(Q)$ has maxima related to the short-range order of the particles; this phenomenon is called 'de Gennes line narrowing' (de Gennes 1959). It can intuitively be understood in the sense that fluctuations corresponding to preferential correlation distances decay with a smaller rate than the others. Obviously, if there is no interaction, the result is symmetric in c and $(1-c)$ which implies that scattering on dilute defects is the same as scattering on dilute vacancies.

For small Q the linewidth Γ_0 (which is the linewidth for $c \to 0$) obeys the Q^2 law, analogously to Eq. (5.88); correspondingly for finite concentration the coherent QENS linewidth is given by

$$\Gamma_{\text{coh}}(\text{small Q}) = \hbar Q^2 \frac{c(1-c)D_{\text{E}}}{c S_{\text{SRO}}(Q)} \quad \text{(mean field)} \tag{6.33}$$

with $S_{\text{SRO}}(\boldsymbol{Q})$ from Eq. (6.7) quoted as 'per interstitial' particle and therefore multiplied by the factor c (in the denominator) in order to refer to 'per site'; we recognize the blocking factor $(1-c)$ in the numerator, compare Eq. (4.14). Apart from the tracer correlation factor (which cannot be expected since a mean field treatment disregards local correlations) the numerator represents the self-diffusion coefficient. Going beyond the mean field approximation we introduce correlation factors and define a Q-dependent collective diffusion coefficient

$$D_{\text{c}}(Q) = \frac{D_{\text{s}}(f_{\text{m}}/f_{\text{t}})}{S_{\text{SRO}}(Q)} = \frac{D_{\text{s}}}{H_{\text{R}}\, S_{\text{SRO}}(Q)} \tag{6.34}$$

where $H_{\text{R}} = f_{\text{t}}/f_{\text{m}}$ is the Haven ratio, see Eq. (4.66), of the tracer and mobility correlation factors. Thus the linewidth of coherent QENS at small Q is given by

$$\Gamma_{\text{coh}}(\text{small } Q) = \hbar D_{\text{c}}(Q) Q^2. \tag{6.35}$$

In the limit $Q \to 0$, which means macroscopic density fluctuations, we remember that $S(0)^{-1}$ equals the thermodynamic factor (4.33) and obtain

$$\Gamma_{\text{coh}}(Q \to 0) = \hbar Q^2 D_{\text{bulk}} \tag{6.36}$$

where the bulk diffusion coefficient, D_{bulk}, is a measure of the rate of mounting and dismantling of long-range density fluctuation, i.e. it is very similar to Fick's chemical diffusion coefficient. In Section 8.1.3 it will be shown that coherence stresses can play an important role in this connection and that for this reason the bulk diffusion coefficient can differ from the chemical one, D_{chem}.

In cases where the system of diffusing defects has a miscibility gap and a coexistence line with a critical point at a temperature T_c, the structure factor $S(0)$ diverges as $\varepsilon^{-\gamma}$ where $\varepsilon = (T - T_c)/T$. Therefore D_{chem} and the quasielastic width Γ_c dissappear for $\varepsilon \to 0$ with the power law ε^{γ} ('critical slowing down'). In the mean field approximation one has $\gamma = 1$ (see Chapter 4.1). At the same time, approaching $\varepsilon = 0$, the structure factor $S(Q)$ is, to a very good approximation, an Ornstein–Zernike function whose width ΔQ is proportional to the inverse correlation length of the fluctuations, $1/\zeta$, which goes to zero as ε^{ν}, where, again in the mean field approximation, $\nu = \frac{1}{2}$. This approximation holds if the interaction of the diffusing particles is of long range. Otherwise, a transition to Ising behaviour is to be expected with other critical exponents (Fisher and Burford 1967; Stanley 1971).

The line narrowing effect described above is related to an intuitive concept which relates the coherent with the incoherent quasielastic width; the latter can be calculated more easily (Sköld 1967). This is achieved by rescaling the scattering vector, namely replacing Q^2 by $Q^2/S_{\text{SRO}}(Q)$, and writing

$$S(Q,\omega) \cong S_{\text{SRO}}(Q)\, S_{\text{i}}\left(\frac{Q}{\sqrt{S_{\text{SRO}}(Q)}}, \omega\right). \tag{6.37}$$

Applying the sum rules, this leads to a recoil $\hbar^2 Q^2/2M S_{\text{SRO}}(Q)$ instead of $\hbar^2 Q^2/2M$, as if the recoiling mass of the atom had been increased by $S(Q)$. With this manipulation, the first and second moments of $S(Q, \omega)$ (in its classical limit) are obeyed (see Section 3.3). This relation is a convenient approximation mainly in the case of liquids. For small Q it agrees with the result of the linear response theory in Eqs. (6.32) and (6.36). Both approximations disregard correlation effects. The experimental verification is discussed in Section 8.1.3 for deuterium interstitially dissolved in niobium.

6.3 ELASTIC DISTORTION SCATTERING

A lattice defect (vacancy, interstitial, substitutional atom) is surrounded by host lattice atoms which are displaced from their regular positions, see Fig. 6.3. Scattering on defects is therefore determined by the interference between the contributions of the defect/atom (vacancy) itself and the displacement field; the latter is scattering from the *difference* between the deformed and undeformed lattice. If the defect

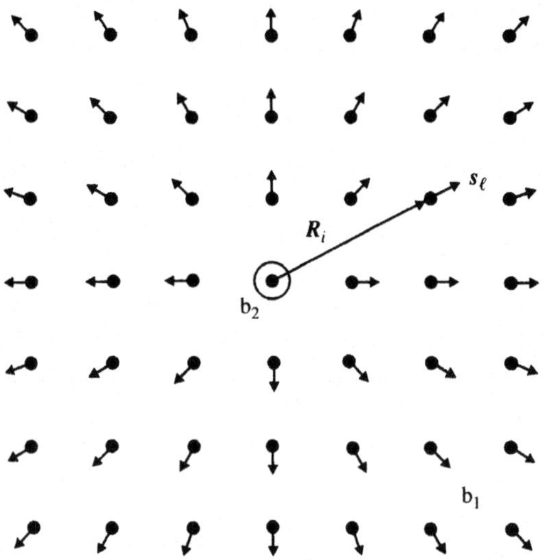

Fig. 6.3 Displacement field around a substitutional atom which is bigger than the host lattice atoms and leads to an outward shift s_l.

is static, the scattering is elastic, and this will be calculated first (see Schmatz 1973; Dederichs 1973; Bauer 1979). If the defect moves by diffusion, it carries the displacement field with it—eventually with retardation effects—such that the defect scattering becomes quasielastic.

The static structure factor (Eqs. (3.45) and (3.46)) is written as

$$S(\boldsymbol{Q}) = I(\boldsymbol{Q}, 0) = N^{-1}\langle|\rho(\boldsymbol{Q}, 0)|^2\rangle. \tag{6.38}$$

For simplicity, we assume vacancy defects, at position $\boldsymbol{R}_s$ which belong to the regular lattice with position vectors $\boldsymbol{R}_i$. This yields

$$\rho(\boldsymbol{Q}, 0) = \sum_{i \neq s} e^{i\boldsymbol{Q}(\boldsymbol{R}_i + \boldsymbol{u}_i^{(s)})} \tag{6.39}$$

where the vacant sites s were left out in the summation. $\boldsymbol{u}_i^{(s)}$ are the displacements of atoms at sites $\boldsymbol{R}_i$, caused by the defects. We introduce the lattice sum over *all* sites without displacements which is zero except at the Bragg points and then subtract the actual sum (6.39). This leads to

$$\rho(\boldsymbol{Q}, 0) = -\sum_{\text{all } s} e^{i\boldsymbol{Q}\boldsymbol{R}_s} + \sum_i e^{i\boldsymbol{Q}\boldsymbol{R}_i}\left(e^{i\boldsymbol{Q}\boldsymbol{u}_i^{(s)}} - 1\right). \tag{6.40}$$

Now we consider the displacement of an atom $\boldsymbol{u}_j^{(s)}$ specifically caused by a single defect, i.e. a vaccancy, at site $\boldsymbol{R}_s$ at a distance vector $\boldsymbol{R}_j = \boldsymbol{R}_i - \boldsymbol{R}_s$ from this defect, yielding

$$\begin{aligned}\rho(\boldsymbol{Q},0) &= -\sum_{\text{all } s} \mathrm{e}^{\mathrm{i}\boldsymbol{Q}\boldsymbol{R}_s} + \sum_{\text{all } j} \mathrm{e}^{\mathrm{i}\boldsymbol{Q}(\boldsymbol{R}_s+\boldsymbol{R}_j)}\left(\mathrm{e}^{\mathrm{i}\boldsymbol{Q}\boldsymbol{u}_j^{(s)}} - 1\right) \\ &= \sum_s \mathrm{e}^{\mathrm{i}\boldsymbol{Q}\boldsymbol{R}_s} f(\boldsymbol{Q})\end{aligned} \tag{6.41}$$

with the introduction of the diffusive structure factor

$$f(\boldsymbol{Q}) = -1 + \sum_{\text{all } j} \mathrm{e}^{\mathrm{i}\boldsymbol{Q}\boldsymbol{R}_j}(\mathrm{e}^{\mathrm{i}\boldsymbol{Q}\boldsymbol{u}_j} - 1). \tag{6.42}$$

We have suppressed the superscript (s) by placing the defect at the origin such that finally, for defects at coordinates $\boldsymbol{R}_s$, $\boldsymbol{R}_{s'}$, one gets

$$I(\boldsymbol{Q},0) = N^{-1} S_{\text{dis}}(\boldsymbol{Q}) \sum_s \sum_{s'} \langle \mathrm{e}^{\mathrm{i}\boldsymbol{Q}\boldsymbol{R}_s} \mathrm{e}^{-\mathrm{i}\boldsymbol{Q}\boldsymbol{R}_{s'}} \rangle \tag{6.43}$$

where

$$S_{\text{dis}}(\boldsymbol{Q}) = |f(\boldsymbol{Q})|^2 \tag{6.44}$$

is the structure factor due to distortion. The result implies that the defects with their surrounding displacemental field are non-interacting, i.e. that they are distributed over the host lattice at random, and that the displacements from different defects are superimposed linearly for a random distribution. Consequently the double sum $\sum_s \sum_{s'}$ is just replaced by $c(1-c)N$ as in Eq. (6.8).

For small displacements the exponential function $\exp(\mathrm{i}\boldsymbol{Q}\boldsymbol{u}_j)$ can be Taylor expanded and truncated after the linear term. The expression thus obtained is mathematically a Fourier transformation yielding the Fourier transform of the displacement field such that the structure factor due to distortions around a vacancy can be written as

$$S(\boldsymbol{Q})^{\text{v}}_{\text{dis}} = |-1 + \mathrm{i}\boldsymbol{Q}\boldsymbol{u}(\boldsymbol{q})|^2 \tag{6.45}$$

where $\boldsymbol{q} = \boldsymbol{Q} - \boldsymbol{G}$ is the distance of the scattering vector $\boldsymbol{Q}$ from a reciprocal lattice vector $\boldsymbol{G}$.

If the (point) defect is not a vacancy but a foreign atom, then the different scattering lengths have to be taken into account, and Eq. (6.43) has to be modified. As an example we consider deuterium interstitially dissolved in Nb. The differential cross-section for the diffuse scattering of the deuterium atoms surrounded by distortion clouds of displaced Nb atoms can, in analogy to Eq. (6.42), be written as

$$\begin{aligned}\left(\frac{\mathrm{d}\sigma}{\mathrm{d}\Omega}\right)_{\text{dis}} &= N_{\text{D}}\left|b_{\text{D}}\mathrm{e}^{-W} + b_{\text{Nb}}\mathrm{e}^{-W_{\text{BM}}} \sum_j \mathrm{e}^{\mathrm{i}\boldsymbol{Q}\boldsymbol{R}_j}(\mathrm{e}^{\mathrm{i}\boldsymbol{Q}\boldsymbol{u}_j} - 1)\right|^2 \\ &= N_{\text{D}} b_{\text{D}}^2 \mathrm{e}^{-2W}\left|1 + \frac{b_{\text{Nb}}}{b_{\text{D}}}\mathrm{e}^{+W_{\text{LM}}} \sum_j \mathrm{e}^{\mathrm{i}\boldsymbol{Q}\boldsymbol{R}_j}(\mathrm{e}^{\mathrm{i}\boldsymbol{Q}\boldsymbol{u}_j} - 1)\right|^2.\end{aligned} \tag{6.46}$$

We have put a deuterium at the origin such that the moduli of $\boldsymbol{R}_j$ are the distances between the deuterium position and the undisplaced niobium positions. The vectors $\boldsymbol{u}_j$ are the displacements of the Nb atoms, but we assume that a given D atom does not displace the next D atom. The Debye–Waller factors (see Eqs. (3.92) and (3.100)) take into account that the D atoms perform localized motions and band motions whereas the Nb atoms perform only band motions. We again perform a Taylor expansion and a Fourier transformation and obtain the structure factor due to distortions around an interstitial D atom in the Nb host lattice:

$$S(\boldsymbol{Q})_{\text{dis}}^{\text{D}} = \left| 1 + \frac{b_{\text{Nb}}}{b_{\text{D}}} \mathrm{e}^{+W_{\text{LM}}} \mathrm{i}\, \boldsymbol{Q}\boldsymbol{u}(\boldsymbol{q}) \right|^2 . \tag{6.47}$$

As the next step, the displacement field $\boldsymbol{u}(\boldsymbol{R}_j)$ of a defect, i.e. an interstitial D atom, is calculated. We characterize the strength of the defect by 'extra' or Kanzaki forces $\boldsymbol{P}_m$ which act on an atom at a distance $\boldsymbol{d}_m$ from the defect position. These forces imitate the action of the defect on the undisturbed host matrix. The forces are introduced only for distances from the defect which are comparable with the interaction range of the defect on the lattice atoms, $d_{\max}$. In order to treat the displacements at larger distances from the defect we introduce the Green's function $\mathbf{G}(\boldsymbol{R}_j - \boldsymbol{d}_m)$ which gives the elastic response of the lattice caused by the extra forces $\boldsymbol{P}_k$, namely

$$\boldsymbol{u}(\boldsymbol{R}_j) = \sum_m \mathbf{G}(\boldsymbol{R}_j - \boldsymbol{d}_m)\boldsymbol{P}_m . \tag{6.48}$$

The components of the tensor $\mathbf{G}$ can be expressed by the elastic constants of the host lattice. The investigation of the long-range interaction implies that we only take into account the values $\boldsymbol{q} \ll \boldsymbol{Q} \approx |\boldsymbol{G}|$, i.e. the region close to a Bragg point. We introduce Cartesian coordinates (s and s' for x, y, z) such that

$$u_s^j = \sum_{s'} \sum_m G_{ss'}(\boldsymbol{R}_j - \boldsymbol{d}_m) P_{s'}^m \tag{6.49}$$

where $G_{ss'}$, are the components of $\mathbf{G}$. Considering the medium as a continuum we write $\boldsymbol{R}$ instead of $\boldsymbol{R}_j$ with continuous Cartesian components $\boldsymbol{R} = (R_1, R_2, R_3)$ and $d_m = (d_1^m, d_2^m, d_3^m)$. An expansion in space coordinates d_s^m yields

$$u_s(\boldsymbol{R}) = \frac{\partial \{G_{ss'}(R) P_{s's''}\}}{\partial R_{s''}}, \tag{6.50}$$

where the first term of the expansion vanishes since it is proportional to the sum over all forces which is zero and where higher-order terms have been neglected. We introduce $\mathbf{P}$, the force dipole tensor with components,

$$P_{s's''} = \sum_m d_{s'}^m P_{s''}^m \tag{6.51}$$

where the sum goes over all components $P_{s''}^{m}$ of the extra forces, multiplied by the coordinate of the distance from the defect $d_{s'}^{m}$. Consequently, this dipole tensor is related to the symmetry and the strength of the defect. For a cubic lattice, $\boldsymbol{P}$ is diagonal with components $P_{s's''} = 2dP\,\delta_{s's''}$ if we assume only forces on next nearest neighbours. For a hydrogen atom on a tetrahedral site in the bcc lattice (see Fig. 6.4 and Section 8.2.1) the set of force vectors has tetragonal symmetry as shown in the figure. One can split the matrix into a cubic and a non-cubic part, namely

$$\mathsf{P} = \begin{bmatrix} A & 0 & 0 \\ 0 & A & 0 \\ 0 & 0 & B \end{bmatrix} \equiv \frac{2A+B}{3}\begin{bmatrix} 1 & 0 & 0 \\ 0 & 1 & 0 \\ 0 & 0 & 1 \end{bmatrix} + \frac{B-A}{3}\begin{bmatrix} -1 & 0 & 0 \\ 0 & -1 & 0 \\ 0 & 0 & 2 \end{bmatrix}. \tag{6.52}$$

The trace of the matrix P amounts to $2A + B$ and is proportional to the experimentally known expansion of the lattice due to the defect and to the elastic constants (Alefeld *et al.* 1970). The non-diagonal components can be determined by Snoek relaxation (Buchholz *et al.* 1973).

Now we calculate the Fourier transform of $\boldsymbol{u}(\boldsymbol{R})$ in order to obtain the diffuse structure factor in Eqs. (6.42), (6.44), and (6.45). The convolution in Eq. (6.48) gives immediately

$$\boldsymbol{u}(\boldsymbol{q}) = \boldsymbol{\phi}^{-1}(\boldsymbol{q})\,\boldsymbol{P}(\boldsymbol{q}) \tag{6.53}$$

with the Fourier transform of the $\boldsymbol{P}_{\mathrm{m}}$, namely

$$\boldsymbol{P}(\boldsymbol{q}) = \sum_{m} \boldsymbol{P}_{m} \mathrm{e}^{\mathrm{i}\boldsymbol{Q}\boldsymbol{d}_{m}}. \tag{6.54}$$

The periodicity of the lattice allows us to replace $\boldsymbol{Q}$ by $\boldsymbol{q}$. The tensor $\boldsymbol{\phi}^{-1}(\boldsymbol{q})$, representing the Fourier transform of the Green's function, is the inverse dynamical matrix of the host lattice, which can be expressed by the dynamical matrix obtained from the phonon dispersion branches $\omega_i(\boldsymbol{q})$ of the host lattice. For instance, in a symmetry direction of a cubic crystal with transverse and longitudinal branches one gets

$$\boldsymbol{\phi}^{-1}(\boldsymbol{q}) = \begin{bmatrix} 1/M\omega_{T1}^2 & 0 & 0 \\ 0 & 1/M\omega_{T2}^2 & 0 \\ 0 & 0 & 1/M\omega_L^2 \end{bmatrix} \tag{6.55}$$

where M is the atomic mass of the host lattice; Eq. (6.54) is valid for small q. The long-range field decays as $1/R^2$ and the corresponding Fourier transform $\boldsymbol{u}(\boldsymbol{q})$ as $1/q$. Consequently, close to Bragg lines the distortion term increases strongly such

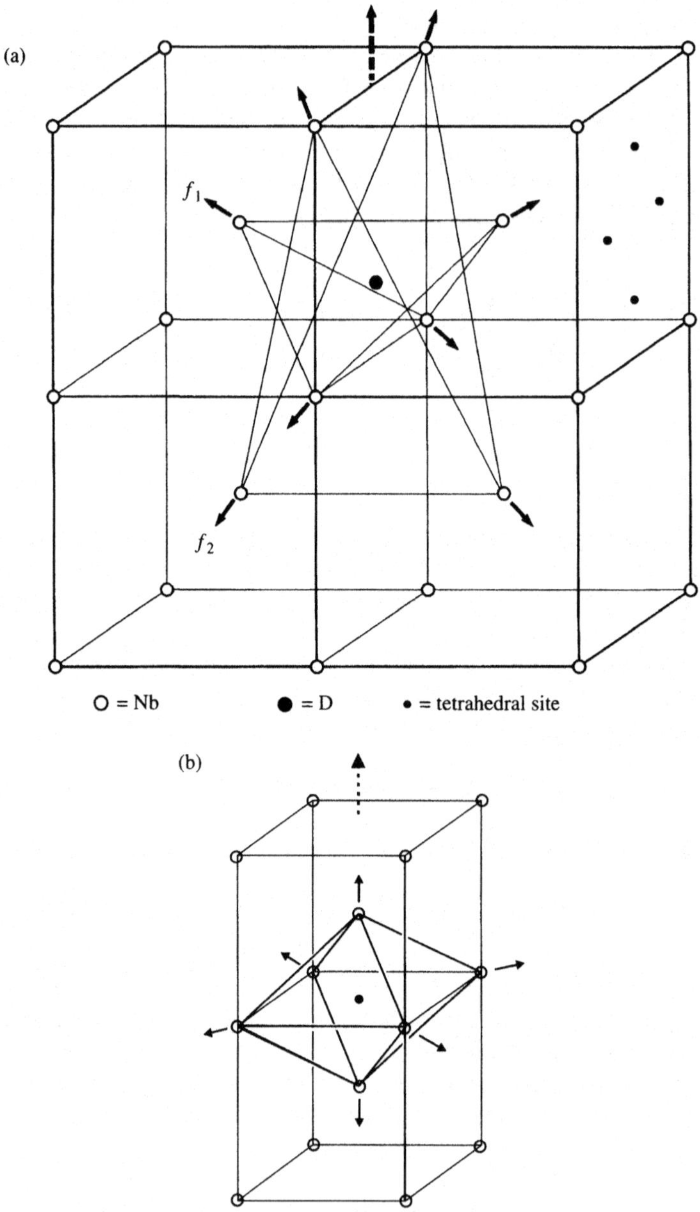

Fig. 6.4 Interstitial atom (a) on a tetrahedral site in a bcc lattice (○ = latticeatoms; • interstitial hydrogen or deuterium). The atoms describe the extra forces, the dashed line is the fourfold axes of the tetragonal displacement field; (b) the same for a octahedral site in a bcc lattice.

that the first term in Eq. (6.47) can be neglected. The resulting so-called *Huang scattering* is then

$$\left(\frac{\mathrm{d}\sigma}{\mathrm{d}\Omega}\right)_{\text{Huang}} = cNb_{\mathrm{h}}^2|\boldsymbol{G}\boldsymbol{u}(\boldsymbol{q})|^2 \tag{6.56}$$

where b_{h} is the scattering length of the host metal atoms. The Huang scattering intensity is proportional to q^{-2}; $\boldsymbol{u}(q)$ can be calculated from the dipole force tensor $\mathbf{P}$ and from the properties of the host lattice by the inverse dynamical matrix $\boldsymbol{\phi}^{-1}(q)$. For a defect with cubic symmetry one obtains a nodal plane through the reciprocal lattice point where the intensity disappears. For defects with lower symmetry this plane can degenerate into nodal lines, or there is nothing of this kind (Dederichs 1973). On the other hand, if $\boldsymbol{Q}$ is not close to $\boldsymbol{G}$, the scattering is strongly influenced by the short-range displacement and by the short-range order term as described, e.g. by Eq. (6.6).

At small Q the inverse dynamical matrix in Eq. (6.53) can be expressed in terms of the elastic constant, c_{11}, and the mean atomic volume of a lattice atom, Ω, yielding

$$u(\boldsymbol{Q}) = \frac{1}{\Omega c_{11}} \frac{1}{Q^2} P(\boldsymbol{Q}). \tag{6.57}$$

As $Q \to 0$ the Fourier transform of the Kanzaki forces, $P(\boldsymbol{Q})$, is given by

$$P(Q) = \mathrm{i}Q\,\mathrm{Tr}(\boldsymbol{P}) \tag{6.58}$$

and the trace of the dipole force tensor by (Peisl 1978)

$$\mathrm{Tr}(\boldsymbol{P}) = \frac{3\Delta a}{ac}\Omega(c_{11} + 2c_{12}); \tag{6.59}$$

$\Delta a/(ac)$ is the lattice expansion per unit concentration. The structure factor Eq. (6.47) can therefore now be evaluated; for small Q we neglect the Q^2 term and obtain

$$S(Q)_{\text{dis}}^{\mathrm{D}} = 1 - 2\frac{b_{\mathrm{Nb}}}{b_{\mathrm{D}}}\frac{c_{11} + 2c_{12}}{c_{11}}\frac{3\Delta a}{ac}\mathrm{e}^{+W_{\mathrm{LM}}}. \tag{6.60}$$

6.4 QUASIELASTIC SCATTERING ON DISTORTIONS COMBINED WITH DIFFUSION

To discuss a *diffusing defect*, and in order to simplify the calculation we consider dilute *vacancies* in a Bravais lattice (Gillan and Wolf 1985; Anderson *et al.* 1987), we assume that the displacement field follows *instantaneously* the diffusive steps of the defect. Consequently, during the rest period τ of the defect on a lattice site, the displacement field reaches equilibrium. Under this assumption we now introduce the time-dependent intermediate scattering function, namely (Eq. 3.44):

$$I(\boldsymbol{Q},t) = N^{-1}\langle \rho(-\boldsymbol{Q},0)\,\rho(\boldsymbol{Q},t)\rangle \tag{6.61}$$

where we again use the reformulation of the lattice sum in Eqs. (6.40) and (6.41), but allow for a time dependence of $\boldsymbol{R}_s$. This leads to

$$\rho(\boldsymbol{Q},t) = \sum_{\text{all } s} \mathrm{e}^{\mathrm{i}\boldsymbol{Q}\boldsymbol{R}_s(t)}\left[-1 + \sum_{\text{all } j} \mathrm{e}^{\mathrm{i}\boldsymbol{Q}\boldsymbol{R}_j}[\mathrm{e}^{\mathrm{i}\boldsymbol{Q}\boldsymbol{u}_j} - 1]\right]. \tag{6.62}$$

Consequently, with $f(\boldsymbol{Q})$ for the square bracket as in Eq. (6.41) and with $S(\boldsymbol{Q})^{\mathrm{v}}_{\mathrm{dis}}$ as in Eq. (6.45) the intermediate scattering function is

$$I(\boldsymbol{Q},t) = N^{-1} S(\boldsymbol{Q})^{\mathrm{v}}_{\mathrm{dis}} \sum_i \sum_j \langle \mathrm{e}^{-\mathrm{i}\boldsymbol{Q}\boldsymbol{R}_i(0)} \mathrm{e}^{\mathrm{i}\boldsymbol{Q}\boldsymbol{R}_j(t)}\rangle. \tag{6.63}$$

For uncorrelated and dilute defects with linear superposition of the displacements, interference terms are negligible and only terms $i = j$ contribute. As for incoherent scattering one obtains

$$I(\boldsymbol{Q},t) = cS(\boldsymbol{Q})^{\mathrm{v}}_{\mathrm{dis}} \langle \mathrm{e}^{-\mathrm{i}\boldsymbol{Q}\boldsymbol{R}_i(0)} \mathrm{e}^{\mathrm{i}\boldsymbol{Q}\boldsymbol{R}_j(t)}\rangle. \tag{6.64}$$

where c is the vacany concentration which comes from the sum over the self-terms $i = j$. If the vacancies perform a jump diffusion process, the Chudley–Elliott model for incoherent scattering is an appropriate treatment; therefore we can express $I(\boldsymbol{Q},t)$ in the usual form $\exp(-t\Gamma(\boldsymbol{Q}))$ such that the energetically broadened diffuse scattering due to the diffusion of vacancies surrounded by their distortion cloud becomes:

$$S^{\mathrm{v}}(Q,\omega) = cS(Q)^{\mathrm{v}}_{\mathrm{dis}} \frac{\Gamma(\boldsymbol{Q})/\pi}{\Gamma^2(\boldsymbol{Q}) + \hbar^2\omega^2} \tag{6.65}$$

where $\Gamma(\boldsymbol{Q})$ in the case of a Bravais lattice is given by Eq. (5.99) for a single diffusing atom (Section 5.3), with τ as the rest time of the vacancy.

In an analogous way we obtain the coherent quasielastic scattering function for diffusing interstitial D atoms with short-range order by a combination of Eqs. (6.32), (6.35), and (6.47):

$$S^{\mathrm{D}}(Q,\omega) = S^{\mathrm{D}}_{\mathrm{dis}}(Q) \cdot S^{\mathrm{D}}_{\mathrm{dif}}(Q,\omega). \tag{6.66}$$

For a Bravais lattice this means

$$S^{\mathrm{D}}(Q,\omega) = S^{\mathrm{D}}_{\mathrm{dis}}(Q)\, S^{\mathrm{D}}_{\mathrm{SRO}}(Q) \frac{1}{\pi} \frac{\Gamma_{\mathrm{coh}}}{\Gamma^2_{\mathrm{coh}} + \hbar^2\omega^2}. \tag{6.67}$$

Generally at small Q one obtains

$$S^{\mathrm{D}}(Q,\omega) = S^{\mathrm{D}}_{\mathrm{dis}}(Q)\, S^{\mathrm{D}}_{\mathrm{SRO}}(Q) \frac{1}{\pi} \frac{\hbar D_{\mathrm{c}} Q^2}{(\hbar D_{\mathrm{c}} Q^2)^2 + (\hbar\omega)^2}. \tag{6.68}$$

Obviously, defects are hopping from site to site, carrying with them instantaneously the surrounding strain field like an elastic polaron, described by the form factor $f(\boldsymbol{Q})$. Also hopping conductivity of electrons, i.e. the diffusion of electrons with their polarons, could thus possibly become visible in this kind of quasielastic scattering (Gillan and Wolf 1985).

As has been shown above, near a Bragg peak the corresponding Huang scattering is strong such that the scattering is amplified and the corresponding coherent quasielastic scattering may thus be measurable. On the other hand, since the width is periodic in the reciprocal lattice, near a Bragg peak we have $\Gamma = \hbar D_c(Q)|Q - G|^2 = \hbar D_c(Q)q^2$. Therefore, if one wants to measure quasielastic Huang scattering one has to compromise between intensity and energetic broadening.

The complicated case that the inverse relaxation rate of the host lattice is larger than the rest time of the interstitial atom was treated by Dosch (1992). The essential modification is a resonance term, the strength of which is related to the force tensor $\mathbf{P}$ of the defect and to the phonon polarization vectors $\boldsymbol{e}(\boldsymbol{q})$. In this case the dynamical form factor for a defect in a single Bravais lattice is more complex and the coherent quasielastic scattering function reads

$$S_D(\boldsymbol{Q}, \omega) = \frac{c\Gamma/\pi}{\Gamma^2 + \hbar^2\omega^2} \left| \sum_s \frac{(\boldsymbol{e}_s \boldsymbol{Q})(\boldsymbol{e}_s \mathbf{P} \boldsymbol{q})}{\omega_s(\boldsymbol{q})^2 - (\omega + i\eta)^2} \right|^2 \tag{6.69}$$

where the sum goes over phonons ω_s which couple to the defect motion, and η is a damping term. This coupling may reduce the displacement scattering.

7

Experimental techniques

7.1 INTENSITY AND BACKGROUND CORRECTIONS

Quasielastic neutron scattering experiments can be performed in direct or inverse scattering geometry. This is schematically displayed in Fig. 7.1 for time-of-flight spectroscopy, the most common method of QENS. In direct scattering geometry a beam of monochromatic neutrons, characterized by a wavevector $\boldsymbol{k}_0$, falls on the sample. As pointed out in previous chapters, the interaction probability is rather small, and in a typical experimental situation most neutrons are transmitted without any interaction. The percentage of scattered neutrons or the scattering probability, respectively, is adjusted to values between 5 and 10% by means of the sample thickness d according to the expression for the transmission,

$$T = \exp\left(-d\sum_i \tilde{N}_i \sigma_{\text{inc}}^i\right)\exp\left(-d\sum_i \tilde{N}_i \sigma_{\text{a}}\right), \tag{7.1}$$

where the first factor correspondingly should amount to 0.90–0.95.

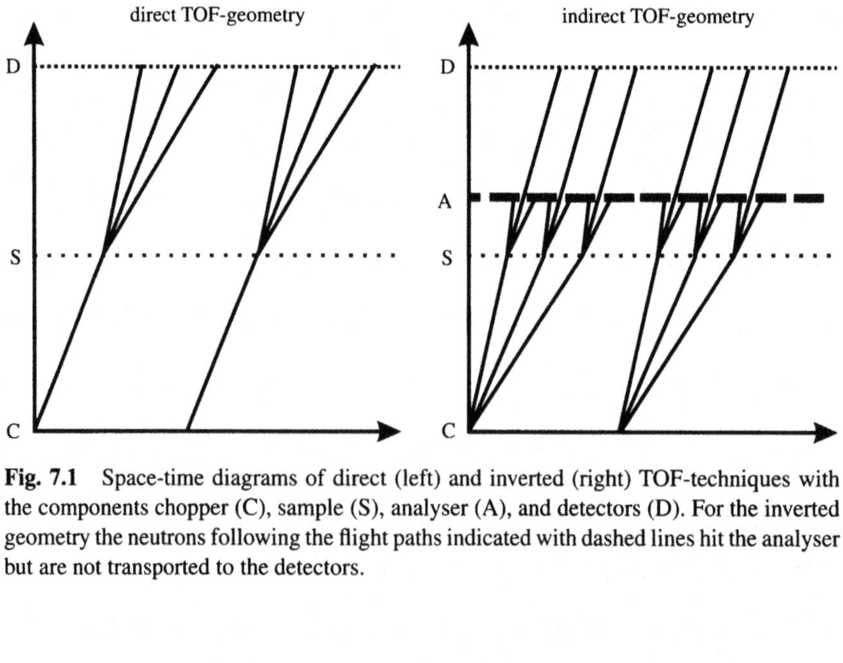

Fig. 7.1 Space-time diagrams of direct (left) and inverted (right) TOF-techniques with the components chopper (C), sample (S), analyser (A), and detectors (D). For the inverted geometry the neutrons following the flight paths indicated with dashed lines hit the analyser but are not transported to the detectors.

Scattering probabilities below 5% imply correspondingly longer measuring times, i.e. in a sense a waste of precious beam time, whereas scattering probabilities above 15% necessitate multiple scattering corrections. If the incident beam is characterized by a uniform flux Φ (strictly speaking, current density Φ), the sample has N identical atoms in the beam, and the detector, placed in direction $\boldsymbol{k}_1$, subtends a solid angle $\Delta\Omega$ and has an efficiency η, then we may expect the count rate C in the detector to be proportional to all these quantities. The constant of proportionality is called the differential cross-section:

$$\frac{\mathrm{d}\sigma}{\mathrm{d}\Omega} = \frac{C}{\Phi N \Delta\Omega\, \eta}. \tag{7.2}$$

In QENS experiments additionally the energy of the scattered neutrons is analysed such that only neutrons with energies between ε_1 and $\varepsilon_1 + \mathrm{d}\varepsilon$ hit the detector while it is counting. Then we obtain the double differential cross-section

$$\frac{\mathrm{d}^2\sigma}{\mathrm{d}\Omega\, \mathrm{d}\varepsilon} = \frac{C}{\Phi N \Delta\Omega\, \eta \Delta\varepsilon}. \tag{7.3}$$

It is connected to the incoherent and coherent scattering functions, respectively. Thus in QENS the double differential cross-section can be considered as the interface between experiment and theory. The detector efficiency (including solid angle) in Eq. (7.3) is usually calibrated by means of a piece of vanadium, of thickness 0.5 mm, which is an incoherent scatterer. In most cases only relative intensities are used, i.e. the relative neutron flux is measured by means of a monitor counter in the incident beam, with typically 1% efficiency, i.e. 1% of the neutrons are destroyed for this purpose. The neutron spectra are then quoted for a certain number of monitor counts, for instance 1000 000.

All neutron spectra contain *background* counts which originate from different sources. Usually there are some neutrons around in the experimental hall—from the neutron source or from other experiments. These neutrons appear at all energy channels of the spectra and thus give rise to a flat background; in order to minimize this effect the spectrometers and particularly the detectors are carefully shielded. A second flat background contribution is air scattering which occurs if the neutrons pass through air on their flight from the sample to the analyser and/or detector. This contribution can be suppressed if on long flight paths air is replaced by argon or, even better, by helium. A vacuum is less common for long flight paths because the constructions become very heavy and expensive. In contrast to these flat backgrounds the sample periphery mostly gives rise to elastic background scattering: neutrons are elastically scattered at the walls of the cryostat or furnace and at the sample container. This contribution is minimized by proper selection of materials and geometry. Al and Nb are best suited because of their extremely small incoherent neutron scattering cross-section. The sample itself usually also gives rise to elastic background scattering. In a metal hydride, for example, in addition to the desired hydrogen scattering, scattering of the host metal lattice also occurs.

In order to correct for these background contributions, a 'background spectrum' $I_B(Q, \omega)$ is recorded and subtracted from the 'sample spectrum' $I_S(Q, \omega)$:

$$I(Q, \omega) = I_S(Q, \omega) - \frac{T_S}{T_B} I_B(Q, \omega). \tag{7.4}$$

The ratio of the transmissions of 'sample' and 'background', T_S/T_B, takes into account self-shielding. The sample in the sample holder scatters (and absorbs) neutrons; therefore the scattering contribution from a sample holder filled with a sample is less than that from an empty sample holder. If in a backscattering spectrometer the sample is passed twice by the beam, the ratio of transmission factors in Eq. (7.4) has to be squared.

In the case of hydrogen in metals the host metal can be measured without hydrogen, and its scattering contribution can be subtracted, together with the contribution of the sample holder. This is of course not possible for instance for solid ionic conductors, with one kind of ion mobile and the other one immobile. In those cases the sample induced background cannot simply be subtracted but has to be handled as an elastic contribution in the data evaluation procedure. Generally, background corrections are not perfect; therefore in most cases a flat background (or sometimes a sloped one) is handled as an adjustable parameter in the data evaluation procedure.

Spallation neutron sources produce primary neutrons with appreciably more intensity at very high neutron energies. These fast neutrons are difficult to shield, but with pulsed sources very low backgrounds are possible, because the fast neutrons are initially concentrated in a short burst, whereas the useful cold neutrons from the moderators arrive later at the sample; the detectors are usually gated during the pulse and start counting somewhat later. For more details on shielding we refer to Windsor (1981).

For inverted scattering geometry, background is a more difficult problem, because in this geometry the full white beam hits the sample whereas otherwise it is only the monochromatic beam, i.e. a small fraction of the total beam. Scintillation counters with a certain γ sensitivity may be sensitive to another kind of sample induced background: for certain elements in the sample gamma radiation is produced by means of (n, γ) reactions; it propagates with light velocity and may fake features apparently in the inelastic part of the neutron spectrum.

7.2 RESOLUTION CORRECTIONS

Each spectrometer exhibits a certain resolution: a theoretical scattering function, $\delta(\hbar\omega)$, appears as a line with a characteristic shape and a finite width in the real measured spectrum. This is called the resolution function $R(Q, \omega)$. In TOF spectrometers of direct geometry, contributions to the resolution originate from the uncertainty in the wavelengths of the incoming monochromatic neutrons, from the uncertainties in the different flight paths, and from the sample and detector thicknesses. For the backscattering spectrometers, however, the resolution is mainly determined by the divergence of the neutron beam, see Section 7.6.

The measured intensity is given by a convolution of the scattering function and the resolution function:

$$J(Q,\omega) = S(Q,\omega) \otimes R(Q,\omega) = \int_{-\infty}^{\infty} S(Q,\omega')R(Q,\omega-\omega')\,\mathrm{d}\omega'. \qquad (7.5)$$

Thus a purely elastic scattering function yields the resolution function as the measured neutron scattering spectrum, whereas, for example, a Lorentzian scattering function yields a corresponding line broadening. In order to correct for resolution effects the resolution function must be known. In high-resolution experiments the change of resolution with energy transfer is usually neglected, and the resolution is measured using a sample which scatters only elastically and with good intensity in order to reach good statistical accuracy. The standard resolution 'material' is vanadium; however, particularly with TOF spectrometers, one has to make sure that the vanadium standard has exactly the same geometry as the sample. It is usually better to determine the resolution function using the sample itself at low temperatures where all motions are frozen-in and the scattering is purely elastic. This simultaneously allows one to check whether intensity is apparently lost at the measuring temperatures, i.e. whether broad scattering components exist with intensity outside the energy window of the spectrometer.

The most common method of data interpretation involves the fitting of an appropriate model for $S(Q,\omega)$ to the measured data taking into account the convolution with the resolution function. In QENS this convolution is usually done numerically, and correspondingly there are some numerical intricacies.

- The numerical resolution function is usually superimposed on a flat background; this background has to be determined, for instance from the first ten and last ten channels, and has to be subtracted, because a flat contribution would yield a sharp peak after convolution.
- If very narrow components occur in the scattering function and if their linewidth is similar or smaller than the width of a single energy channel, then there is the risk that intensity is apparently lost between two adjacent points.

A possible remedy for the latter problem is to interpolate between the data points of the resolution function and in this way to create (artificially) very narrow channel widths. But since the convolution, performed in the central loop of the fitting routine, thus requires a lot of CPU time, the above procedure slows down fitting routines appreciably. A preferable way of solving is as follows: the components of the scattering function due to diffusive motions are Lorentzians of the general shape

$$S_{TQ}(\omega) = \frac{1}{\pi} \frac{\Gamma_{TQ}}{\Gamma_{TQ}^2 + \hbar^2(\omega - \omega_0)^2}. \tag{7.6}$$

Γ_{TQ} denotes the HWHM (for given values of T and Q) and ω_0 the shift on the energy axis (ω_0 should be zero, but actually, due to experimental misalignments, is slightly different). If we define, as the value of the scattering function at the position ω_i, the integral of the function from $\omega_i - \Delta/2$ to $\omega_i + \Delta/2$, where Δ is the channel width, we obtain

$$S_{TQ}(\omega) = \frac{1}{\pi}\left[\arctan\left(\frac{\omega_i + \Delta/2 - \omega_0}{\Gamma_{TQ}}\right) - \arctan\left(\frac{\omega_i - \Delta/2 - \omega_0}{\Gamma_{TQ}}\right)\right]. \tag{7.7}$$

In this way even very narrow components of the scattering function are completely registered, without any loss of intensity. This procedure is also very well suited to handling δ-functions. A δ-function is described as a Lorentzian with an extremely narrow, not variable linewidth, say $\Gamma = \Delta/20$. The numerical convolution with the resolution function using the procedure in Eq. (7.7) then reproduces the resolution function.

An alternative approach for the resolution correction is that of Fourier transformation (Howells 1996) using the fast Fourier transform (FFT) technique. The measured data are the convolution of the scattering function $S(Q, \omega)$ with the resolution function $R(Q, \omega)$, so the transformation becomes $F(Q, t) = I(Q, t)\, R(Q, t)$. As this is just a product, the required intermediate scattering function $I(Q, t)$ is obtained by simple division. In principle, the back transform will then provide the deconvoluted $S(Q, \omega)$. However, in some cases theory actually provides $I(Q, t)$ so it would be more logical to keep the data as $I(Q, t)$.

If the scattering function comprises several components, i.e. if $I(Q, t)$ is a superposition of several exponential functions, this is, like the superpositon of Lorentzians in frequency space, an ill-posed problem, and FFT does not help too much. If, however, the theory yields a stretched exponential as intermediate scattering function, then a temporal Fourier transformation to $S(Q, \omega)$ is not possible analytically, and it is more appropriate to fit the data in the time domain. Stretched exponential relaxation functions are a common feature of polymer dynamics (Richter 1998) and of glass transitions (Mezei *et al.* 1987).

7.3 SIMULTANEOUS FITS AND BAYESIAN ANALYSIS

The physical model of diffusive motions in condensed matter consists of a single Lorentzian in the simplest case, that of translational diffusion, but in many cases there are several Lorentzians or there is an elastic peak with one or more Lorentzians, and it is often hard to decide how many components are involved.

The most common approach of data evaluation is a least squares fit comprising the convolution of the resolution function with the scattering function; one starts with the simplest possible model and, if the fit is not satisfactory, continues with models of increasing complexity. According to experience it is already difficult to fit a two-component scattering function to an individual QENS spectrum at a given Q value and given temperature, $S_{TQ}(\omega)$, but it is usually impossible to fit more components. A possible remedy is a simultaneous fit of the ω and Q dependences of a set of spectra, $S_T(Q, \omega)$, at a given temperature, or even a simultaneous fit of the ω, Q, and T dependences of all recorded spectra. This is illustrated for the simple example of translational diffusion. Let us assume we have recorded QENS spectra, each consisting of 256 energy channels, at 10 different Q values and at five different temperatures, comprising altogether 12 800 data points. We assume a Lorentzian ω dependence and, in a first step, perform corresponding single spectra fits of $S_{TQ}(\omega)$. This reduces the data to 50 linewidths $\Gamma(T, Q)$. The second step is to assume a Q dependence of the linewidths, say according to the isotropic Chudley–Elliott model, Eq. (5.103), and to fit the five sets of linewidths $\Gamma_T(Q)$ correspondingly; this reduces the data to five diffusion coefficients, $D(T)$, and five jump lengths; the latter should all be equal, but usually exhibit some fluctuations, so the mean value is formed. The third step would then be an Arrhenius fit of the diffusion coefficients $D(T)$ yielding the prefactor D_0 and the activation E_a.

The alternative method of data fitting is the simultaneous fitting of all 12 800 data in one step:

$$S(T, Q, \omega) = \frac{1}{\pi} \frac{\Gamma(T, Q)}{[\Gamma(T, Q)]^2 + \hbar^2[\omega - \omega_0(Q)]^2} + B(T, Q) \qquad (7.8)$$

with

$$\Gamma(T, Q) = \frac{\hbar D(T)}{6l^2}\left(1 - \frac{\sin Ql}{Ql}\right) \qquad (7.9)$$

and with

$$D(T) = D_0 \exp\left(-\frac{E_a}{kT}\right). \qquad (7.10)$$

The relevant physical parameters l, D_0, and E_a are common to all spectra. The misalignment parameters ω_0 are, according to experience, different for each detector (Q value), but for a given detector do not depend on the temperature; usually they are taken from previous individual fits and are no longer allowed to vary. The background parameters B are also without physical relevance; they somehow contain part of the phonon density of states, and thus may change with temperature; therefore these parameters have to be considered as individual to each spectrum. Thus in this example the simultaneous fit implies 53 variable parameters, requires good starting values, and some CPU time, but the requirements on the statistical quality of the data are considerably less stringent than for individual spectra fits.

For this simple example both methods of data fitting are feasible, as are some intermediate methods, for instance individual spectra fits $S_{TQ}(\omega)$ and a simultaneous fit of the linewidths, $\Gamma(T, Q)$. But for complex situations where the scattering function consists of two or more components, individual spectra fits are usually not successful.

The decision whether an assumed model is appropriate and the data evaluation is satisfactory is taken by the responsible scientist by considering (i) whether the data are described well or at least satisfactorily by the model, and (ii) whether the resulting parameters make sense physically. This 'decision' requires experience and is to a certain extent subjective. The Bayesian analysis of QENS data analysis (Sivia *et al.* 1992) puts this on a firmer ground and allows one, on the basis of maximum entropy, to decide for instance how many Lorentzians are most probable. However, since QENS in this respect is an ill-posed problem, the probability difference between two, three, or more components is usually not significant, whereas the distribution between one or more than one components is easy in most cases: an experienced neutron scatterer is able to do this just by visual inspection of the data. Therefore, the common wisdom nowadays seems to be that with respect to QENS Bayesian 'probability theory is nothing but common sense reduced to calculations' (in the words of Laplace 1812). In some cases a Bayesian analysis of QENS data might be helpful but conventional data fitting is considerably more commonly used and the 'common sense' of an experienced neutron scatterer should not be undervalued.

7.4 MULTIPLE SCATTERING CORRECTIONS (MSCs)

The distortion of the widths and shapes of QENS spectra due to multiple scattering is particularly pronounced at small scattering angles, i.e. at small Q values. At small scattering angles most of the twice-scattered neutrons (the largest fraction of multiple scattering) have been scattered twice at large scattering angles (i.e. at large Q values) where the scattering function is broad. This makes the linewidths at small Q appear too broad. At large Q values, however, the influence of multiple scattering is not so substantial. Most of the twice-scattered neutrons have been scattered once at small and once at large scattering angles, or twice at medium scattering angles. In both cases the resulting linewidths do not differ too much from those due to single scattered neutrons. A correction of multiple scattering effects (MSC) can be done using the Monte Carlo program DISCUS (Johnson 1974) in an iterative manner. The simulation of the scattering process in DISCUS requires knowledge of the scattering at large Q, which, however, in some investigations is just the aim of the data evaluation and not known *a priori*. In other investigations, it is considered too complex and the diffusion coefficient is the only aim; then a pragmatic procedure is to take the isotropic Chudley–Elliott model as an approximation. For the purpose of MSC this is often reasonable and sufficient; the resulting scattering function is a single Lorentzian with linewidth (5.103). An iterative correction procedure is performed as follows. First all measured QENS spectra are fitted with a Lorentzian function (including a convolution with the resolution function). In the second step the resulting linewidths $\Gamma(T, Q)$ are simultaneously fitted with Eq. (7.9) and the Arrhenius law for the temperature dependence of the diffusion coefficient; in this way zeroth-order values of the jump length l as well as of the prefactor D_0 and of the activation energy E_a of the diffusion coefficient are determined. Using these three parameters (and, of course, with the necessary information about the geometry and the scattering properties of the sample) DISCUS simulates the scattering process and calculates correction factors $f_Q(\omega)$ for the different Q values,

$$f_Q(\omega) = \frac{I_1(\omega)}{I_1(\omega) + I_2(\omega) + I_3(\omega)}, \tag{7.11}$$

i.e. the ratio of the flux of once-scattered neutrons, $I_1(\omega)$, over the total flux of scattered neutrons. Neutrons which are scattered more than three times in the sample are very rare and can be neglected. Now within the fitting routine the scattering function for once-scattered neutrons, $S_Q(\omega)$, is divided by the correction function, $f_Q(\omega)$, channel-by-channel, and thus the scattering function for multiple scattering is obtained. Corresponding fits of the QENS spectra yield first-order values for the linewidths and subsequently first-order values for l, D_0, and E_a. These 'output' values have to agree with the 'input' values of DISCUS, otherwise the cycle is repeated. Usually after three iterations self-consistency is reached.

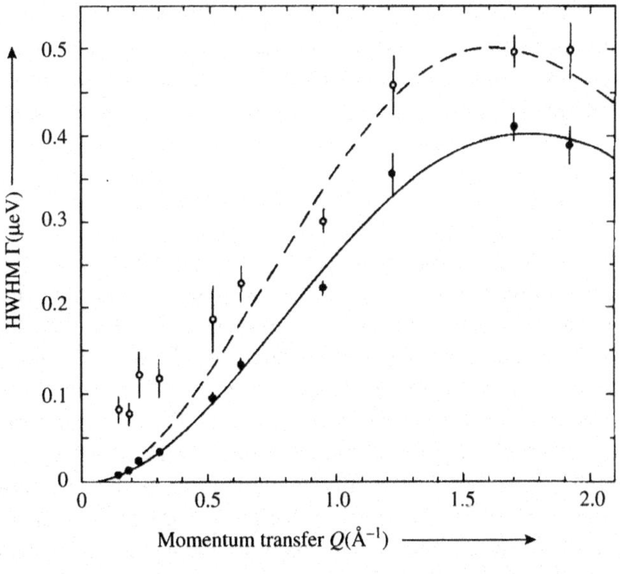

Fig. 7.2 Quasielastic linewidths of $LaNi_5H_6$ at $T = 303$ K as a function of Q: dashed line and open symbols: fit of Eq.(5.103) to the datapoints without multiple scattering correction; solid line and solid symbols: after multiple scattering corrections (from Richter *et al.* 1982, with permission from Elsevier Science).

Finally, the corrected linewidths at small Q can be fitted with the Q^2 law in order to determine a model-independent diffusion coefficient. Figure 7.2 shows, for the example of H diffusion in $LaNi_5H_6$, the effect of MSC. The linewidths at small Q suffers an appreciable correction whereas those at large Q values are 'only' corrected by about 25%. Of particular interest are the error bars: for the data points at small Q the error bars are smaller than the size of the symbols: obviously the lineshape is perfectly Lorentzian-like, which proves that the scattering function at small Q is a single Lorentzian, as it should. Multiple scattering not only broadens the scattering function but also changes its shape as is evident from the large error bars. At large Q, on the other hand, the scattering function obviously is not a single Lorentzian, because corresponding fits—with or without MSC—yield large error bars. The apparent line broadening due to multiple scattering is larger, the narrower the lines are. Therefore, without MSC the apparent diffusion coefficients at low temperatures (where the linewidths are narrow) are more increased than those at higher temperatures; in this way the apparent activation energy is smaller than the correct one.

7.5 TIME-OF-FLIGHT SPECTROMETERS

Neutron time-of-flight (TOF) spectroscopy is a powerful method for the study of dynamic processes over a wide dynamical range for polycrystalline and disordered solids or for liquids. In contrast to a triple axis spectrometer, specialized to single-crystal samples, a TOF experiment simultaneously detects several points in (Q, ω) phase space by time-resolved neutron detection within a large solid angle by the use of typically several hundreds of detectors. TOF spectroscopy uses the relation between the energy of the neutron, ε_i or ε_f, respectively, and its velocity v, which reads

$$\varepsilon = \tfrac{1}{2}mv^2 \rightarrow \varepsilon[\mu\text{eV}] = 5.227v^2[\text{m ms}^{-1}]. \tag{7.12}$$

Typical neutron velocities used in a scattering experiment range from 0.2 to 3.0 m ms^{-1}. Therefore the flight time for experimentally relevant distances of several metres are conveniently accessible. By measuring the time the neutron needs to pass a known distance one directly calculates the energy ε. Thus in a TOF experiment a facility is needed to give an electronic signal starting the clock that counts the neutron's necessary time to the detectors. This time marker chops up (and is therefore called a chopper) the beam into several bunches by opening and closing in defined intervals. Thus usually more than 98% of the beam is lost due to the closed chopper. The loss in intensity, however, is more than compensated for in a TOF experiment by the simultaneous detection of many points in (Q, ω) space.

TOF experiments can either have direct or inverted geometry. Direct geometry means that the sample is illuminated by a monochromatic neutron beam. After the scattering process the final energies ε_f of the detected neutrons are calculated from the flight time they need to pass the known distance from the sample to the detectors. From the known incident neutron energy ε_i one is able to calculate the energy transfer $\hbar\omega$ at the sample, the quantity one is interested in. Inverted TOF spectrometers operate in the opposite way. Here a polychromatic neutron beam hits the sample. After the scattering process crystal analysers reflect only neutrons with a certain final energy ε_f into the detectors. From the measured time of flight one is able to calculate the initial energy ε_i and thus again the energy transfer. Figure 7.1 shows the respective space-time diagrams for both geometries. The main advantage of the inverted TOF method is the fixed geometry for every experiment. The disadvantage is that the white beam illuminating the sample usually provides larger background signals than the monochromatic beam of a direct geometry TOF instrument. The prototype inverted-geometry TOF spectrometer for high-resolution neutron spectroscopy is IRIS at the spallation source ISIS. Since this instrument represents a backscattering instrument as well, it will be discussed in the next section. For direct-geometry TOF spectroscopy nowadays there are two different principles of monochromatizing the beam. The first type of spectrometer uses several disk choppers separated from each other, i.e. the multichopper TOF spectrometer. The first instrument and still the best-known example is the IN5 instrument of the ILL (Lechner *et al.* 1973); other and younger instruments of this type are MIBEMOL at LLB (Saclay), NEAT at HMI (Berlin), and DCS at NIST (Gaithersburg). The other possibility of monochromatizing and pulsing a neutron beam is the use of a crystal monochromator in combination with a Fermi chopper. This is actually the original method of TOF spectroscopy, and numerous thermal instruments of this type exist. For cold neutrons this method is nowadays again competitive with multichopper TOF spectroscopy, because by the use of focusing techniques an enormous intensity gain has been achieved. The first instrument of this type is the IN6 (Scherm *et al.* 1978), again at the ILL; another example is FOCUS at PSI (Villigen) (Mesot *et al.* 1996). Since the famous ILL TOF spectrometers have already been described many times (see, e.g. Bée 1988, p. 81–91) we will concentrate in the following on new developments outside the ILL.

As a new example of a multichopper TOF spectrometer we mention the NEAT instrument at the HMI Berlin (Lechner *et al.* 1996). Although its basic conception is derived from the ILL spectrometer IN5, a number of modifications have been implemented, giving NEAT an improved performance. As a result, for a given energy resolution, the duration of experiments with the new instrument is comparable to that of IN5, in spite of the 10 times smaller flux of the host reactor. The spectrometer set-up and its essential components are shown in Fig. 7.3. The primary part of the spectrometer creates a pulsed monochromatic neutron beam by means of four phased disk chopper systems, three of which consist of pairs of disks, closely spaced, and rotating in opposite direction. Basically the pulse production and the monochromatization works with only two disk choppers. As one can see from the left part of Fig. 7.1, it might happen that the fast neutrons of a later pulse overtake the slow neutrons from the preceeding one (frame overlap). Therefore one uses a third chopper rotating with a lower frequency to enlarge the

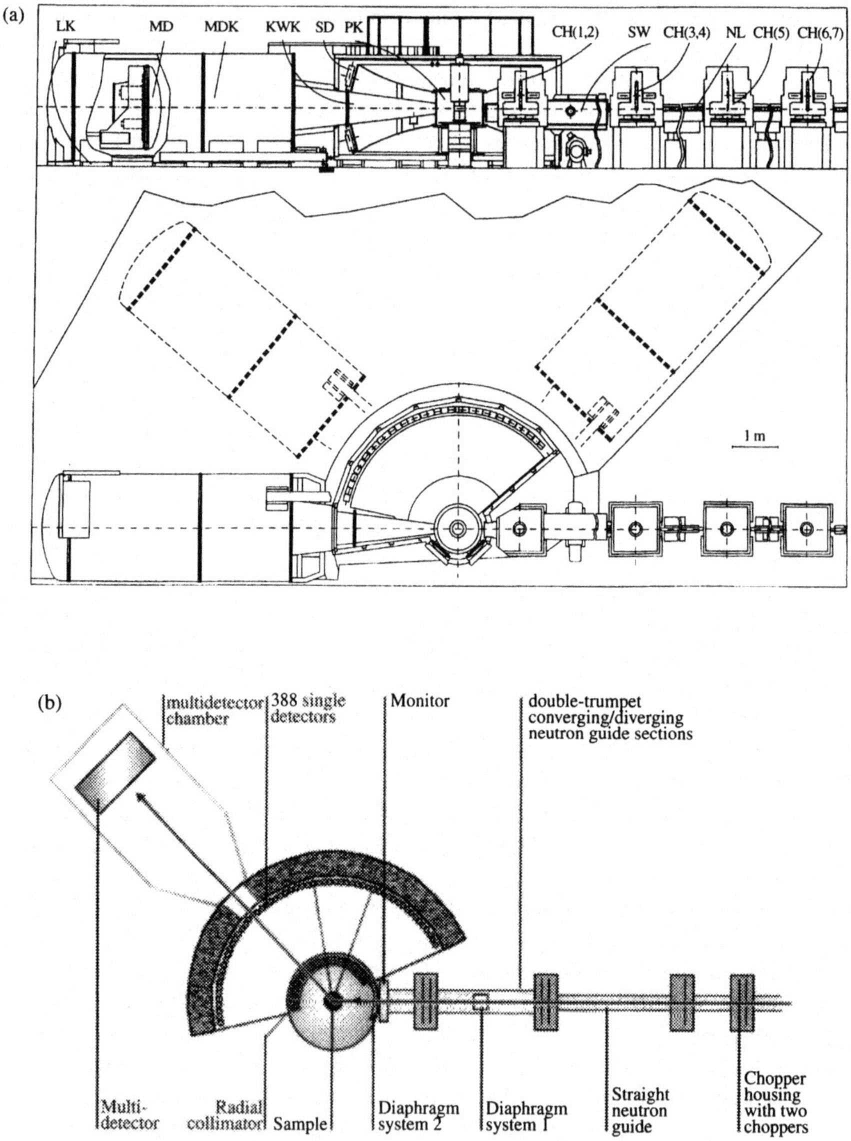

Fig. 7.3 Time-of-flight spectrometer NEAT, schematic sketch showing the essential components of this instrument. The sample, mounted in a chamber (PK) filled with the gas or evacuated, receives an incident pulsed monochromatic neutron beam created by seven synchronized disk choppers (right side of the figure). The two detector systems, shown on the left side of the figure, are mounted at the periphery of the detector chamber (DK: 388 single detectors) and in the interior of the multidetector chamber (MDK: two-dimensional position-sensitive detector), respectively (from Lechner *et al.* 1994).

temporal distance between the pulses. The fourth chopper is then used to suppress neutrons that travel with integer multiples of $\frac{1}{2}$, $\frac{1}{3}$, $\frac{1}{4}$, etc. of the desired velocity that would pass a two- or three-chopper system.

Counter-rotating chopper pairs represent a better compromise between intensity and resolution than single choppers rotating at the same speed (Lechner 1991; Copley 1992). Burst time, an essential contribution to the resolution, is proportional to W/v where W is the beam width and v is the mean linear velocity of the disk. On the other hand transmitted intensity varies as W^2/v, being proportional to the product of the burst time, W/v, and the beam width. Thus v (and W) should be maximized in order to achieve the highest intensity at a given resolution. Given engineering and safety constraints, which limit v for any given disk, the effective chopping speed of a counter-rotating chopper pair is twice that of a single disk.

A second and very important reason for using counter-rotating chopper pairs is that they allow a choice of beam widths so that the burst time of a chopper pair can be (discretely) changed without changing its speed. This is achieved by placing several windows on a pair of disks.

Let us first consider the hypothetical case that the two disks of a pair are mounted at the same place on the guide (which is of course not possible in reality). The synchronously rotating disks are then phased in such a way that the desired windows of both disks cross the neutron beam simultaneously, thus creating a periodic sequence of neutron pulses. All other windows are placed on the disks at phase angles, which for the given disk phasing do not allow a simultaneous opening of the beam by two different windows. Thus unwanted combinations of windows ('cross-talk' within a chopper pair) are excluded. In reality the two disks of a pair have to be mounted at a finite distance of about 50 mm. Due to this finite distance the transmission through the disk pair of neutron pulses due to unwanted combinations of neutron windows is no longer completely forbidden. The challenge is to determine where to place the windows on the disks in order to minimize cross-talk and find useful chopper configurations where cross-talk does not disturb the beam.

As a new example of a focusing TOF spectrometer we mention the FOCUS instrument at PSI (Mesot *et al.* 1996). It has been equipped with two interchangeable monochromators (pyrolithic graphite and mica) that continuously cover the range of initial energies $0.27\,\text{meV} < \varepsilon_0 < 20\,\text{meV}$. High flexibility is achieved by operating either in time- or monochromatic focusing mode. In Fig. 7.4 a schematic horizontal cut through the spectrometer and its main components is given. At the end of the neutron guide a disk chopper is located that acts as an anti-overlap chopper. The crystal monochromator has a variable curvature and focuses the beam horizontally as well as vertically on the sample. The main TOF chopper is a Fermi chopper; interchangeable rotors are available with different collimator packages; in this way the collimation of the Fermi chopper can be changed.

In the time-focusing mode the *end of neutron guide–monochromator* distance amounts to 3 m, and the *monochromator–sample* distance amounts to 1.5 m; the curvature of the monochromator is tuned such that neutrons emerging from

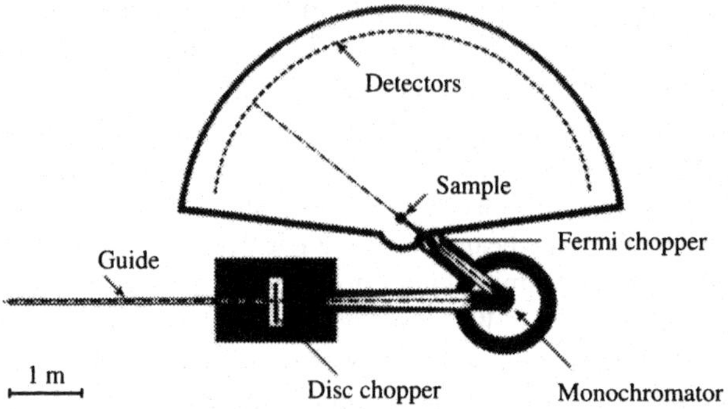

Fig. 7.4 Horizontal cut through the TOF spectrometer FOCUS at the Swiss spallation neutron source SINQ (from Janssen *et al.* 1997, with permission from Elsevier Science).

different sides of the monochromator have slightly different wavelengths. Now the Fermi chopper scans the monochromator in such a way that it allows the slower neutrons of each pulse to pass earlier than the faster ones. Neutrons having a wavelength distribution around a certain λ_0 then simultaneously arrive at the detectors within a sharp pulse. The time-focusing mode is well suited for applications which are interested either in a single inelastic excitation at a certain energy transfer or for quasielastic scattering (around energy transfer $\hbar\omega = 0$), because the energy resolution gets rather poor out of the focal condition.

In the monochromatic focusing mode the *end of guide–monochromator* distance and the *monochromator–sample* distance are equal, and the curvature of the monochromator is tuned such that there is no wavelength distribution (besides that arising from local divergences). This configuration reveals a good resolution over a wide range of energy transfers and is attractive since it provides high flexibility: with both the above-mentioned distances as either 3.0 m or 1.5 m one can choose between a high-resolution or high-intensity mode.

7.6 BACKSCATTERING SPECTROMETERS

The basic principle of backscattering spectroscopy was invented and proposed by Maier-Leibnitz in about 1965 (unpublished); according to this principle Alefeld *et al.* (1968 and 1969) built the first backscattering spectrometer. Meanwhile, the technique has reached a very high stage of development, and is being applied very succesfully in many fields of physics and chemistry (for a review see Alefeld *et al.* 1992). The backscattering principle is based on Bragg reflection of a perfect single crystal (free of dislocations) at a scattering angle 2Θ at or near to 180°. The Bragg law can be formulated as

$$|\mathbf{G}_{hkl}| = 2k \sin\Theta \qquad (7.13)$$

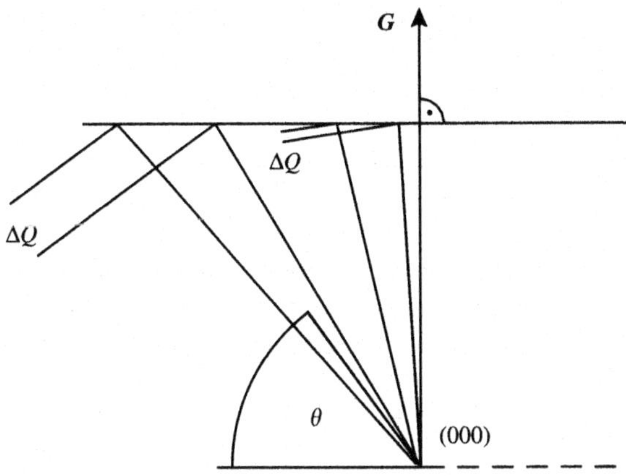

Fig. 7.5 Ewald construction in reciprocal space. All end points of the $\boldsymbol{k}$ vectors lie on the bisectrix B perpendicular to the reciprocal lattice vector $\boldsymbol{G}$. For backscattering and a beam divergency $\Delta\Theta$ centred at $\Theta = \pi/2$ the width of $\boldsymbol{k}$ is $\delta k_{\mathrm{div}}/k = (\Delta\Theta)^2/8$.

where $\boldsymbol{G}_{hkl}$ is a reciprocal lattice vector with modulus $|\boldsymbol{G}_{hkl}| = 2\pi/d_{hkl}$, i.e. proportional to the reciprocal interplanar spacing of the scattering lattice plane. This is visualized in Fig. 7.5. The uncertainity or resolution width is obtained from the angular uncertainity by differentiating:

$$\delta k_{\mathrm{div}}/k = \cot\Theta\,\Delta\Theta. \tag{7.14}$$

In this way resolutions of about 10^{-4} eV are usually obtained in triple axis spectroscopy. In the case of backscattering ($\Theta = \pi/2$) the linear term disappears and only a second-order term has to be taken into account:

$$\delta k_{\mathrm{div}}/k \approx (\Delta\Theta)^2/8. \tag{7.15}$$

For small beam *divergences* this leads to resolution widths of 10^{-7} eV. The count rates are still reasonable because to first order resolution and intensity are decoupled.

In addition to δk_{div} there is a contribution to the resolution width due to *primary extinction*. This is related to the fact that only a finite number of lattice planes contributes to the Bragg-reflected beam:

$$\delta k_{\mathrm{ex}}/k = 16\pi N_{\mathrm{c}} F_G/G^2. \tag{7.16}$$

F_G is the structure factor for $\boldsymbol{Q} = \boldsymbol{G}$; N_{c} is the number of lattice cells per unit volume. Both contributions together yield:

$$\delta\varepsilon/\varepsilon = 2\delta k/k \approx 2\left[\frac{\Delta\Theta^2}{8} + \frac{16\pi N_{\mathrm{c}} F_G}{G^2}\right]. \tag{7.17}$$

For a monochromator placed at the end of a nickel-covered neutron guide the divergence is given by the critical wavevector for nickel, $\Delta \boldsymbol{k}_{\mathrm{Ni}}$. Considering a dislocation-free single crystal of silicon and a (111) reflection one obtains $\delta\varepsilon = (2.4 + 0.8) \cdot 10^{-7}$ eV, i.e. a resolution much better than that of a triple axis or time-of-flight spectrometer. Experimentally values between 0.2 and 0.3 μeV have been found.

Since the Bragg angle has to be fixed at 90°, the scan of a scattering spectrum is obtained by rapid motion of the monochromator crystal, thus shifting the energy by the Doppler effect. This, of course, restricts the energy range covered by the spectrometer. To first order, with a ratio of the crystal speed v_D perpendicular to the reflecting lattice planes and the neutron speed v_0, the energy shift $\Delta\varepsilon = \hbar\omega$ by the Doppler effect is to first order (Shull and Gingrich 1964)

$$(\Delta\varepsilon_0/\varepsilon_0)_D = 2(v_D/v_0). \tag{7.18}$$

For $v_D = 2.5\,\mathrm{m\,s^{-1}}$, which can easily be achieved, this leads to an accessible energy range of ±15 μeV for incident neutrons of 6.27 Å. We point out that the resolution does not in practice depend on the analysed energy transfer; this is an advantage of this type of spectrometer.

Figure 7.6 shows the backscattering spectrometer operated at the liquid hydrogen cold source of the FRJ-2 reactor in Jülich. Neutrons are conducted from the source to the monochromator by a ^{58}Ni covered neutron guide of 70 cm^2 cross-section. The monochromator is a silicon wafer of $10 \times 10\,\mathrm{cm^2}$ area and 1 mm thickness which leads to practically 100% reflectivity. The Doppler motion is produced by a hydraulic system with a maximum speed of $3\,\mathrm{m\,s^{-1}}$. It is possible to carry out different velocity profiles as a function of time; usually a triangular profile is chosen. The velocity is measured by an inductive pick-up coil placed on the crystal holder. In an older instrument a sinusoidal crystal motion was used with the disadvantage that the measuring time is unnecessarily long at the turning point of the Doppler drive. The maximum energy range to be covered is ±18 μeV. The monochromized beam hits an array of focusing pyrolytic graphite crystals, which deflect the beam onto the sample. After scattering, the neutrons are analysed again in (or close to) the backscattering geometry by large spherical shells covered with 0.7 m^2 silicon wafers in the (111) orientation. The beam is focused by the curved crystals into several detectors covering different regions of scattering angle. The total resolution with contributions from ε_0 and ε_1 is 1 μeV FWHM. A typical Q resolution is 0.1 Å^{-1}. In most quasielastic scattering this resolution is sufficient since the spectrum depends only smoothly on Q. By applying focusing techniques the new backscattering spectrometer IN16 of the ILL improves both the neutron flux and the energy resolutions over existing backscattering spectrometers (Frick *et al.* 1997).

In cases where larger energy transfers are to be investigated the Bragg energy can be shifted by means of silicon doped with germanium with an off-centre combination, having undoped Si (111) as analysing crystals. Instead of making use of a Doppler drive, the lattice parameter $d(T)$ can be changed by heating (Heidemann and Alefeld 1992; Cook *et al.* 1992; Randl and Johnson 1997), and

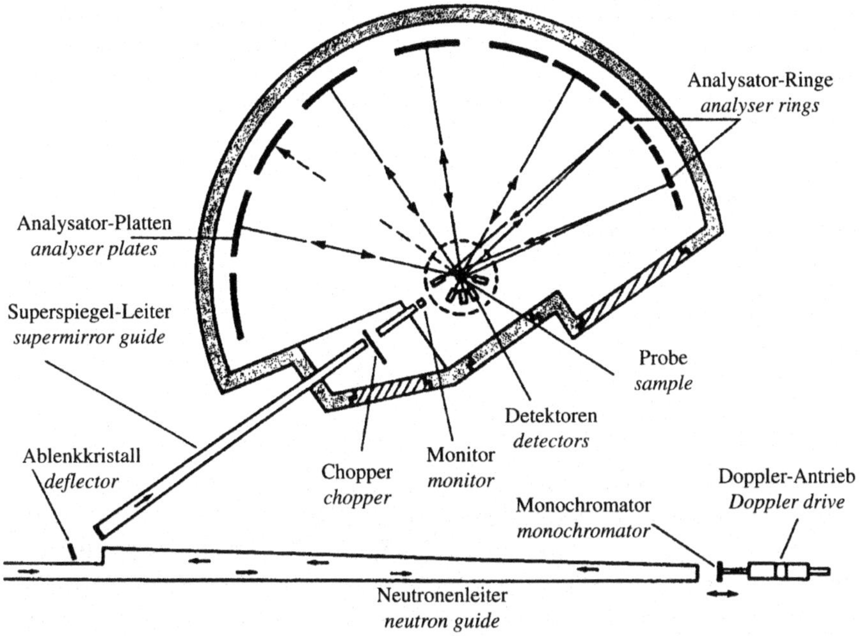

Fig. 7.6 The Jülich backscattering instrument at the FRJ-2 reactor. Neutrons from the hydrogen cold source arrive at the monochromator, a plate of ideal Si crystals, and are back reflected. The crystals move linearly and follow a given velocity profile. The monochromatized beam is separated from the incident beam by a graphite crystal, and is deflected onto the sample. After scattering on the sample, the neutrons are again back reflected on spherical monochromator shells covered with small Si wafers. The backscattered neutrons are focused into several detectors corresponding to different Q values. The chopper prevents neutrons from being scattered directly from the sample to the detector.

the energy change is given by

$$\varepsilon_0 = \hbar^2\pi^2/2md(T)^2. \tag{7.19}$$

For a temperature change from 80 to 700 K a scanning range of several per cent can be achieved except for Si which allows only 0.3%. This principle has already been used for the medium-resolution instrument IN13 at ILL in Grenoble with a heated or liquid nitrogen cooled CaF_2 (422) crystal as monochomatizer, and at room temperture CaF_2 crystals as analyser. This instrument works at $E_0 = 16$ meV (2.23 Å) and covers larger Q values than IN10. The ω range goes from -125 to 300 μeV.

Table 7.1 (Randl and Johnson 1997) summarizes the energy range for a number of monochromator crystals with a temperature shift between 80 and 700 K, paired with Si (111) analysers. In addition, this range can also be combined with the

Table 7.1 Monochromators available at the backscattering spectrometer IN10B of the ILL. Si* (111), used only on IN10A, is given for comparison (from Randl and Johnson 1997).

Crystal	Reflection	Minimum energy (μeV)	Maximum energy (μeV)	Flux (a.u.)
Si*	111	−14	+14	1.0
NaF	111	−813	−602	0.3
NaCl	200	−534	−339	1.7
NaBr	200	−252	−80	0.4
BaF_2	200	−62	+19	0.25
KCl	200	−16	+128	1.3
NaI	200	+92	+226	0.5
NaCl	111	+119	+266	

conventional Doppler shift. Also crystals with heat gradients were investigated in order to broaden the resolution width and to adapt it to a larger energy window, with a corresponding gain of intensity. So far this has not yet been applied practically. Also the effect of longitudinal sound waves in the MHz range produced in a Si crystal by a $LiNbO_3$ transducer and its effect on reflectivity has been studied (Hock *et al.* 1993; Remhof and Magerl 1997).

A new feature in neutron backscattering spectroscopy is the possibility of transforming, by means of a phase-space transform chopper, the shape of the incoming neutron beam in phase space such that the neutron flux of the diffracted beam is enhanced at the backscattering energy E_0. The well-collimated energy dispersive incoming beam is transformed into an angle dispersive monochromatic beam. This idea is due to Schelten and Alefeld (1984), and was developped to address the significant mismatch in divergence between primary and secondary spectrometers. A backscattering spectrometer exploiting this effect has recently been built at the Cold Neutron Research Facility of the National Institute of Standards and Technology (Gehring *et al.* 1995).

7.7 BACKSCATTERING WITH TIME-OF-FLIGHT

Instead of using backscattering for monochromatization before scattering and for energy analysis after scattering, one can monochromatize by time-of-flight analysis. For this purpose, the incident beam is chopped into pulses of length $\Delta\tau$ as obtained from a pulsed spallation source. The corresponding repetition rate of chopping is $\nu = 1/\tau_R$ where τ_R is the pulse distance. Figure 7.7 shows the IRIS time-of-flight backscattering spectrometer at the ISIS pulsed source in a schematic representation.

For the benefit of the intensity the width $\Delta\tau$ should be adapted to the resolution available in the backscattering geometry in the analyser. We take $\Delta\varepsilon_1 \approx \Delta\varepsilon_0$

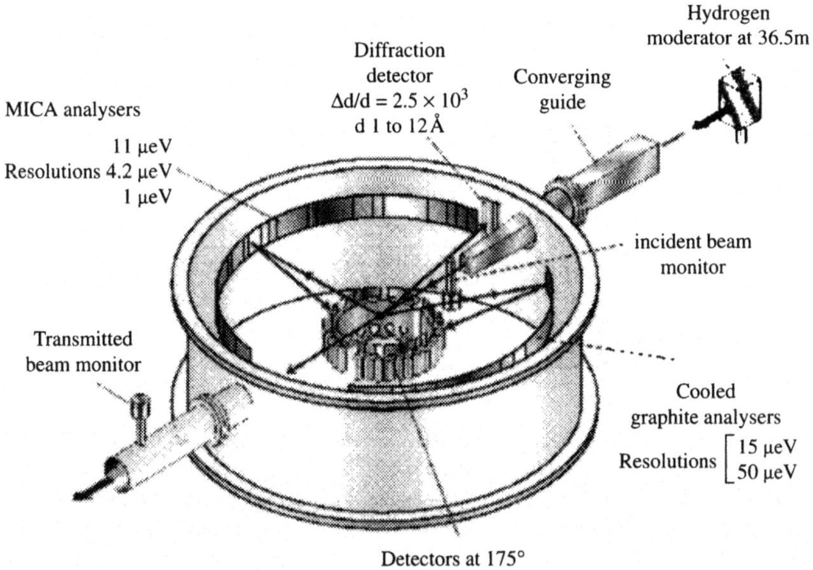

Fig. 7.7 The IRIS time-of-flight backscattering spectrometer at the ISIS pulsed source. The neutron beam starts in the liquid hydrogen moderator at the target. The flight path is 36.5 m between source and detector. The scattered neutrons are back reflected on graphite or on high resolution mica analysers into a two-dimensional detector. The resolution is determined by the pulse length $\Delta\tau$ and by the natural width of the backscattering process. (With kind permission of the Rutherford Appleton Laboratory.)

which means

$$\Delta\tau/t_0 = \Delta\varepsilon_0/2\varepsilon_0 \tag{7.20}$$

where $t_0 = L/v_0$ is the flight time over the flight path between the chopper and detector of length L. These leads to an optimized path length, matched to the monochromator width:

$$L = v_0 t_0 = v_0 \Delta\tau/(\Delta\varepsilon_0/2\varepsilon_0). \tag{7.21}$$

The time schedule of neutrons for this instrument is shown in Fig. 7.8. A chopper between the pulsed source and the sample cuts out a time window which is sufficiently broad to cover the observed scattering spectrum of interest. A quasielastic spectrum, resulting from diffusing particles, is centred at $\hbar\omega = 0$. The width of this window is given by

$$t_{\mathrm{w}} = L/v - L_{\mathrm{s}}/v_0 - L_s/v_1 \approx L/v \tag{7.22}$$

since the energy transfer is small or $v_0 \cong v_1$, which is the variable velocity before scattering. v_1 is the velocity after scattering fixed by the energy of the

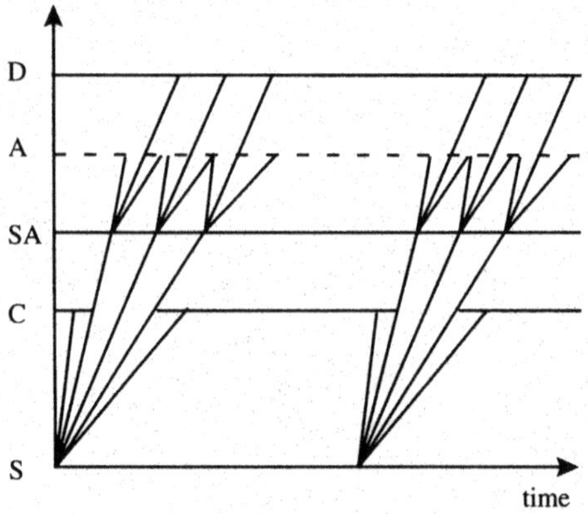

Fig. 7.8 Time-of-flight diagram of neutrons in a pulsed backscattering spectrometer. S = source position, SA = sample position, and D = the detector plane. C indicates the position of the chopper which periodically opens a certain velocity window synchronized with the source pulses. The velocity after scattering, v_1, is fixed and determined by the backscattering analyser. The velocity range covered by this spectrometer is $v_{max} - v_{min}$. The abscissa is a time or a wavelength scale. $\Delta\tau$ is the pulse length of the source.

backscattering analyser. In order to reduce the frame overlap, i.e. the superposition of neutrons during a given pulse due to the previous pulses, a 'window chopper' has to be introduced with an appropriate time window. Consequently the time between pulses, t_0, must be sufficiently large with respect to t_W. For the IRIS spectrometer the flight path is 36.5 m and the chopper sits 6.4 m from the source.

7.8 QUASIELASTIC MÖSSBAUER SPECTROSCOPY AS A RELATED METHOD

Quasielastic Mössbauer spectroscopy (QMS) is closely related to QENS in that it also measures the double Fourier transform $S(Q, \omega)$ of the self-correlation function, which however, is not deduced from scattering, but from resonant emission or absorption of nuclear γ radiation (Vogl 1996; Vogl and Feldwisch 1998). The scattering vector for QENS is the difference between the incident and scattered wavevector $\boldsymbol{Q} = \boldsymbol{k}_i - \boldsymbol{k}_f$, whereas for QMS in nuclear absorption mode $\boldsymbol{Q}$ equals the incident γ wavevector ($\boldsymbol{k}_f = 0$). Correspondingly, while for QENS both the direction and the length of Q is variable, for QMS it is only the direction of the photon wavevector which can be varied with respect to the crystallographic directions of the system to be investigated. In general, this is a limitation for obtaining maximum information on the diffusion mechanism. But since (for ^{57}Fe) the value of Q is 7.3 Å^{-1} the sensitivity of QMS for details of the jump process may sometimes be comparable to that of presently available neutron spectrometers.

In any case the energy resolution for QMS with ^{57}Fe is considerably better than QENS (one or two orders of magnitude). In terms of the diffusion coefficient D, the energy window corresponds to $10^{-12}\,m^2\,s^{-1}$ with no upper limit for QENS and to $10^{-14}\,m^2\,s^{-1} < D < 10^{-10}\,m^2\,s^{-1}$ for QMS. This means that in both methods sufficiently high diffusivities are needed to see diffusional line broadening. With respect to the best-suited element (nuclei) both methods are different, of course. For QENS, hydrogen is by far the best choice because of both the nuclear properties (large incoherent neutron scattering cross-section) and the dynamic properties (high mobility), whereas QMS needs 'Mössbauer nuclei', and rests on ^{57}Fe and marginally ^{119}Sn, and thus has a rather limited range of applications.

Concerning diffusion in cubic Bravais lattices the key experiments in QMS were performed by Mantl *et al.* (1983) and Steinmetz *et al.* (1986) for ^{57}Fe diffusion in Al and Cu single crystals. Figure 7.9 shows a typical result of these investigations. The dashed line in this figure is the theoretical curve for tracer diffusion by jumps into nearest neighbour sites. Jump frequency and jump mechanism agree fully with expectations from tracer diffusion experiments. The full line in Fig. 7.9 demonstrates that a so-called five-frequency model which takes into account the complicated correlated motion of Fe atom and vacancy, provides a better fit.

As an example of diffusion in a non-Bravais lattice we mention Fe diffusion in Fe_3Si (Sepiol and Vogl 1993). Fe_3Si is an ordered intermetallic compound with DO_3 structure, a cubic superstructure consisting of four sublattices; three of the sublattices are occupied by Fe atoms.

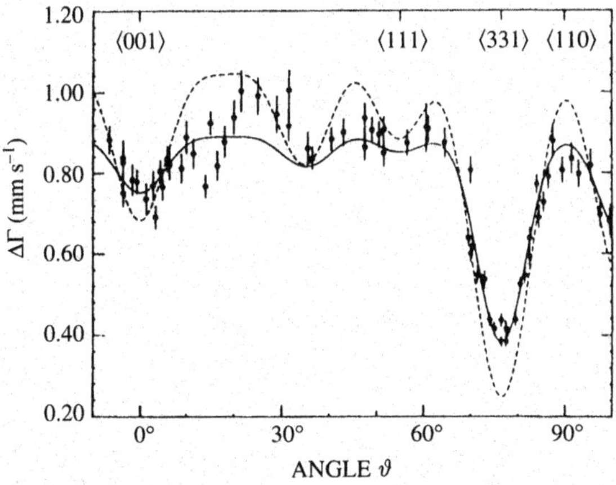

Fig. 7.9 Diffusional line broadening 2Γ of the ^{57}Fe Mössbauer spectrum as a function of the observation direction in Cu at 1313 K. θ denotes the angle between the [001] crystal axis and the observation direction in a plane slightly tilted from the (110) crystal plane. For the interpretation of full and dashed lines, see text (from Steinmetz *et al.* 1986, copyright by the American Physical Society).

Correspondingly the jump matrix $\mathbf{\Lambda}$ (see Eq. (5.119)) is a 3×3 matrix, and the Mössbauer absorption probability is a superposition of three Lorentzians:

$$S(Q, \omega) = \sum_{p=1}^{3} w_p(\boldsymbol{Q}) \frac{\Gamma_p(\boldsymbol{Q})}{\Gamma_p(\boldsymbol{Q})^2 + \omega^2} \tag{7.23}$$

where the weights w_p are given by the eigenvectors and the widths Γ_p by the eigenvalues of $\mathbf{\Lambda}$. Figures 7.10(a) and (b) show predictions of these calculations; it is evident that for the Mössbauer γ radiation parallel to certain single crystal directions *one* Lorentzian line dominates the Mössbauer spectrum: in [111] it is the narrowest of the three Lorentzians, for [113] it is the broadest one. Measurements

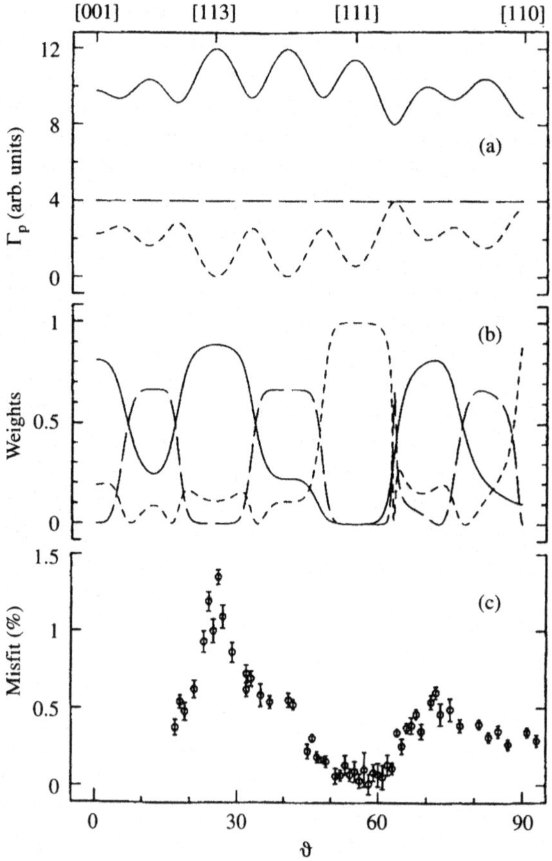

Fig. 7.10 Jump diffusion model for Fe in Fe_3Si (from Sepiol and Vogl 1993, copyright by the American Physical Society).

have therefore been performed parallel to these directions and thus the most discriminative information has been obtained. It was confirmed that for stoichiometric Fe_3Si the elementary diffusion jumps of the Fe atoms are nearest neighbour jumps to the adjacent sublattice.

In summary we emphasize the striking similarities between both quasielastic methods, QMS and QENS.

Part II

Special classes of materials

8

Hydrogen in metals

Many metals dissociatively dissolve hydrogen (Alefeld and Völkl 1978; Schlapbach 1988; 1992; Fukai 1993; Wipf 1996). At low H content the host lattice is unchanged (apart from a slight lattice expansion) and the hydrogen atoms occupy random certain sites in the interstitial lattice (e.g. octahedral interstices in Pd or tetrahedral interstices in Nb). In the metal/H phase diagram this regime is called the α phase; from a theoretical point of view it is called a lattice-gas. At higher H concentration stoichiometric hydride phases appear, in which the hydrogen atoms form an interstitial lattice with long-range order and in which the host lattice structure may differ from the 'empty' host lattice. In some cases phases are observed without long-range order but with pronounced H–H short-range order (sometimes called α') which can be thought of as lattice-liquids.

For QENS studies metal/hydrogen systems are particularly suited:

(i) Due to the large neutron scattering cross-section of hydrogen, QENS spectra exhibit high statistical accuracy; in host metals with low incoherent cross-section like Nb, hydrogen concentrations can still be studied as low as 0.1 at.%.
(ii) Scattering from light hydrogen (hydrogen isotope H) is incoherent (the coherent contribution is only about 2%); thus QENS spectra of metal hydrides reveal the single-particle diffusional behaviour of hydrogen, for which many theoretical approaches exist.
(iii) Deuterium scatters both coherently and incoherently; consequently in deuterides both the collective and the single-particle diffusional behaviour can be studied, if it succeeds—either by software (data fitting) or by hardware (polarization analysis)—to separate coherent QENS from incoherent QENS; see Table 5.1.

It is appropriate to distinguish between the small Q range, $Q < 1\,\text{Å}^{-1}$ (Section 8.1), where macroscopic diffusion coefficients are measured, and the large Q range, $Q > 1\,\text{Å}^{-1}$ (Section 8.2), where individual diffusional jump processes can be observed and the atomistic diffusion mechanism can be elucidated. Section 8.3 deals with the nature of the single jump event. While at high temperatures information can be obtained by QENS in the form of the temperature and Q dependence of classical jump processes or diffusion coefficients (Section 8.2), at low temperatures in bcc hydrides tunnelling is directly evident from high-resolution inelastic neutron spectra, but only for the trapped state (Section 8.3). For the positive muon which chemically can be considered as a light hydrogen isotope, quantum effects in low-temperature diffusion are very pronounced. Because of its close relation to H diffusion these effects will be explained briefly.

8.1 MACROSCOPIC HYDROGEN DIFFUSION COEFFICIENT

8.1.1 Intermetallic hydrides

Macroscopically, hydrogen diffusion coefficients in metals have been determined for many metal/hydrogen systems, mainly by electrochemical methods (Züchner 1970; Kirchheim and McLellan 1980) and by means of Gorsky effect measurements (Völkl and Alefeld 1978) yielding the chemical diffusion coefficient D_{chem}. Microscopic methods like NMR and QENS yield the self-diffusion coefficient D_s. In the following we present examples of hydrogen diffusion in intermetallic hydrides which are used as hydrogen storage materials like $TiFeH_x$ (Töpfer *et al.* 1978), $LaNi_5H_x$ (Richter *et al.* 1983), $TiMn_{1.5}H_x$ (Hempelmann *et al.* 1983a), and $Ti_{0.8}Zr_{0.2}CrMnH_3$ (Hempelmann *et al.* 1983b). More recent examples are $HfV_2H_{0.1}$ (Campbell *et al.* 1998), $ZrTi_2H_{3.6}$ (Fernandez *et al.* 1997), $ZrCr_2H_x$ (Skripov *et al.* 1999), and TaV_2H_x (Skripov *et al.* 1997; 1998).

Intermetallic compounds are brittle and disintegrate into a fine powder on hydrogenation; therefore the common macroscopic methods cannot be applied. Intermetallic compounds often contain paramagnetic impurities, particularly technical storage materials; this is a severe obstacle for NMR measurements.

Figure 8.1 displays some QENS spectra of $Ti_{0.8}Zr_{0.2}CrMnH_3$ taken at the Jülich backscattering spectrometer (Hempelmann *et al.* 1983b). In spite of the limited count rates a broadening of the spectra with increasing Q and T is clearly visible. At sufficiently small Q values the Q^2 relation is valid and can be quantitatively evaluated. However, in Fig. 8.2 the linewidths versus Q^2 immediately determined from the QENS spectra do not exactly extrapolate to the origin but seem to intersect the ordinate at positive values. This is indicative of multiple scattering: neutrons which have twice been scattered at about $2\Theta = 90°$ reach the small Q detectors and cause appreciable line broadening. In order to correct for multiple scattering (Chapter 7.4) the scattering function at large Q has to be known. For the purpose of multiple scattering correction it is often sufficient to apply the isotropic Chudley–Elliott model, as shown in Fig. 7.2 for the example of $LaNi_5H_6$. After these scattering corrections, the linewidths in Fig. 8.2 obey the Q^2 law quite well, and from their slopes the self-diffusion coefficients can be determined. Since narrow linewidths are affected more than broad ones, multiple scattering gives too large absolute values and too small activation energies of the diffusion coefficient.

Hydrogen diffusion in TiFeH is comparatively slow, whereas in $ZrCr_2H_3$ it is particularly fast. For $Ti_{0.8}Zr_{0.2}CrMnH_3$ it was estimated whether bulk hydrogen diffusion could be the rate-determining step of the complex absorption and desorption kinetics for a $Ti_{0.8}Zr_{0.2}CrMnH_2$ powder. If bulk diffusion is the slow step, the hydrogen content n_t at time t of spherical powder particles with radius r is

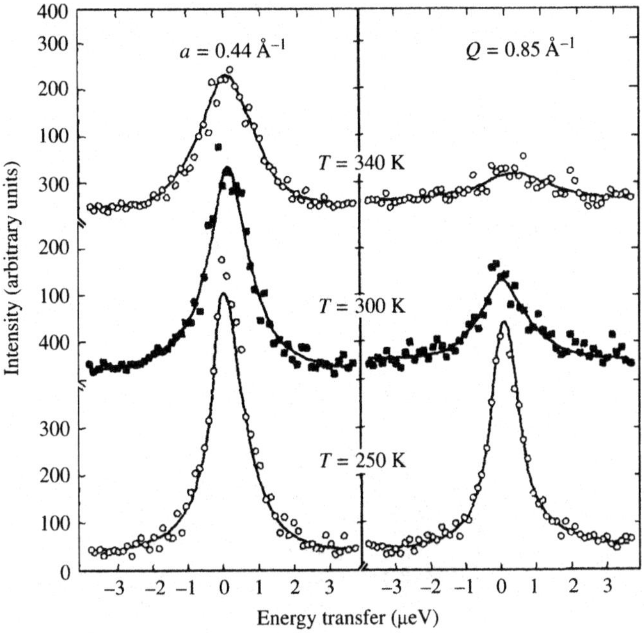

Fig. 8.1 Typical quasielastic spectra for hydrogen in an intermetallic compound ($Ti_{0.8}Zr_{0.2}CrMnH_3$), as measured with a backscattering spectrometer. Solid lines show Lorentzians convoluted with the resolution spectrum (from Hempelmann *et al.* 1983b, with permission from IOP Publishing Ltd).

obtained as a solution of Fick's second law (Crank 1975)

$$\frac{n_t}{n_\infty} = 1 - \frac{6}{\pi^2}\sum_{m=1}^{\infty}\frac{1}{m^2}\exp\left(-\frac{m^2\pi^2 Dt}{r^2}\right). \tag{8.1}$$

When t is not too small the higher-order terms can be neglected and the reaction rate constant is given by

$$k^{\text{diffusion}} = \pi^2 D/r^2. \tag{8.2}$$

For $T = 300\,\text{K}$, with $D = 6.0 \cdot 10^{-8}\,\text{cm}^2\,\text{s}^{-1}$ from QENS and $r \approx 0.5\,\mu\text{m}$ from a scanning electron micrograph, the diffusion-controlled reaction-rate constant would amount to $k^{\text{diffusion}} = 200\,\text{s}^{-1}$. A comparison with the experimental values from powder absorption measurements, $k^{\text{absorption}}_{300\,\text{K}} = 0.10\,\text{s}^{-1}$, and also $k^{\text{absorption}}_{300\,\text{K}} = 0.06\,\text{s}^{-1}$ for the closely related hydride $Ti_{0.8}Zr_{0.2}Cr_{0.8}Mn_{1.2}H_x$ (Suda *et al.* 1980), clearly indicates that the diffusion-controlled rate is orders of magnitude larger than the macroscopic absorption and desorption rates. The hydrogen atoms in the hexagonal multicomponent Laves phase hydride $Ti_{0.8}Zr_{0.2}CrMnH_3$ are so mobile that hydrogen diffusion is not the elementary

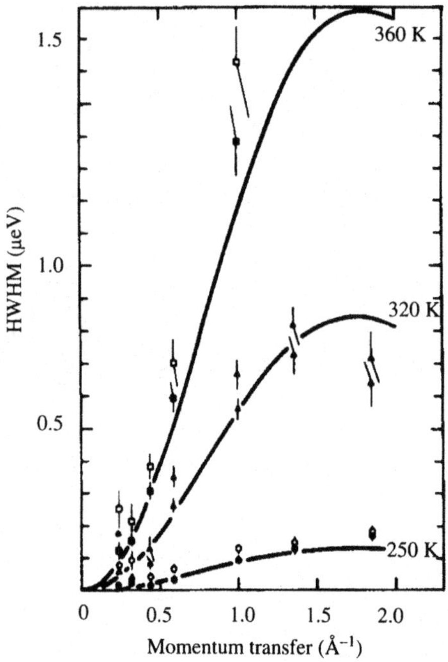

Fig. 8.2 The widths from Fig. 8.1, uncorrected (○) and corrected (•) for multiple scattering, plotted versus Q (from Hempelmann *et al.* 1983b, with permission from IOP Publishing Ltd).

step which regulates the overall kinetics of hydrogen absorption and desorption in this technical storage system.

The opposite situation was found for $TiFeH_x$ (Lebsanft *et al.* 1979) where the self-diffusion is very slow and the reaction is diffusion controlled. However, if one calculates $k^{\text{absorption}}$ the resulting value is smaller than the measured one. Using metallographic pictures this was explained by microcracks in the grains which bypass the diffusion path. Obviously, neutron quasielastic scattering does not 'see' this effect since the distance between microcracks is much larger than $2\pi/Q$.

8.1.2 Face-centred cubic TiH_x and YH_x as examples of diffusion in a concentrated lattice-gas

The self-diffusion coefficient of hydrogen in the dihydrides TiH_x and YH_x has been investigated by Stuhr *et al.* (1992) using QENS. In both cases the metal atoms form fcc lattices. In TiH_{2-y} hydrogen occupies the simple cubic sublattice of the tetrahedral sites, and accordingly the maximum hydrogen concentration amounts to $x = 2$. In YH_x, on the other hand, a small fraction of the hydrogen (typically below 10%) is also located on octahedral sites, so that x can reach values up to 2.1.

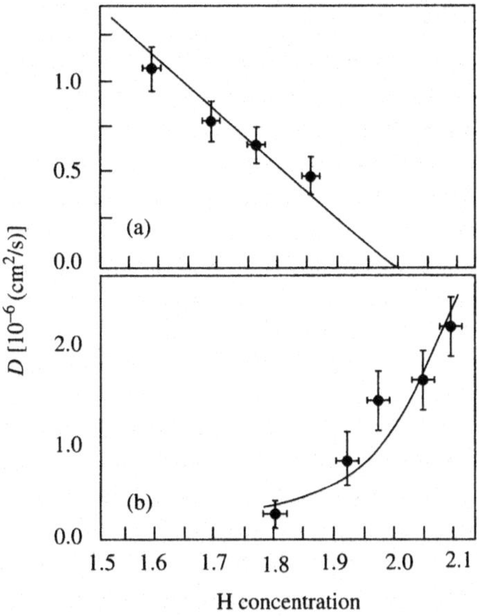

Fig. 8.3 H diffusion coefficient in a plot versus H concentration (a) TiH_x at 550 °C, (b) YH_x at 450 °C (from Stuhr *et al.* 1992).

The QENS spectra taken for $Q \leq 0.41\,\text{Å}^{-1}$ were evaluated in terms of the Q^2 law. Multiple scattering effects were thereby accounted for as described in Section 7.4. Figure 8.3 presents the concentration dependence of the hydrogen self-diffusion coefficient in TiH_x and YH_x, which exhibit entirely different features. In the case of TiH_x (Fig. 8.3(a)) the self-diffusion coefficient decreases continuously with rising x, whereas it is found to increase in the case of the YH_x samples. The figure shows also that the x range of the YH_x samples exceeds the value of 2.

The concentration dependence of D_s observed for TiH_x is in agreement with the behaviour of a concentrated lattice-gas on account of the factor $V(c)$ in Eq. (4.14). In particular, the value of D_s goes to zero as a consequence of the decreasing number of vacant sites into which diffusive jumps can occur, if the occupation probability c_t of the tetrahedral sites approaches the limiting value of 1 (for TiH_x, $c_t = x/2$). The concentration dependence of the tracer correlation factor f_t, see Eq. (4.65), can be neglected (it changes between $f_t = 1$ at $c = 0$ and $f_t = 0.78$ at $c = 1$ for an fcc lattice). In the case of the YH_x system in Fig. 8.3(b) the self-diffusion coefficient increases with rising x by a factor of 8 between $x = 1.80$ and 2.09. This behaviour is attributed to the additional occupation of energetically higher octahedral sites, over which the hydrogen atoms can essentially freely diffuse, because this sublattice is nearly empty. It would be possible to prove such an interpretation by means of QENS if single crystalline samples were available.

8.1.3 Collective diffusion in NbD_x

Deuterium as an interstitial particle offers the possibility of studying both collective and single-particle diffusional behaviour simultaneously because the coherent thermal neutron scattering cross-section, σ_{coh}, is comparable with the incoherent neutron scattering cross-section, σ_{inc} ($\sigma_{coh} = 5.6$ barn, $\sigma_{inc} = 2.0$ barn), in contrast to hydrogen where 98% of the total scattering cross-section is incoherent.

As has been outlined in Section 6.2, coherent QENS line broadening at small Q yields the Q-dependent collective diffusion coefficient $D_c(Q)$. In the macroscopic limit, $Q \to 0$, the so-called bulk diffusion coefficient is obtained (Wipf *et al.* 1989):

$$\lim_{Q\to 0} D_c(Q) = D_{bulk}. \tag{8.3}$$

The coherent QENS intensity is given by the structure factor $S_{tot}(Q)$ of coherent diffuse neutron scattering which can be approximately factorized into a short-range order term and a lattice distortion term, see Eq. (6.67):

$$S_{tot}(Q) = S_{SRO}(Q) \cdot S_{dis}(Q). \tag{8.4}$$

In the macroscopic limit, $Q \to 0$, the short-range order structure factor equals the reciprocal thermodynamic factor, see Eq. (4.33):

$$\lim_{Q\to 0} S_{SRO}(Q) = S_{SRO}(0) = \phi_{bulk}^{-1}. \tag{8.5}$$

The subscript 'bulk' indicates that both the chemical diffusion coefficient and the thermodynamic factor, when determined by means of QENS, contain effects of coherence stress, details of which follow below at the end of this section, and are thus different from the corresponding quantities measured macroscopically. If we go beyond the mean field approximation of Chapter 6, i.e. if we take into account local correlations by the appropriate correlation factors (Eqs. (4.26) and (4.59)) we can determine the ratio of the correlation factors, often called Havens' ratio, from a comparison of the coherent and incoherent linewidths:

$$\Gamma_{coh} = \frac{\Gamma_{inc}}{S_{SRO} \cdot f_t/f_m}. \tag{8.6}$$

Figure 5.8 displays the structure of $\alpha(\alpha')$-NbD_x: whilst at low concentration the D atoms are distributed at random over the tetrahedral interstices of the bcc host lattice (α phase), at large concentration (α' phase)—due to multiple site blocking—short-range order develops. It can be expressed in terms of the so-called Cowley short-range order parameters α_i (Cowley 1950), see Eq. (6.1); for the spatially averaged structure factor per D atom we thus obtain (compare Eq. (6.6)):

$$S_{SRO}(Q) = \left(1 - \frac{c}{6}\right)\left\{1 + \sum_{i=1}^{\infty} z_i \alpha_i \frac{\sin(QR_i)}{QR_i}\right\} \tag{8.7}$$

where the sum runs over the coordination shells (from Fig. 5.8 it is evident that the coordination numbers z_i are 4, 2, and 8 for the first, second, and third coordination shells, respectively). We denote by R_i the radius of the ith coordination shell; the term $c/6$ appears because we use the atomic ratio $c = N_D/N_{Nb}$, as is conventional in the field of metal–hydrogen systems, instead of the ratio of interstitial particle per accessible site. For a lattice-gas or lattice-liquid the short-range order parameters α_i describe deviations from a random mixture of interstitial particles (here D atoms) and interstitial vacancies:

$$\alpha = \begin{cases} -c/(6-c) & \text{complete blocking} \\ 0 & \text{random} \\ +1 & \text{complete clustering.} \end{cases} \tag{8.8}$$

Note that for a non-interacting system (a random mixture, consequently $f_m = 1$) the sum in Eq. (8.7) vanishes and thus if we insert Eq. (5.88) and Eq. (4.59) with $V(c) = 1 - c/6$ into Eq. (8.6), we obtain

$$\Gamma_{coh} = \frac{\hbar D_s Q^2}{(1-c/6)\cdot f_t} = \hbar D_E Q^2 \quad \text{(non-interacting system)} \tag{8.9}$$

i.e. without D–D interactions the width of the coherent quasielastic Lorentzian is independent of the concentration (in contrast to the width of the incoherent Lorentzian!) and equals the incoherent value at $c = 0$. Therefore at large c the coherent width is considerably larger than the incoherent width, and this offers the possibility of separating them in a data fitting procedure, if both incoherent QENS and coherent QENS are measured simultaneously.

In the coherent diffuse neutron scattering experiment the deuterium atoms in the centre of their respective tetrahedral interstitial sites in the Nb lattice give rise not only to D–D interference effects but also to D-Nb interference. The former effect is described by the short-range order term $S_{SRO}(Q)$; the latter effect is due to small polaron formation: D atoms (assumed to be in undisplaced positions) displace the neighouring Nb atoms and thus distort the long-range Nb order. This distortion field travels together with the D atoms and thus yields the same line broadening, but due to the additional Nb scattering the intensity of the coherent QENS increases. At very dilute $NbD_{0.03}$, where $S_{SRO}(Q) = 1 - c \approx 1$, Bauer *et al.* (1977) studied (integrated) diffuse scattering only due to lattice distortions and found that for $Q \le 1.5\,\text{Å}^{-1}$ the structure factor due to distortion, $S_{dis}(Q)$, does not differ appreciably from its value at $Q = 0$. So for the purpose of coherent QENS data evaluation at small Q, the zero-Q result can be taken, and this has been outlined in Section 6.3.

In summary, the scattering intensity of D in polycrystalline NbD_x was evaluated (Hempelmann *et al.* 1988) by means of the following equation

$$\frac{\partial^2\sigma}{\partial\Omega\,\partial\omega} = \frac{R^*_{NbD}}{R^*_V}\frac{I_V(Q)}{N_V\sigma_V^{inc}/(4\pi)} N_{Nb} e^{-2W(Q)}$$

$$* \left\{ c\frac{\sigma_D^{inc}}{4\pi}\Lambda_{inc}(Q,\omega) + b_D^2 c S_{tot}(Q)\Lambda_{coh}(Q,\omega)\right\}, \tag{8.10}$$

where the linewidth of the coherent Lorentzian, Λ_{coh}, contains the short-range order structure factor according to Eq. (8.7). The R^* values can be calculated by means of DISCUS and concern the different absorption properties; the subscript V means vanadium: absolute intensities are evaluated with reference to a vanadium standard.

There are two possible approaches to the problem of separating the incoherent and coherent components of the total scattering. The first was employed by Hempelmann *et al.* (1988) and involved a theoretical separation of the two parts by fitting the model described above to the total scattering. The second approach, used by Cook *et al.* (1990), is an experimental separation by neutron polarization analysis. Both techniques have their relative merits and disadvantages. The former requires a significant line broadening difference and good statistics (in order to separate reliably the two components by fitting procedures) and—of course—reliable models. The latter, for supermirror polarizer and analyser elements, suffers from a rather large reduction of the cold neutron flux by spin polarization and analysis. This is due to the inevitable collimation imposed by these components and by the fact that only one spin state is measured at any time. But polarization analysis has the advantage of giving an entirely unambiguous separation of any combination of coherent and incoherent line broadenings.

A QENS study with polarization analysis has been performed by Cook *et al.* (1990). Figure 8.4 displays the separated incoherent and coherent scattering around the elastic peak of single crystalline $NbD_{0.70}$ at 600 K. The spectra show the expected qualitative features: the incoherent peak narrows as $Q \to 0$, and its intensity slightly decreases with Q according to the Debye–Waller factor. The

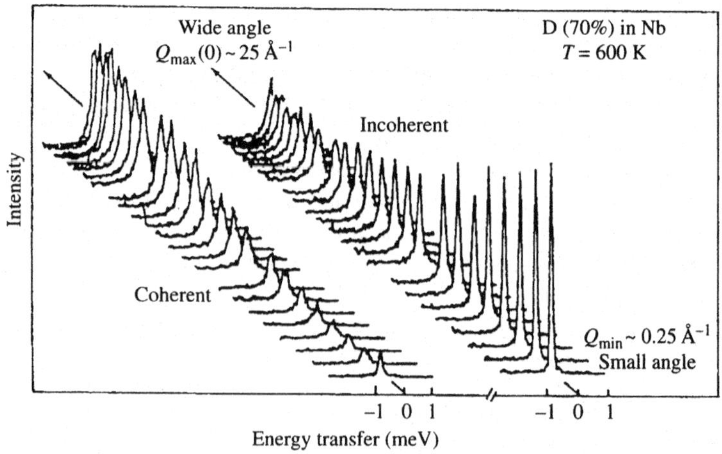

Fig. 8.4 The separated coherent and incoherent parts of the total D scattering around the elastic position of $NbD_{0.70}$ at 600 K, as a function of detector angle (from Cook *et al.* 1990, with permission from IOP Publishing Ltd).

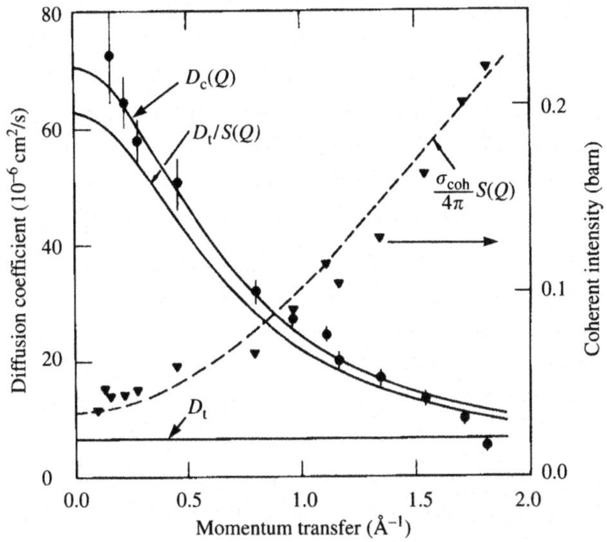

Fig. 8.5 Q dependence of the deuterium diffusion coefficients in $NbD_{0.85}$ at 581 K (left ordinate); the black dots result from a simultaneous fit of coherent and incoherent scattering with $D_c(Q)$ free to vary individually for each Q value; the upper solid line is the result of a simultaneous fit with the Q dependence of $D_c(Q)$ as $D_c(Q) = \text{Const.}/S(Q)$. The lower solid line represents the de Gennes narrowing result $D_c(Q) = D_t/S_{SRO}(Q)$. The black triangles display the coherent deuterium intensities (right coordinate). The dashed line results from the simultaneous fit.

coherent peak, on the other hand, is low in intensity and broad at low Q; the intensity maximum coincides with a narrowing of the width.

In Fig. 8.5 (Hempelmann *et al.* 1988) the result of the 'software' separation of incoherent and coherent scattering following Eq. (8.10) is summarized for polycrystalline $NbD_{0.85}$ at 581 K. Diffusion coefficients (widths divided by Q^2, left ordinate) and coherent intensity (structure factor, right ordinate) are plotted versus momentum transfer. The results of individual data fits for each Q are represented as circles (collective diffusion coefficient) and triangles ($b_D S(Q)$), whereas the result of a simultaneous fit according to Eq. (8.10) is indicated by the dashed line ($b_D S_{tot}(Q)$), straight solid line (D_s), and the upper solid curve ($D_c(Q)$).

From this data evaluation a large variety of diffusional quantities was derived. We start with the short-range order derived from $S_{SRO}(Q)$. The resulting Cowley short-range order parameters indicate a complete blocking of the first and second coordination shells (see Fig. 5.8) whereas the occupation of the third shell varies depending on temperature and concentration. This is in accordance with two very simple arguments:

(i) Maximum D content: theoretically, for self-site blocking only NbD_6, for one-shell blocking $NbD_{2.1}$, for two-shell blocking $NbD_{1.35}$, for three-shell blocking

$NbD_{0.8}$; upper limit of observed D-content is NbD_1. Thus two shells are blocked.

(ii) Separation distance between neighbouring tetrahedral sites: first shell 1.16 Å, second shell 1.65 Å, third shell 2.01 Å(disregarding the lattice expansion due to deuterium absorption); according to Switendick's 2.1 Årule there cannot be a D atom in the first or second coordination shell, whereas, taking account of lattice expansion, occupation of the third shell is just possible.

The pronounced short-range order derived from the liquid-like shape of the structure factor causes us to consider the interstitials in α'-NbD_x as a *lattice-liquid* with strongly repulsing D–D interaction.

Next we consider single-particle diffusion, i.e. the straight solid line in Fig. 8.5. We compare, in Fig. 8.6, the concentration dependence of the self-diffusion coefficient D_s with Monte Carlo simulations by Faux and Ross (1987) for self-site blocking (index '0') and one/two/three-shell blocking (indicies '1'/'2'/'3'). The concentration dependence of the self-diffusion coefficient suggests a scenario between two- and three-shell blocking.

Now we consider the thermodynamic factor, derived from the reciprocal extrapolated $S_{SRO}(Q)$, see Eq. (4.13) or Eq. (4.33):

$$S_{SRO}(0)^{-1} = \phi = \frac{c}{RT}\frac{\partial \mu}{\partial c}. \tag{8.11}$$

This is a thermodynamic quantity and is to be compared with a thermodynamic measurement. Kuji and Oates (1984) have measured pressure composition

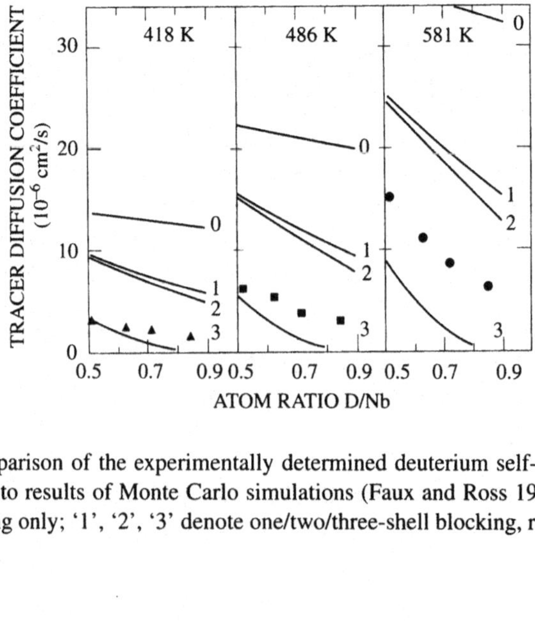

Fig. 8.6 Comparison of the experimentally determined deuterium self-diffusion coefficient (♦, ■, •) to results of Monte Carlo simulations (Faux and Ross 1987): '0' denotes self-site blocking only; '1', '2', '3' denote one/two/three-shell blocking, respectively.

isotherms of NbH_x and from these measurement they have determined the chemical potential for H in Nb. Disregarding the isotope effect the derivative of these quantities can be calculated; Fig. 8.7 shows the comparison with the neutron results explained above and with Gorsky-effect results by Bauer *et al.* (1978). Whilst the agreement of the Gorsky-effect data is excellent, the neutron data semiquantitatively exhibit the correct trends.

Finally, Fig. 8.8 shows a comparison between Gorsky-effect results and QENS results on the diffusion and spatial fluctuations of D in $NbD_{0.5}$. As was pointed out by Wipf *et al.* (1989) the two experimental techniques probe concentration eigenmodes of a different type where the short-wavelength modes studied by neutrong scattering are influenced by coherence stresses which raise the elastic energy of the short-wavelength D fluctuations. Gorsky-effect measurements, on the other hand, probe long-range fluctuations which are coherence-stress free. As Fig. 8.8 shows, the elasticity-theoretical concepts of Cahn (1961; 1962), Krivoglaz (1969), and Wagner and Horner (1974), Wagner (1978) allow a quantitative calculation of neutron scattering results from Gorsky-effect data. This demonstrates for the first time the applicability of these theories to diffusion coefficients.

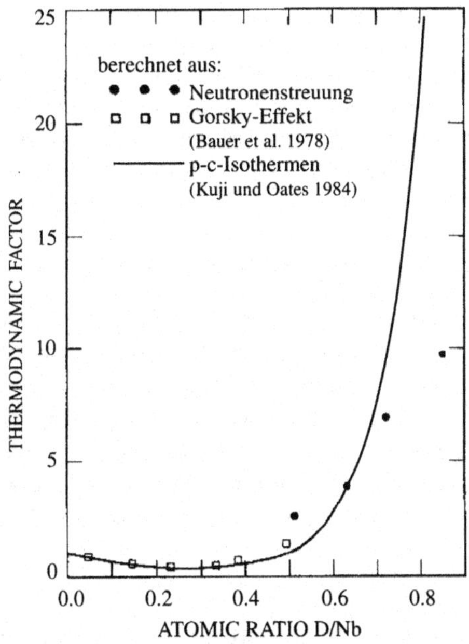

Fig. 8.7 Comparison of deuterium thermodynamic factors derived from the CQENS experiment of Hempelmann *et al.* (1988) (•), from Gorsky-effect measurements of Bauer *et al.* (1978) (○), and from pressure-composition isotherms of Kuji and Oates (1984) (solid line).

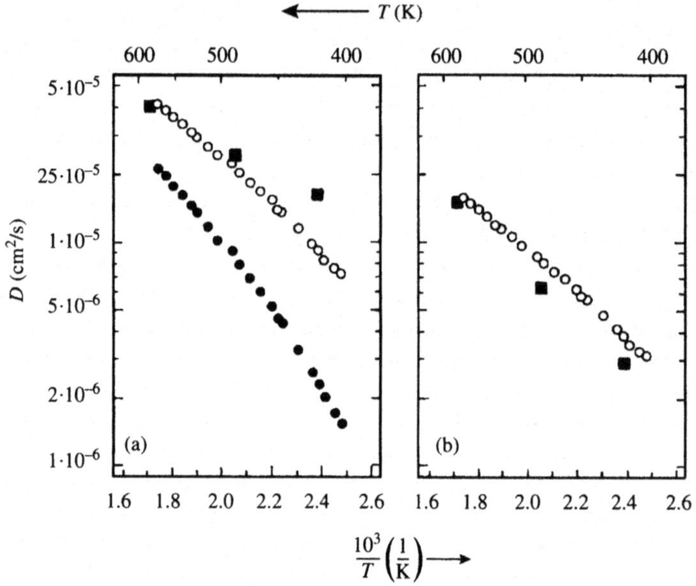

Fig. 8.8 Various deuterium diffusion coefficients in $NbD_{0.5}$ as reported from Gorsky-effect (Bauer *et al.* 1978) and QENS (Hempelmann *et al.* 1988) measurements in an Arrhenius representation. The left hand side shows Gorsky-effect results for D_{chem} (full circles •) and D_{bulk} (open circles ∘). The full squares (■) are neutron scattering data for D_{bulk}. The right hand side shows Gorsky effect results for D_s (open circles ∘) obtained under the assumption of a correlation factor $f_t = 1$. The full squares (■) are the QENS data for D_t; the ratio represents the tracer correlation factor (from Wipf *et al.* 1989, copyright by Springer-Verlag).

8.2 THE HYDROGEN JUMP-DIFFUSION MECHANISM

The investigation of quasielastic scattering at small Q, as discussed in the previous section, yields the self-diffusion constant. At larger Q, comparable with the inverse jump distances, more detailed atomistic information of the elementary process of diffusion can be obtained. The cubic fcc and bcc hydrides of Pd, Nb, Ta, and V were investigated very early, and quite detailed results are now available, in particular dealing with the coupling of diffusive jumps and lattice vibrations. Later on, more complicated hydrides were also studied in this sense, and two examples will be presented in detail.

8.2.1 Face-centred cubic and body-centred cubic α-phase hydrides

The very first quasielastic experiments to study the diffusive motion of hydrogen were performed by Sköld and Nelin (1967) on α-palladium hydride in its disordered fcc phase where the hydrogen atoms occupy octahedral sites. More recent results of a series of single-crystal experiments in two symmetry crystal directions of palladium hydride by Rowe *et al.* (1972) are presented in Fig. 8.9. The quasielastic width is normalized by the self-diffusion constant D_S, replacing $\Gamma(\boldsymbol{Q})$ by $\Gamma(\boldsymbol{Q})\, a^2/D_S$, where a is the lattice parameter. The dashed line in the figure is the result of the Chudley–Elliott model based on independent or uncorrelated jumps between the octahedral sites (see Chapter 5.3). The agreement is excellent. For reasons of comparison the mean rest times $\tau(0)$ taken from a number of quasielastic scattering experiments and from 'macroscopic' measurements like NMR, permeation, Gorsky relaxation are extrapolated to a hydrogen concentration $c = 0$, namely

$$D(0) = a^2/6\tau(0) \tag{8.12}$$

where $\tau(0)$ for the dilute system was expressed by $\tau(c)$ as taken from $D(c)$ with the relation $1/\tau(0) = (1 - c)/\tau(c)$ (see Eq. (4.60)), i.e. with the blocking factor. The comparison shows that there is fairly good agreement and consistency of these data.

Figure 8.9 also includes results from molecular dynamics calculation by Li and Wahnström (1992). The motion of the hydrogen and the palladium atoms is described classically, with calculated interaction energies $E(R_1, R_2, \ldots)$ of the host lattice atoms in the neighbourhood of the interstitial hydrogen. The calculations introduce the possible non-adiabatic behaviour of the conduction electrons expressed in terms of electronic correlations. This effect is included in the equation of motion by an electronic stopping power, acting on the hydrogen moving in the

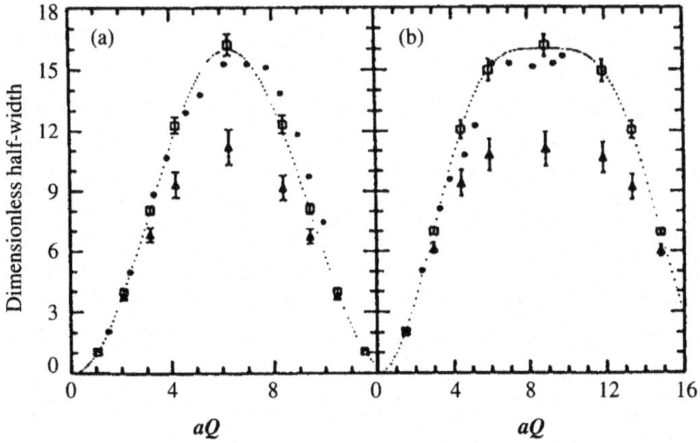

Fig. 8.9 Normalised quasielastic halfwidths for hydrogen in Palladium $\Delta(a\boldsymbol{Q}) = \Gamma(\boldsymbol{Q})a^2/D_s$ as a function of $a\boldsymbol{Q}$ where $\boldsymbol{Q}$ is the scattering vector, a is the lattice parameter, and D_s the self-hydrogen diffusion constant (a) $\boldsymbol{Q} \parallel \langle 100 \rangle$ and (b) $\boldsymbol{Q} \parallel \langle 100 \rangle$. • Experiments from Rowe *et al.* (1972). Δ Computer simulation without electronic friction ($\eta = 0$) and □ friction $\hbar\eta = 4.54\,\text{meV}$. Dotted line: Chudley–Elliott jump model. Only a significant value of η leads to agreement (from Li and Wahnström 1992, copyright by the American Physical Society).

electron gas. This is described by a term $m_H \eta \dot{R}_H$. The classical equation for the hydrogen coordinate $\mathbf{R}_H$ then reads

$$\frac{m_H \, d^2 R_H(t)}{dt^2} = -\nabla E(R_1, R_2, \ldots) - m_H \eta \dot{R}_H + F(t) \qquad (8.13)$$

where m_H is the hydrogen mass, and F is a fluctuating stochastic force of the host lattice. The motion of the host lattice coordinates $R_1, R_2, \ldots$ is taken into account knowing the force constants of the lattice. Surprisingly, the results of the adiabatic behaviour of the electrons in the friction-free case ($\eta = 0$ in Fig. 8.9) leads to a quasielastic width which is too small in the middle of the Brillouin zone as compared to the experimental data and, as well, to the Chudley–Elliott model. Fig. 8.10 illustrates what happens. Hydrogen trajectories as found from earlier computer simulations by Gillan (1986) projected in a (100) plane containing the octahedral interstitial site show strong correlations along the H trajectories, and the concept of well-separated jumps of Chudley–Elliott type is not at all applicable. Sometimes, the atom spends a relatively long time interval in the vicinity of an equilibrium octahedral site carrying out vibrations, then it passes quickly over other 'sites' performing only one oscillation; see also simulations by Culvahouse and Richards (1988), and Li and Wahnström (1992). However, in order to produce simulation results consistent with experiments and the Chudley–Elliott model, Li and Wahnström (1992) introduced a relatively large friction into the equation of motion to avoid these correlations (see Fig. 8.9), i.e. a strong hydrogen–electron interaction had to be considered (see Section 4.4.2). It is not yet clear to what extent the effect of non-adiabatic electron–hydrogen interactions must be assumed to be responsible for the large discrepancy between computer simulations without friction and experiments.

In contrast to the fcc palladium hydride as discussed before, with hydrogen on octahedral sites, striking discrepancies were found for $\Gamma(\mathbf{Q})$ in bcc metal hydrides, in particular for niobium where hydrogen sits on tetrahedral sites. This was observed by Rowe *et al.* (1974) for α-tantalum hydride and also for α-niobium hydride by Rowe *et al.* (1972), Rush and Rowe (1974), and Lottner *et al.* (1979a, b). Since there are six eigenvalues, the quasielastic line is a sum of (at most) six contributions. Figure 8.11 shows the effective linewidth of the composite spectrum as measured for different crystal orientations, as a function of Q^2. Beyond the region of small Q, where Γ goes as $Q^2 D$, there are discrepancies. The experimental points $\Gamma(\mathbf{Q})$ are considerably lower than expected for the Chudley–Elliott model. This could be formally related to a larger effective jump distance in the diffusive process. Lottner *et al.* (1979b) have proposed a two-state model (see Section 5.5) introducing jump sequences over the tetrahedral sites in the following sense. The hydrogen atom alternates between a state of high mobility where it performs rapid jump sequences between adjacent sites with a rate $1/\tau_1$ and during a period τ_e, and an immobile state; here the hydrogen stays well localized on a site for a relatively long time τ_R. This model leads to good agreement between experiment and measurements. At 580 K the sequences have a length of about four jumps with

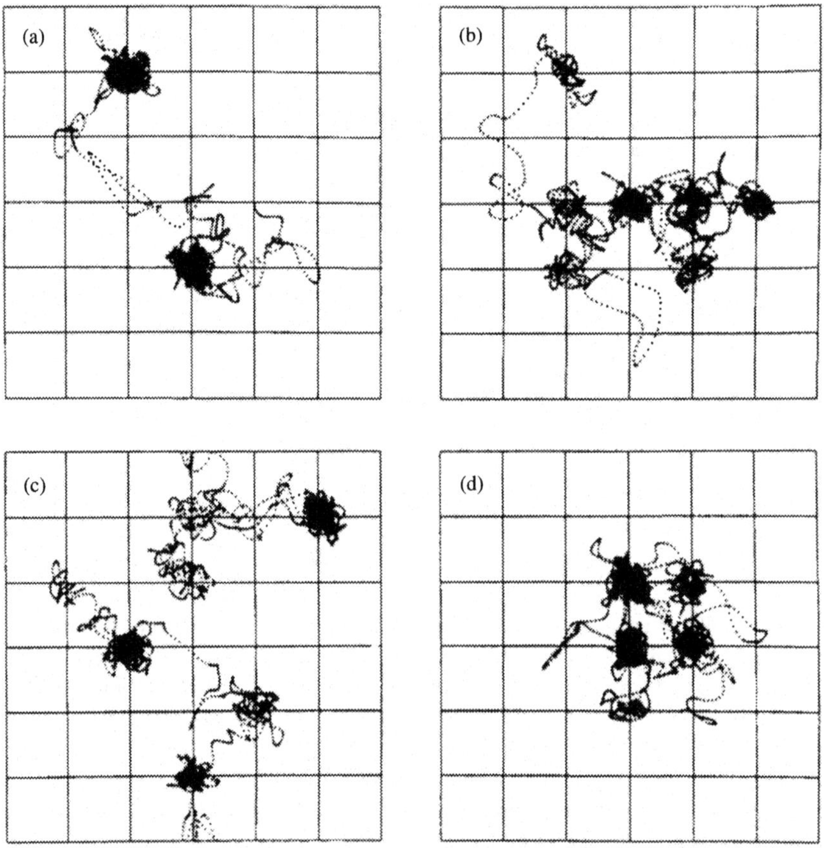

Fig. 8.10 Typical hydrogen trajectories in Pd as obtained from molecular dynamics at 967 K. The grid is the projection of the lattice of octahedral interstitial sites (from Gillan 1986, with permission from IOP Publishing Ltd).

a jump rate of $1/\tau_1 = 1.0 \cdot 10^{12}\,s^{-1}$. Also the alternative concept of correlated jump pairs between nearest neighbours (including back and forth motion) leads to fair agreement as can be seen from the Fig. 8.11 (Lottner *et al.* 1979a). In view of the six components in the spectrum, a more sophisticated analysis of the experimental data is adequate (Lottner *et al.* 1979b). From a fit at different $\boldsymbol{Q}$, the apparent jump rate of the model was calculated individually for various values of $\boldsymbol{Q}$. For high temperatures, only the model of correlated jump sequences gives the same characteristic time τ for all $\boldsymbol{Q}$ values. The other models fail as can be seen from the Q dependence of τ^{-1} in Fig. 8.12 which is an artifact. At low-temperature, however, the simple Chudley–Elliott jump model still works.

We qualitatively interpret the existence of multiple jumps in the following way. A hydrogen atom causes a distortion of the host lattice which has a tetragonal

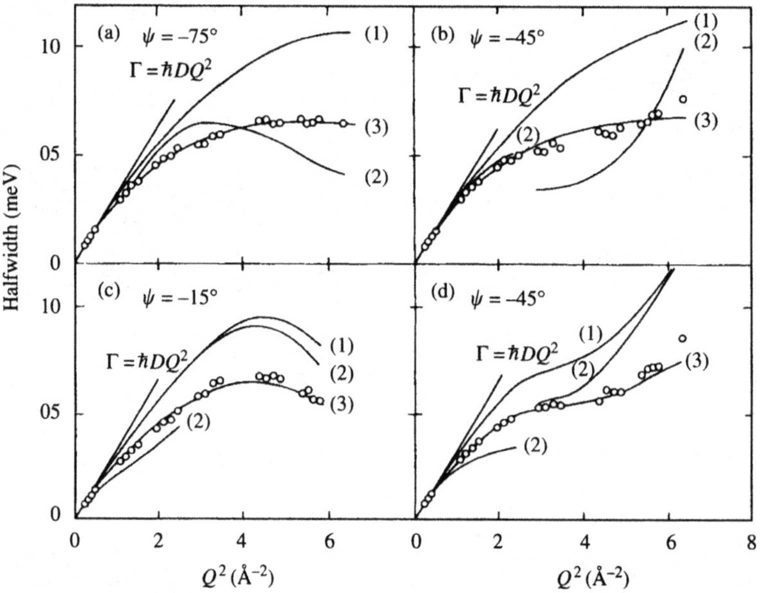

Fig. 8.11 Halfwidth at half maximum of the composite spectrum for hydrogen diffusion in α-$NbH_{0.02}$ at 580 K versus Q^2; $\cdot\psi$ = angle between k_0 and $\langle 110 \rangle$ in the scattering plane. Solid curves: (1) nearest neighbour tetrahedral jumps, following the Chudley–Elliott model; (2) as before, allowing jumps to second nearest neighbours; (3) correlated double jumps (see text) (from Lottner *et al.* 1979a, with permission from Elsevier Science).

axis of symmetry (Section 6.3 and Fig. 6.4). Obviously, a jump from a tetrahedral site to an adjacent site reorients the axis of the tetragonal field and leads to a rearrangement of the displacements for all surrounding atoms. A certain time τ_R is needed until the new equilibrium configuration has been reached. Consequently the activation barrier opposing the next hydrogen jump for leaving the site is—at first—lower than for a fully relaxed site. Therefore, the probability of leaving this site before it is fully relaxed will increase with temperature. This makes jump pairs or correlated jump sequences increasingly probable as temperature increases. The corresponding relaxation time of the lattice can be estimated on the basis of small-polaron hopping (Emin 1971) which leads to

$$\tau_R \simeq (12/\pi\omega_D)(2E_{el}/kT)^{2/3} \simeq 5 \cdot 10^{-12}\,\text{s} \tag{8.14}$$

where $E_{el} \simeq 170$ meV is the energy of elastic lattice deformation for the hydrogen atom on its site, and ω_D is the Debye frequency (for a detailed discussion see Dosch *et al.* 1987). This time is comparable with the rest time of the hydrogen atom as obtained from the self-diffusion constant which shows that this picture makes sense.

There are two remarkable observations related to the diffusion of hydrogen in niobium which could also be connected with correlated jumps. As explained in

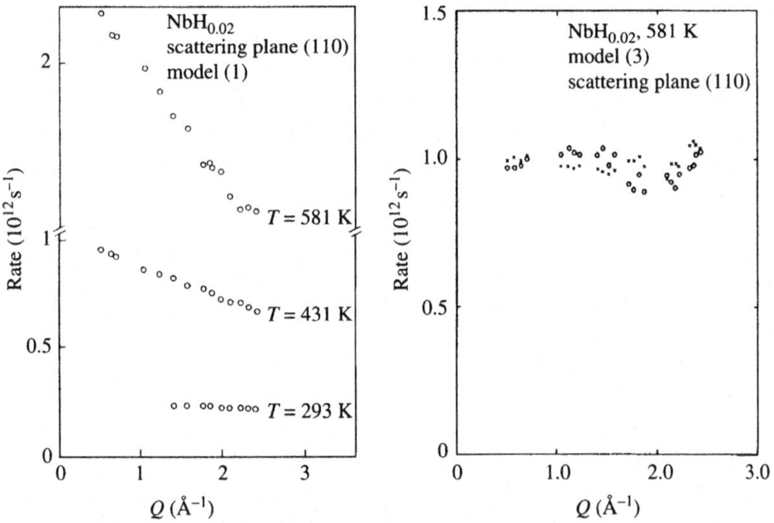

Fig. 8.12 *Left side*: apparent jump rate $1/\tau$ for H in $NbH_{0.02}$ from individual fits for different $\boldsymbol{Q}$ using the model of nearest neighbour jumps. At 293 K the fit consistently yields the same $1/\tau$ for all $\boldsymbol{Q}$. The model obviously fails at larger temperatures. *Right side*: the same for the model with correlated double jumps. The model is consistent with experiments also at high temperature (from Lottner *et al.* 1979c, copyright by Springer-Verlag).

Fig. 6.4 the strain field around a hydrogen on a tetrahedral site has tetragonal symmetry. Huang scattering of X-rays around the Bragg lines (Section 6.1) probes the symmetry of the long-range strain field around a defect; unexpectedly, Huang scattering behaves as if the strain field were cubic (Bauer *et al.* 1975; Metzger *et al.* 1976). Secondly, a striking anomaly is the missing Snoek effect of hydrogen in niobium and also in tantalum (Buchholz *et al.* 1973). Snoek relaxation is sensitive to the tetragonal field created by the hydrogen atoms in the bcc lattice. Its strength is proportional to the quantity $(A - B)^2$ from the double force tensor (see Eq. (6.52)). Experimentally, Snoek relaxation was unobservable within the experimental accuracy which again indicates that the long-range field is practically cubic.

Dosch *et al.* (1992) have worked out a model which considers the coupling of the host lattice to the proton jumps via phonons, i.e. applying the frequency-dependent host lattice Green's function. Thus, this work goes beyond the theory of instantaneous relaxation (Gillan and Wolf 1986) and takes into account the finite relaxation times of the various parts of the lattice distortion field. It then turns out that the coherent neutron scattering from the fluctuating strain fields of the host lattice exhibits, in analogy to incoherent neutron scattering, a quasielastic broadening governed by the six eigenvalues associated with the rate equations. In addition, there appears a phonon-coupling term which is responsible for the relaxation behaviour. One of the eigenvalues,

$$\Gamma_1 = D_s q^2 \equiv D|\boldsymbol{Q} - \boldsymbol{G}|^2 \tag{8.15}$$

($\boldsymbol{G}$ = reciprocal lattice vector) is related to those jump components which contribute to long-range diffusion and, thus, yields the narrowest quasielastic component. In the coherent neutron scattering cross-section this Γ_1 is coupled to the cubic part of the strain field. This very small energy broadening of the associated Huang diffuse neutron scattering was indeed observed for the first time in $NbD_{0.02}$ by Dosch *et al.* (1992). The other five eigenvalues $\Gamma_2, \ldots, \Gamma_6$ lead to broader components whose widths are proportional to the jump rate of the hydrogen. They are coupled to the tetragonal part of the strain field. Consequently, they are strongly affected by the phonon-coupling term and exhibit a noticeable relaxation behaviour. For a sufficiently high jump rate the tetrahedral contributions become small, and consequently, the long-range part of the strain field may approach cubicity. The calculation indicates that the sluggish tetragonal components in Huang diffuse scattering should survive as long as the jump rate is not faster than $5 \cdot 10^{11}\,\mathrm{s}^{-1}$ (note that $\tau_{\mathrm{R}}^{-1} \approx 2 \times 10^{11}\,\mathrm{s}^{-1}$), while they should start to fade away for a jump rate between $10^{12}\,\mathrm{s}^{-1}$ and $10^{13}\,\mathrm{s}^{-1}$. As intuitively expected, the critical jump frequency, where this decay starts, increases with decreasing wavenumber q: the long-range components of the displacement (associated with small q) fluctuate more slowly than the local lattice distortions (observable at large q). This is consistent with the above observation that Huang diffuse X-ray scattering exhibits virtually no tetrahedral components up to $q = 0.1\,\text{Å}^{-1}$. Within the framework of the theory it was concluded that a local jump rate of 10^{12}–$10^{13}\,\mathrm{s}^{-1}$ is required to explain this surprising observation.

In order to visualize these ideas qualitatively, we once more refer to Fig. 6.4. There are three kinds of sites, of type x, y, and z, each of which leads to a certain axis of this tetrahedral displacement field. Let us assume that the proton performs rapid jumps between a sequence x, y, and z of the potential minima. This induces fluctuating lattice distortions in the local neighbourhood. At a sufficiently large distance from the mobile proton, the associated displacement field exhibits a relaxation time which is larger than the inverse jump rate of the defect and, thus, starts to average over the displacement which follows from the occurrence of this x-y-z sequence of jumps. Consequently, the tetragonal part of the long-range displacement field fluctuates only very weakly and, if the rate is sufficiently high, cancels out entirely. This effect will increase when the jump rate (i.e. the temperature) increases. It is tempting to assume that these fast jumps are related to the rapid jump sequences as inferred from the incoherent quasielastic scattering experiments described above. However, it should be noted that the jump rate as quoted by Lottner *et al.* (1979a, b) have characteristic times longer than 10^{-13} s, thus the connection between this picture and the correlated jumps from the quasielastic experiments may not be so straightforward.

In this context we also mention other observations related to the coupling between proton jumps and the lattice phonons. Neutron spectroscopy has shown that the TA_1 [110] branch for bcc tantalum deuteride and for niobium deuteride (Magerl *et al.* 1977; 1978) show an anomalous change at a certain phonon frequency ω. This phonon branch is related to the elastic sheer coefficient c_{12} and it is observed that the slope $d\omega(q)/dq$ of this phonon branch deviates from the value obtained from ultrasonic c_{12} measurements in the region of 10^7 Hz (Magerl *et al.* 1976). The change of sound velocity occurs at a frequency of about $1/\tau$, the corresponding jump rate of hydrogen diffusion in this metal matrix. This observation, and, as well, the strong isotope effect observed for c_{12} comparing NbH and NbD (Magerl *et al.* 1976) could also be related to the coupling between diffusive jumps and phonons. A consistent explanation of all these phenomena is still missing.

Apart from numerous experiments on the quasielastic line shape and its width there are also investigations of the line intensities. This can be investigated from the ratio of the integrated quasielastic spectra, namely

$$I(Q)/I(0) = \int_{-\Delta\omega/2}^{\Delta\omega/2} S_i(Q,\omega)\,d\omega = \int_{-\Delta\omega/2}^{\Delta\omega/2} S_i(Q \to 0,\omega)\,d\omega \qquad (8.16)$$

where $S_i(Q,\omega)$ is the incoherent scattering function, and $\Delta\omega$ is the integration window. For a harmonic oscillator or a crystal with harmonic lattice waves this ratio is the Debye–Waller factor which gives the intensity of the infinitively sharp elastic line, namely

$$I(Q)/I(0) = e^{-Q^2 u^2(T)} \qquad (8.17)$$

where u^2 is the temperature-dependent mean square amplitude of the hydrogen oscillations on its interstitial site (see Section 3.5). If the temperature or Q value is sufficiently small, the quasielastic line is narrow and the intensity ratio or u^2 can be easily obtained with Eq. (8.17). This is valid under the condition that no relevant intensity contribution is hidden in the wings of the line where the hydrogen vibration frequency is comparable with $1/\tau$, the diffusive jump rate. Here, the vibrational line becomes overdamped and merges into the diffusive motion.

In order to deal with this problem, two integration procedures can be used. First, extend the integral in Eq. (8.16) only over a fixed energy window $\Delta\omega/2$ which is sufficiently narrow to exclude the spectral part which is determined by the oscillatory motion [say $\Delta\omega < (5,\ldots,10)/\tau$]. If the window is chosen too narrow, this falsifies the result and leads to a spurious increase of $\langle u^2\rangle$. Secondly, one can fit the quasielastic spectrum in a region of small ω with a certain diffusion model; then one integrates the fitted quasielastic spectrum over all energy transfers up to infinity. This procedure is more reliable, in particular for small Q where the shape of the quasielastic spectrum is model independent.

Figure 8.13 shows results of the quasielastic spectrum as obtained by Lottner *et al.* (1979a) as a logarithmic plot vs. Q^2, whose slope yields the mean square amplitude. Similar results were also reported for hydrogen in tantalum. One recognizes that the quasielastic intensity, as obtained from the second procedure described above, follows the Debye–Waller formula (8.17), with differences in the slope $\langle u^2 \rangle$, depending slightly on the choice of the model.

Mean square amplitudes were calculated by Schober and Lottner (1979) for several metal hydrides in the harmonic approximation. This quantity contains two contributions. (i) The first is due to the high-energy modes of the hydrogen atom which are well separated from the Debye spectrum (this does not hold for PdH_x); therefore, these modes can be treated as localized Einstein oscillators. (ii) There is a contribution caused by the host lattice waves. We could expect that the dissolved hydrogen simply mirrors the spectrum of these host lattice vibrations, with a certain amplitude coupling factor (Kley *et al.* 1970; see also Blaesser *et al.* 1968). We notice that for hydrogen in Nb a strong resonance enhancement of the hydrogen vibrations appears for certain modes thus yielding an especially large contribution to $\langle u^2 \rangle$ (Lottner *et al.* 1979b). For a more detailed review dealing with the vibrational modes of hydrogen in simple metals see Springer (1978). Figure 8.14 summarizes calculated mean square amplitudes as a function of temperature for H in Nb and Pd, together with experimental values. There is fairly good agreement.

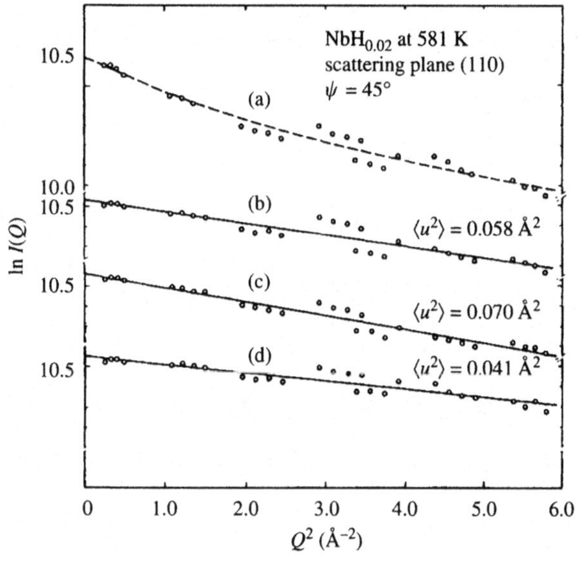

Fig. 8.13 Debye–Waller factor for α-$NbH_{0.02}$: intensity of the quasielastic line in a logarithmic plot versus Q^2 (a) integration of the line intensity over a fixed energy window of 2.2 meV halfwidth; (b) fit of the experimental points with the simple Chudley–Elliott model, then extrapolated and integrated to $\hbar\omega \to \pm\infty$; (c) same with model of correlated double jumps. The H mean square amplitude $\langle u^2 \rangle$ depends on the choice of the procedure (from Lottner *et al.* 1979c, copyright by Springer-Verlag). (d) fit with a two-state-model.

An early controversy dealt with an anomalously large value of $\langle u^2 \rangle$ for H in Nb; later, however, this discrepancy was explained by a too small integration range. However, there are recent results which indicate strong deviations from Eq. (8.17) which may be related to rapid local proton motion (Dosch *et al.* 1992).

For hydrogen in vanadium, the quasielastic intensity ratio reveals strong deviations from the usual Debye–Waller factor, as shown in Fig. 8.15. The origin of this deviation is not yet clear. Hydrogen in vanadium has an extremely high diffusion constant which approaches, at elevated temperatures, a value between 10^{-5} and 10^{-4} $cm^2\,s^{-1}$ like for atoms in a liquid. Consequently, the time needed for a diffusive jump ($\tau_j \cong s/v_{th} \cong s(k_B T/2M_H)^{-1/2} \cong 10^{-13}$ s, where v_{th} is the thermal velocity of a free H atom and s is the distance) is comparable with the mean rest time τ on a site during the diffusive process. This motion during the jump leads to scattering features like those of a nearly free particle. The corresponding width from such scattering is much larger than $1/\tau$, namely of the order of $(\hbar^2 Q^2 k_B T/2M_H)^{1/2}$. With increasing Q, the fraction $\tau_j/(\tau + \tau_j)$ of scattering due to the jump process itself is supposed to decrease, which removes intensity from

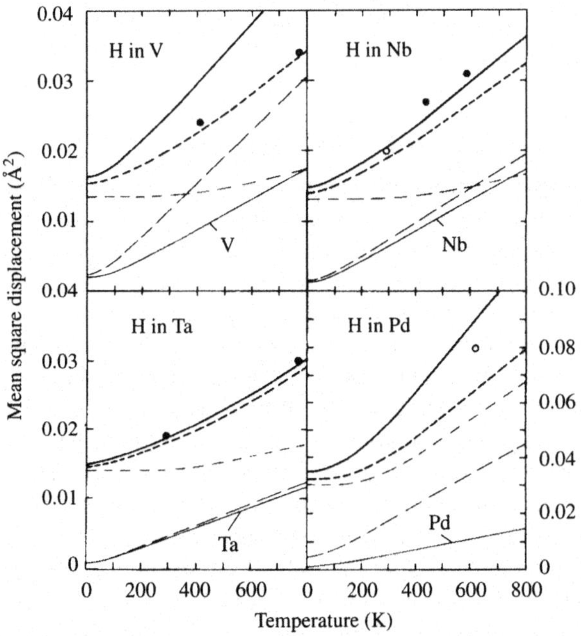

Fig. 8.14 Mean square amplitudes $\langle u^2 \rangle$ of hydrogen in Nb and Pd, obtained from the intensity of the quasielastic line. *Solid line*: harmonic theory. *Dashed line*: contribution of the localised (optic) modes only. *Thin line*: contribution from vibrations of host lattice atoms. Experiments: For V, Nb, and Ta: • Lottner *et al.* (1979a) and ○ Gissler *et al.* (1973). For Pd: ○ Rowe *et al.* 1972 (calculation from Schober and Lottner 1979).

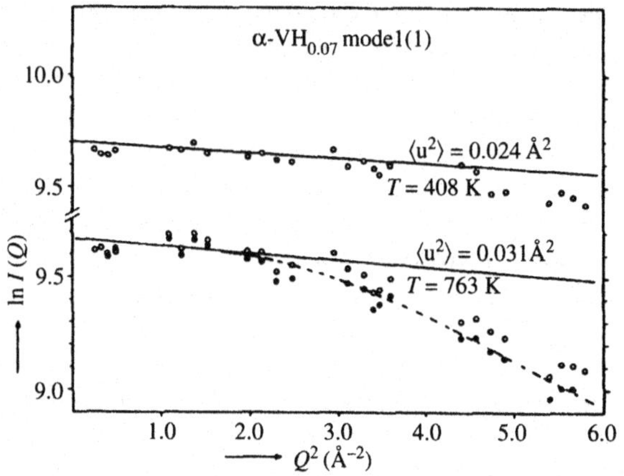

Fig. 8.15 Same as in Fig. 8.13 for α-$VH_{0.07}$ using the simple Chudley–Elliott model for integration. Deviations from the usual Debye–Waller factor behaviour at $Q^2 \geq 3$ Å^{-2} (from Lottner *et al.* 1979c, copyright by Springer-Verlag).

the quasielastic line; this effect may be responsible for the anomalous decrease of the quasielastic line intensity.

8.2.2 Hexagonal close packed α-phase hydrides

Yttrium is a transition metal with an hcp structure at room temperature. Like its lighter homologue Sc and like several rare earth metals (Tm, Lu) it can dissolve hydrogen or deuterium in solid solution up to relatively high concentrations (Bonnet and Daou 1974). This solid solution phase (α phase) extends up to 24 at.% H, and the phase boundary is virtually independent of temperature below 500 K. Above this temperature the maximum extent of the α phase increases with temperature. At higher concentrations the α phase coexists with the cubic (fcc) β phase which has the CaF_2 structure, see Section 8.1.2.

In the hexagonal solid solution phase there are two types of interstitial sites: hydrogen preferentially occupies tetrahedral sites, and there is hardly any indication from neutron vibrational spectroscopy (Udovic *et al.* 1997) that octahedral sites are simultaneously occupied to an appreciable extent. But since the T–T pairs are located on the c-axis it is nevertheless unlikely that diffusive jumps can directly occur between separate pairs of T sites because of geometrical hindrance by the metal atoms; therefore the microscopic proton diffusion mechanism must involve T $\rightarrow$ O jumps.

The QENS experiments were performed on the IN13 thermal backscattering spectrometer at the ILL Grenoble using polycrystalline samples in the temperature range 300–600 °C. The data could be described by a single Lorentzian component for incoherent quasielastic scattering from the proton and background terms.

A jump diffusion model was developed analogous to that described in Section 5.4.2. Due to the six sublattices present in α-YH_x (two O sites and four crystallographically different T sites) a superposition of six Lorentzians results. Since the sample used in the experiment was a polycrystal, the calculated scattering function had to be orientationally averaged. Although much information on individual jump patterns is lost in this way, some general features remain. For given values of residence times the model scattering was calculated, numerically averaged over all directions, and fitted with a Lorentzian. This Lorentzian was compared with that from a fit of the data; thus some conditions for the residence times could be established. In any case the self-diffusion coefficient could be extracted from the line broadening at small Q. For the interesting low-temperature diffusional behaviour of H in hcp α-ScH_x we refer to Section 8.3.2.

8.2.3 β-V_2H as an example of an ordered metal hydride

In dilute metal/hydrogen systems (α phase) the hydrogen diffusion mechanism can be considered as a vacancy mechanism on the interstitial sublattice; this sublattice contains few hydrogen atoms and many vacancies, so the hydrogen jumps are hardly impeded, which is the basic reason for the high hydrogen mobility in such a '*lattice-gas*'.

At elevated hydrogen concentration and higher temperatures some metal/hydrogen systems form so-called '*lattice-liquids*' refered to as α' or β phases. In these phases the hydrogen atoms on their sublattice exhibit short-range order, see Section 8.1.3 on α'-NbD_x. As another example we mention a study by Anderson *et al.* (1978) of a single crystal of β-$PdH_{0.73}$. If the residence time τ is replaced by $\tau' = \tau(1-c)$ where $(1-c)$ is the blocking or site availability factor, then the data could satisfactorily be described by the standard Chudley–Elliott model. By means of Monte Carlo simulations, Ross and Wilson (1978) have compared the low and finite concentration versions of $S_i(Q, \omega)$ at large and small ω. For high hydrogen content at large ω, i.e. short time, they showed that the scattering law corresponds to the Chudley–Elliott model but using τ' instead of τ, whereas at small ω, i.e. large time, and at small Q the same functional shape is obtained but with $\tau'' = \tau/((1-c) f_t)$ where f_t is the tracer correlation factor. Experimentally, such behaviour could not be proven unambiguously, however.

At high hydrogen concentration metal/hydrogen systems form *ordered hydrides*, particularly at lower temperatures. β-V_2H is an example, and its structure is shown in Fig. 8.16. The hydrogen atoms occupy sets of octahedral sites with nearly tetragonal point symmetry which form sheets in (110) directions in pseudocubic notation. The actual symmetry is monoclinic (Somenkov *et al.* 1972). In this way occupied sheets alternate with empty layers, see Fig. 8.16. A vacancy diffusion mechanism on the hydrogen sublattice as mentioned above is ineffective in this case. Actually, as will be explained below, hydrogen diffusion in β-V_2H proceeds via an interstitial mechanism on two interstitial sublattices: some hydrogen atoms

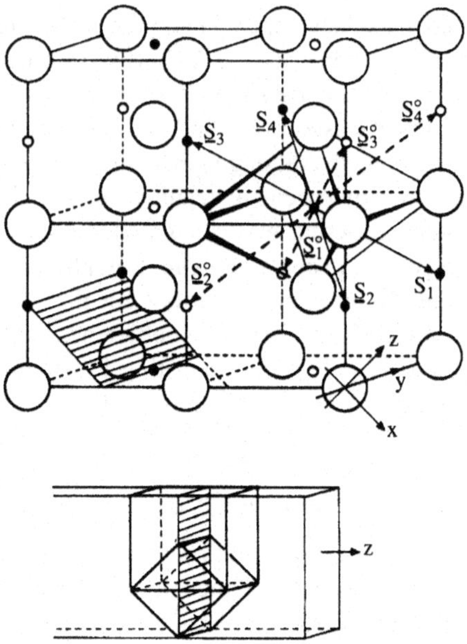

Fig. 8.16 Structure of β-V_2H, explanation in the text.

occupy sites in the forbidden layer, i.e. anti-structure sites or, in other words, they form Frenkel defects. These are the hydrogen atoms which are significant for the diffusion process; usually, however, they cannot be investigated by means of QENS since their number is too small. In β-V_2H, however, the situation is fortunate: with increasing temperature more and more sites in the 'forbidden' layers are occupied, and at 448 K an order–disorder transition to the ε phase occurs in which the hydrogen atoms are randomly distributed over both types of layers. At 390 K $\leq T \leq$ 440 K, the temperatures of the QENS study of Richter *et al.* (1989), the number of H atoms on anti-structure sites turned out to be sufficient to be detectable by means of QENS.

The jump model employed by Richter *et al.* (1989) comprises four different jump rates:

(i) the rate within the occupied plane, τ^{-1}, along the jump vector $\boldsymbol{S}_1$ to $\boldsymbol{S}_4$;
(ii) the jump rate within the empty plane, τ_u^{-1}, also along the jump vector $\boldsymbol{S}_1$ to $\boldsymbol{S}_4$;
(iii) the rate of change-over from the occupied to the empty plane, τ_0^{-1};
(iv) the inverse rate of change-over, τ_1^{-1}; the latter jumps proceed along the jump vector $\boldsymbol{S}_1^0$ to $\boldsymbol{S}_4^0$, see Fig. 8.17.

If $P(\boldsymbol{r}, t)$ denotes the probability of finding a proton in the occupied plane and $U(\boldsymbol{r}, t)$ the probability for a proton to be in the empty layer, then the set of master

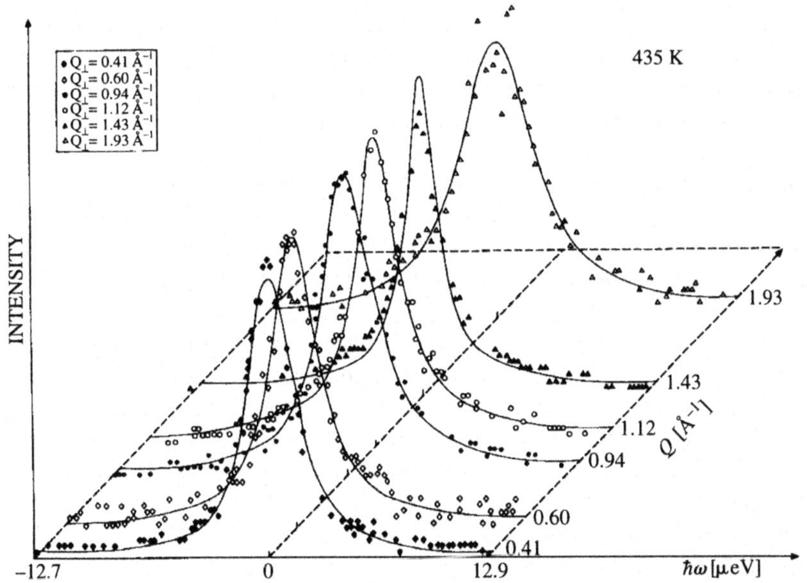

Fig. 8.17 Experimental QENS spectra of β-V_2H at 435 K, measured at a backscattering spectrometer, for $\boldsymbol{Q}$ mainly oriented perpendicular to the hydrogen sheets. The solid lines are the result of a fit with the jump model as explained in the text (from Richter *et al.* 1989).

equations for hydrogen diffusion is given by

$$\begin{aligned} \frac{\partial P(\boldsymbol{r},t)}{\partial t} &= -\left(\frac{1}{\tau}+\frac{1}{\tau_0}\right)P(\boldsymbol{r},t)+\frac{1}{4\tau}\sum_{i=1}^{4}P(\boldsymbol{r}+\boldsymbol{S}_i,t)\\ &\quad+\frac{1}{4\tau_1}\sum_{i=1}^{4}U(\boldsymbol{r}+\boldsymbol{S}_i^0,t)\\ \frac{\partial U(\boldsymbol{r},t)}{\partial t} &= -\left(\frac{1}{\tau_u}+\frac{1}{\tau_1}\right)U(\boldsymbol{r},t)+\frac{1}{4\tau_u}\sum_{i=1}^{4}U(\boldsymbol{r}+\boldsymbol{S}_i,t)\\ &\quad+\frac{1}{4\tau_0}\sum_{i=1}^{4}P(\boldsymbol{r}+\boldsymbol{S}_i^0,t). \end{aligned} \tag{8.18}$$

After Fourier transformation a 2×2 matrix equation is obtained. The incoherent scattering function, therefore, consists of the superposition of two Lorentzians; their widths are the eigenvalues of the above jump matrix, and their weights are related to the eigenvectors. The jump rates τ_0^{-1} and τ_1^{-1} are interconnected through the thermal occupation numbers of the two layers. These occupation numbers are directly related to the Bragg intensity $I(T)$ of the superlattice reflection from the layered structure. Therefore, such a neutron diffraction measurement was performed in order to eliminate one of the four rate parameters in the problem.

The essential prerequisite for the QENS experiment was the preparation of a single-domain β-V_2H single crystal (Hempelmann 1986), because on poly- or multidomain-crystals QENS cannot fully develop its capabilities. The QENS experiments were performed, using IN10 and IN6 at ILL, for two sample orientations. In the orientation $Q_\perp$ the sample was placed into the beam such that the sheet orientation was mainly perpendicular to the scattering plane. Figure 8.17 presents a set of spectra thus obtained at the backscattering spectrometer IN10 at 435 K. The data exhibit considerable quasielastic broadening which generally increases towards higher Q but passes through an intermediate minimum around $Q = 1.43\,\text{Å}^{-1}$, where $\boldsymbol{Q}$ is oriented exactly perpendicular to the H sheets. At the TOF-spectrometer IN6 with $\boldsymbol{Q}$ parallel to the sheets a small quasielastic component is clearly discerned; its relative intensity and width increases with increasing temperature.

The quantitative data evaluation in terms of the jump model outlined above allowed the derivation of the four jump rates and led to the following picture. Let us consider a hydrogen atom starting in the filled layer. Then, depending on temperature (temperature range $390\,\text{K} \leq T \leq 440\,\text{K}$), the jump probability to change into the empty layer is 6 to 3 times higher than to move among the occupied sites. If the hydrogen has changed into the empty layer, then on average it performs 5 to 2 jumps before it drops back into a vacancy in one of the adjacent filled layers. Thus diffusion parallel to the sheets consists in a repeated trapping and release process between the filled and the empty layer. In particular at lower temperatures where the ratio between the fast jump rate in the empty layer and all the other rates is largest, hydrogen diffusion is strongly anisotropic.

8.2.4 $LaNi_5H_\alpha$ as an example of an intermetallic hydride

The interstitial lattice of intermetallic compounds usually consists of several crystallographically and energetically different sites. As an example we consider $LaNi_5H_\alpha$ (Schönfeld *et al.* 1992; 1994). Figure 8.18 shows the structure; there is a low-energy octahedral 3f site located in the basal plane, and an energetically higher tetrahedral 6m site in the (002) lattice plane. The former is characterized by equal distances to each neighbouring site of the same type in the basal plane, thus forming infinite layers perpendicular to the c-axis. The latter exhibits two distances, a very short one which always connects six of the sites to a regular hexagon, and a long one that represents the shortest distance between two hexagons in this layer. 3f and 6m sites are interconnected by jump vectors such that the diffusing H atom can reach eight 6m sites from every 3f site, and four 3f sites from every 6m site. With these interlayer jumps a translation parallel to the hexagonal c-axis is

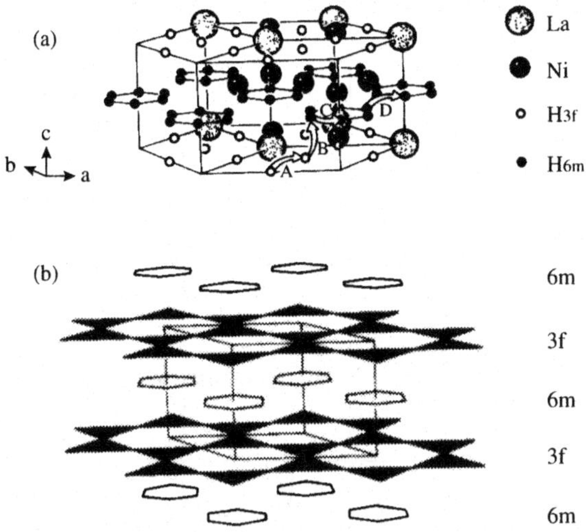

Fig. 8.18 (a) Crystal structure of α-$LaNi_5$ hydride. A–D denote four possible jumps between neighbouring interstitial sites. Host metal atoms are partially omitted for clarity. (b) Geometrical representation of hydrogen sublattice. The 3f sites form infinite two-dimensional layers in the basal plane. The 6m sites are grouped together to regular hexagones in the ($z = \frac{1}{2}$) plane (from Schönfeld *et al.* 1994, copyright by the American Physical Society).

achieved. The QENS jump model based on these considerations has been described in Section 5.4.2.

Figure 8.19 shows the eigenvalue and their relative weights as a function of Q for three selected directions in the α-$LaNi_5$/H system. x, y, and z are laboratory fixed axes corresponding to the three Cartesian axes. x and z point along the a- and c-axes of the hexagonal crystal. This figure uses the experimentally determined jump rates in order to display the eigenvalue spectra.

The time constants of the involved jump processes differ by two orders of magnitude. For this reason the nine eigenvalues (modes) split into two groups, one in the range of a few μeV and the other in the meV regime. In each direction one *diffusive mode* appears, which is the link to the macroscopic property of long-range diffusion in the limit of very small Q (λ_1 in Fig. 8.19). As $Q \to 0$ it is the only component with non-vanishing weight, and it is proportional to Q^2, the proportionality constant being the self-diffusion coefficient $D_s^{\perp}$ or $D_s^{\parallel}$, respectively. The modes λ_2 and λ_3 in Fig. 8.19 can be attributed to diffusion among 3f sites. In the special case of the z-direction, only two out of the nine eigenvalues (λ_1 and λ_4) have non-vanishing weight, which can therefore be attributed to interlayer jumps. The five higher-energy modes (λ_5–λ_9 in Fig. 8.19) can be assigned to jumps over the hexagons of the 6m tetrahedral sites. The sample was a $La^{60}Ni_5H_{0.08}$ single

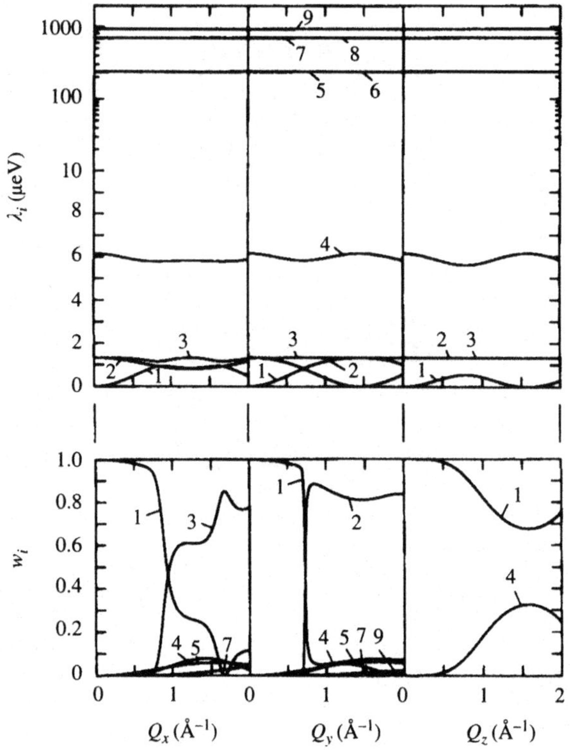

Fig. 8.19 Eigenvalues and weights of the jump model in the three Cartesian directions for $LaNi_5H_\alpha$ (from Schönfeld *et al.* 1994, copyright by the American Physical Society).

crystal. The Ni isotope was chosen in order to avoid the large incoherent neutron scattering of natural Ni.

Using a high-resolution backscattering spectrometer four QENS spectra with $\boldsymbol{Q} \parallel \boldsymbol{c}$ were recorded. From a simultaneous fit of the above model the λ_1 and λ_4 modes could be determined; these imply the jump rates Γ_{mf} and Γ_{fm} and, due to the detailed balance condition, the energetic difference between 3f and 6m sites. With knowledge of these parameters the spectra with $\boldsymbol{Q}$ perpendicular to the c-axis have been evaluated and thus the jump rate Γ_f between two 3f sites has been determined.

The investigation of the fast jumps within the hexagons in the (002) plane required a multichopper time-of-flight spectrometer with an energy resolution of about 100 μeV. Spectra were recorded at 423 K on $La^{60}Ni_5H_{0.16}$ with $\boldsymbol{Q} \perp \boldsymbol{c}$ and $\boldsymbol{Q} \parallel \boldsymbol{c}$, respectively. The additional quasielastic broadening in the ($\boldsymbol{Q} \perp \boldsymbol{c}$) spectrum confirms the confinement of the fast motion to the plane perpendicular to the c-axis as described earlier. From the linewidth, which is nearly independent

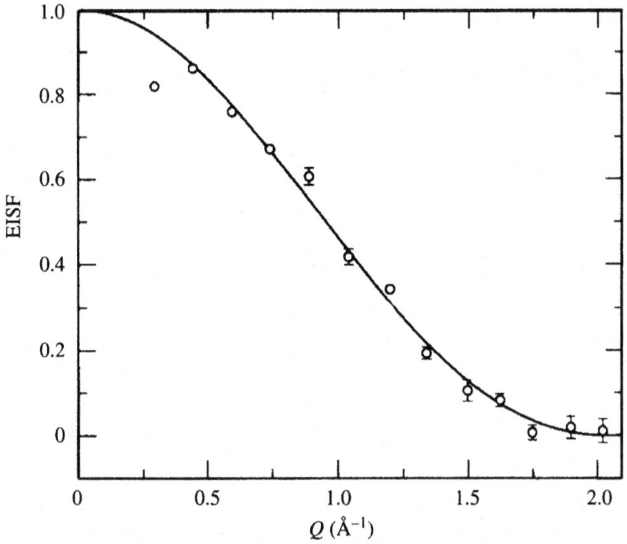

Fig. 8.20 Elastic-incoherent structure factor of the localized hydrogen diffusion in $La^{60}Ni_5H_{1.2}$. The open circles represent the weighted average from four measurements between 380 and 440 K. The solid line is $A_0(Q)$ from Eq. (8.19) with the jump length $l = 1.18$ Å obtained from the structure (from Schönfeld *et al.* 1994, copyright by the American Physical Society).

of Q and of T, the jump rate within the hexagon could be derived. In Fig. 8.20 the apparently elastic intensity divided by the total intensity is plotted versus Q. The solid line is the EISF of localized motion on a hexagon:

$$A_0(Q) = \tfrac{1}{6}[1 + 2J_0(Ql) + 2J_0(Ql\sqrt{3}) + J_0(2Ql)] \qquad (8.19)$$

where $J_0(x)$ are the cylindrical Bessel functions. The jump length $l = 1.18$ Å was taken from the structure and is not used as an adjustable parameter. The agreement between theory and experiment is perfect and supports the model of localized diffusion on the 6m hexagons.

8.3 QUANTUM MOTION OF H AND μ^+

8.3.1 Hydrogen tunnelling in Nb doped with O, N, C

Low-temperature heat capacity measurements of Nb doped with hydrogen exhibit an anomaly at about $T = 2$ K (Morkel *et al.* 1978) which is due to the simultaneous presence of small amounts of both H and O.

If the hydrogen isotope ^{1}H is replaced by ^{2}H, the excitation energy is reduced by about one order of magnitude (Wipf and Neumaier 1984). Because of this unusually large isotope effect this anomaly has been assigned to quantum mechanical tunnelling of hydrogen atoms.

High-resolution neutron spectroscopy allows experimental access to the dynamics of hydrogen over a large range. This comprises free diffusion at room temperature and extends to coherent quantum tunnelling at temperatures below 1 K. The analysis of the resulting neutron scattering structure factors gives detailed information about the geometry of the tunnelling system. Most systems investigated so far can be described as a tunnelling motion within an asymmetric double minimum potential. With decreasing temperature the damping effect of the phonons on the hydrogen dynamics decreases and, eventually, does not obscure the coupling of the tunnelling system to the conduction electrons. This interaction leads to a broadening of the quasielastic linewidth with decreasing temperature. For Nb as host material the coupling to the conduction electrons can be demonstrated in a particularly convincing way at temperatures below the superconducting transition temperature, $T_c = 9.2$ K, by applying a magnetic field. A double minimum potential can be modelled by two approaching harmonic potentials, see Fig. 8.21, such that the wavefunctions of a particle oscillating in the respective minimum overlap. Then the wavefunctions of the double minimum system are given by linear combinations of the wavefunctions of the respective harmonic oscillators located at $x = \pm a/2$ if a is the distance between the minima:

$$\Psi_s = \alpha|-a/2\rangle + \beta|+a/2\rangle$$
$$\Psi_a = \alpha|-a/2\rangle - \beta|+a/2\rangle. \tag{8.20}$$

The normalization factors, $\alpha^2 + \beta^2 = 1$, in the case of a symmetrical double minimum potential are given by

$$\alpha^2 = \beta^2 = \tfrac{1}{2}, \tag{8.21}$$

i.e. the particle is delocalized and equally distributed over both potential minima.

In analogy to the binding and anti-binding orbitals of the H_2 molecule the symmetric and antisymmetric wavefunctions ψ_s and ψ_a exhibit slightly different energy eigenvalues, and this tiny energy difference is called the tunnelling splitting J.

A distortion of the double minimum potential by an amount ε (see Fig. 8.21c) increases the splitting of the energy levels. Then the tunnelling excitation of the ground state is given by

$$E = \sqrt{J^2 + \varepsilon^2}. \tag{8.22}$$

According to Imry (1969) the normalization factors, $\alpha^2 + \beta^2 = 1$, in the case of a distorted double minimum potential, are interrelated by

$$\left(\frac{\alpha}{\beta}\right)^2 = \frac{J^2 + \varepsilon^2}{J^2}. \tag{8.23}$$

Correspondingly, if $\varepsilon > J$, the particle is preferentially localized in the energetically lower potential minimum, see Fig. 8.21c.

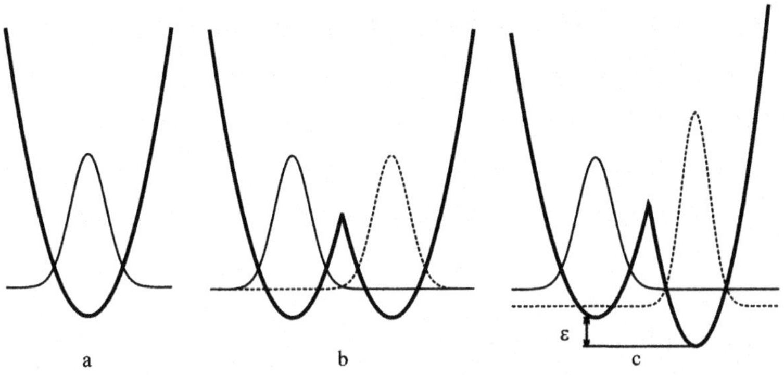

Fig. 8.21 Schematic representation of the energy levels and wavefunctions (a) for a harmonic potential, (b) for a symmetric double minimum potential, (c) for an asymmetric double minimum potential. A typical value for an oscillator excitation energy in metal/hydrogen systems is $\hbar\omega = 100\,\text{meV}$, for the tunnelling splitting of the ground state $J_0 = 0.2\,\text{meV}$. The asymmetry is characterised by the distortion ε.

For the transition $\psi_s \rightarrow \psi_a$ the double differential neutron scattering cross-section is given by Fermi's golden rule:

$$\frac{d^2\sigma}{d\omega\, d\Omega} \propto |\langle \Psi_a| \exp(i\boldsymbol{Q}\boldsymbol{r})|\psi_s\rangle|^2. \tag{8.24}$$

Consequently, with increasing localization of the particle in the energetically lower potential minimum, the inelastic scattering intensity decreases. Stronger distorted double minimum potentials exhibit higher excitation energies, but with very low inelastic scattering intensity: the scattering is dominantly elastic.

The distortion discussed above originates from elastic interactions between the tunnelling systems. Thus the distortion fields depend on the concentration of the point defects. It appears reasonable to assume a Lorentzian shaped distribution function $Z(\varepsilon)$ for the distortions (Stoneham 1969),

$$Z(\varepsilon) = \frac{1}{\pi} \frac{\Delta\varepsilon}{(\Delta\varepsilon)^2 + \varepsilon^2} \tag{8.25}$$

where $\Delta\varepsilon$ is the width of the distribution function and ε characterizes a typical energy difference between the two sites. A Lorentzian distribution function can also be introduced for the tunnelling splitting J:

$$Z(J) = \frac{1}{\pi} \frac{\Delta J}{(\Delta J)^2 + (J - J_0)^2}. \tag{8.26}$$

After orientational averaging the inelastic scattering function (Wipf *et al.* 1981; Magerl *et al.* 1986) is given by

$$S_{\mathrm{i}}^{\mathrm{inel}}(Q,\omega)$$
$$=\left(\frac{1}{2}-\frac{\sin Qa}{Qa}\right)\frac{O(\hbar\omega)}{1+\exp(\hbar\omega/k_{\mathrm{B}}T)}\int_{-\infty}^{\infty}\mathrm{d}J\,Z(J)\int_{-\infty}^{\infty}\mathrm{d}\varepsilon\,Z(\varepsilon)\frac{J^2}{J^2+\varepsilon^2} \tag{8.27}$$

with the damped oscillator function

$$O(\hbar\omega)=\frac{1}{\pi}\frac{\Gamma E^2}{((\hbar\omega)^2-E^2)^2+(\Gamma(\hbar\omega))^2}. \tag{8.28}$$

The corresponding elastic scattering function is given by

$$S_{\mathrm{i}}^{\mathrm{el}}(Q\omega)=\frac{1}{\Delta\varepsilon+J}\left[\Delta\varepsilon+J\left(\frac{1}{2}-\frac{\sin Qa}{2Qa}\right)\right]\delta(\hbar\omega). \tag{8.29}$$

It contains a Q-independent term, which increases with increasing distortion $\Delta\varepsilon$, and a Q-dependent term, which corresponds to the EISF and contains information about the geometry of the tunnelling system. For vanishing $\Delta\varepsilon$ Eq. (8.29) equals the EISF of a dumb-bell, see Eq. (5.76).

In Eq. (8.27) the dynamical coupling of the tunnelling system to the environment (i.e. to phonons and/or conduction electrons) is taken into account by the damped oscillator function $O(\hbar\omega)$ which is determined by the resonance frequency E and by the damping constant Γ. For small Γ the response function approaches δ functions located at positions $\pm E$. In terms of similar models (tunnelling in asymmetric double minimum potentials) low-temperature anomalies in the heat capacity of amorphous solids are also discussed (Anderson *et al.* 1972; Phillips 1972) although in those systems the identity of the tunnelling system is a matter of debate.

Figure 8.22 displays examples of neutron spectra of $Nb(OH)_{0.002}$ in the temperature range 145 K to 10 K (Steinbinder *et al.* 1988); the experimental data can be described by a superposition of an elastic and a quasielastic scattering component. The ratio I_{qu}/I_{tot} does not deviate identifiably from a linear relationship with Q^2, which indicates a distance d that is in the range of 1 Å or less in view of the maximum investigated Q of about 2 Å^{-1}. Figure 8.23 shows the temperature dependence of the quasielastic width $\Gamma(T)$: the striking feature is the minimum at about 70 K; below this temperature, the width increases with decreasing temperature! Below $T_c = 9.2$ K one has to distinguish between the superconducting state and the normal-conducting state in Nb (superconductivity can be suppressed by an external magnetic field!). This is demonstrated in Fig. 8.24: in the superconducting state a tunnelling excitation is observed both at 0.2 K and

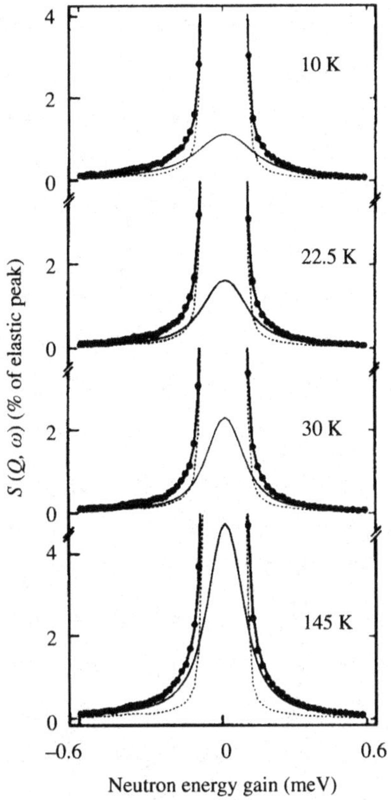

Fig. 8.22 QENS spectra of a $Nb(OH)_{0.002}$ sample at four temperatures. The spectra comprise data taken in the Q range from 1.25 to 2.04 $Å^{-1}$. The thin and thick solid lines are fit curves for the quasielastic and the total scattering intensity. The broken lines indicate the measured resolution function (from Steinbinder *et al.* 1988).

at 4.3 K whereas a tunnelling splitting is not observed at 4.3 K in the normal-conducting state (with magnetic field) (Magerl *et al.* 1986; Wipf *et al.* 1987). The existence of tunnelling excitations directly confirms a coherent tunnelling process. In the normal-conducting state the low-temperature value of J is slightly lower than in the superconducting state. The damping Γ, on the other hand, is distinctly larger in the normal-conducting than in the superconducting state. Both effects prove the direct influence of the conduction electrons.

At low temperatures in the superconducting state the inelastic line exhibits a residual intrinsic width. This width clearly depends on the defect concentration x (Wipf *et al.* 1981; Magerl *et al.* 1986). For $x = 0.000\,15$ a well-defined excitation is observed; hardly any width $\Delta\varepsilon$ of the distortion ε of the double minimum potential and no width ΔJ of the distribution of tunnelling splitting is necessary

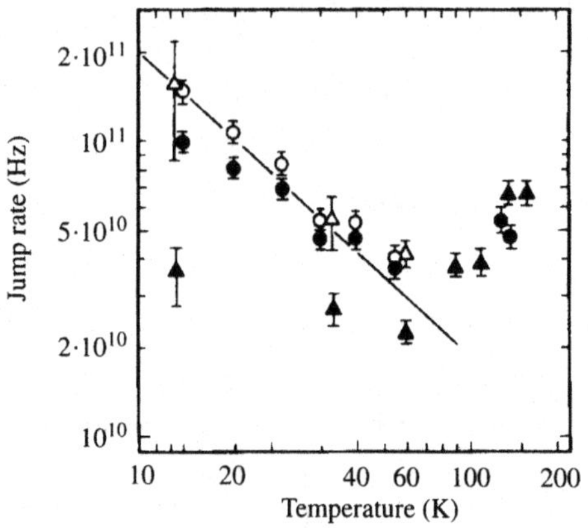

Fig. 8.23 Jump rates $\nu(T)$ (quasielastic widths) of hydrogen in two $Nb(OH)_x$ samples with $x = 0.002$ (circles) and 0.011 (triangles) (from Steinbinder *et al.* 1988).

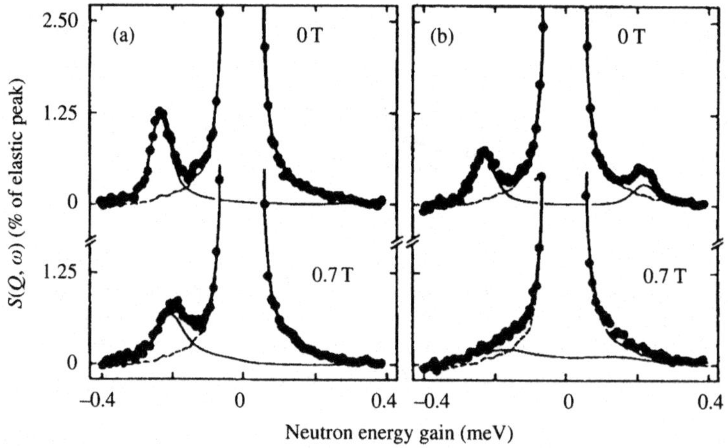

Fig. 8.24 Neutron spectra of a $Nb(OH)_{0.0002}$ sample at 0.2 K (a) and 4.3 K (b). For both temperatures, the spectra were taken in the superconducting (0 T) and normal-conducting (0.7 T) electronic state. The solid lines represent fit curves for the inelastic scattering intensity. The broken lines are for the elastic intensity (from Wipf *et al.* 1987).

in order to describe the data quantitatively in terms of Eq. (8.27). At less minute concentrations, $x = 0.0022$ and $x = 0.011$, the tunnelling system is already distorted and distributions both in ε and in J have to be assumed in order to describe the data.

Neutron vibrational spectroscopy on H in Nb doped with O, and other experimental evidence, show that the double minimum potential in Nb doped with O is formed by two hardly distorted tetrahedral interstitial sites at a distance of about 3.5 Å from the oxygen atom and with a mutual distance of 1.2 Å. A hydrogen atom on this double minimum potential interacts with the conduction electrons, but this interaction which retards the mobility decreases with decreasing temperature, particularly in the superconducting state, as was first pointed out by Kondo (1984). In the case of weak coupling ($K \ll 1$) the temperature dependence of the linewidth is given by $\Gamma(T) \propto T^{2K-1}$. Experimentally a $T^{-0.89}$ dependence is found for Nb (Magerl *et al.* 1986; Steinbinder *et al.* 1988) which confirms the electronic coupling as the dominant mechanism of damping. The resulting coupling parameter is $K = 0.055$.

In the superconducting state $J_0^{\mathrm{s}} = 0.226 \pm 0.004\,\mathrm{meV}$ was found experimentally; in the normal-conducting state $J_0^{\mathrm{n}} = 0.206 \pm 0.003\,\mathrm{meV}$ at very low temperatures. According to

$$\Gamma^{\mathrm{n}}(T) = \pi K (J_0^{\mathrm{n}}/2k_{\mathrm{B}}T) \tag{8.30}$$

for the normal-conducting state, and according to

$$\Gamma^{\mathrm{s}}(T) = \frac{4\pi K k_{\mathrm{B}} T}{1 + \exp(E_\delta(T)/k_{\mathrm{B}}T)} \tag{8.31}$$

for the superconducting state ($E_\delta(T)$ denotes the electronic gap in the superconductor), both tunnelling matrix elements are in accordance with Kondo's theory. For more details we refer to the recent review by Wipf (1997).

8.3.2 Rapid low-temperature hopping of H in α-ScH$_x$

Hydrogen dissolved in hcp rare-earth metals exhibits unusual short-range ordering which apparently stabilizes the α phase to much higher concentration at low temperatures than in other metal/hydrogen systems. Neutron diffraction (Khatamian *et al.* 1981) and diffuse scattering (Blaschko *et al.* 1985; McKergow *et al.* 1987) studies show that the hydrogen atoms are in tetrahedral sites and are arranged in next-nearest neighbour pairs bridging the metal atoms. Quasielastic spectra of a $ScH_{0.16}$ single crystal have been recorded between 300 K and 10 K (Anderson *et al.* 1990), see Fig. 8.25. The data were fitted with a superposition of an elastic and a quasielastic component. The analysis of the Q dependence of the elastic incoherent structure factor gave a hopping distance of 1.0 Å, in agreement with results from diffuse scattering experiments. Figure 8.26 shows the quasielastic linewidth Γ versus temperature with hopping rates exceeding $7 \cdot 10^{10}$ s, indicative of very rapid motion compared to bulk diffusion in this system.

The remarkable upturn of Γ below the minimum at about 100 K is approximately $\propto T^{-1}$ in the range shown, and is similar in appearance to the behaviour of hydrogen trapped at oxygen impurities as described in the previous section. Also the interpretation by the authors is in terms of Kondo's theory, i.e. the low-temperature dynamical behaviour is determined by weak coupling to the conduction electrons.

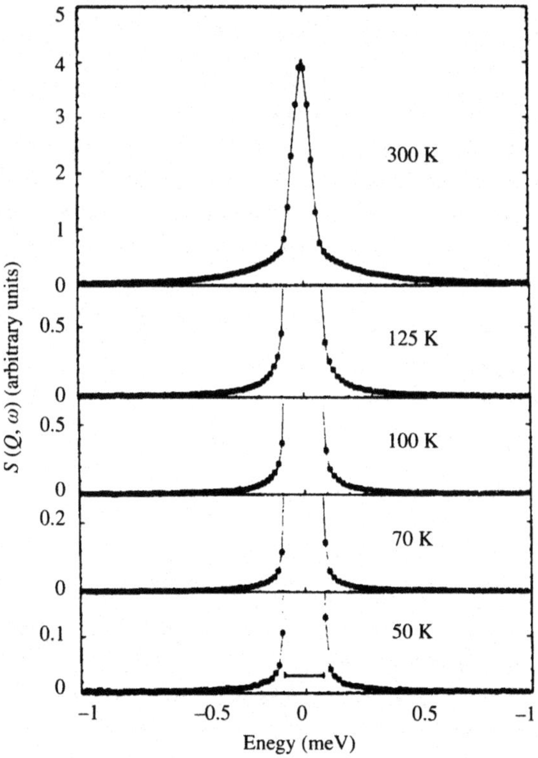

Fig. 8.25 QENS spectra for $ScH_{0.16}$ at several temperatures (from Anderson *et al.* 1990, copyright by the American Physical Society).

The essential difference to the system of the previous section is that in Sc no impurity atom is necessary to create a double minimum potential but the double minimum potential is intrinsic to the structure of the pure-metal/hydrogen system.

8.3.3 Muon diffusion in Al and Cu

Since the positive muon in metals behaves like a light hydrogen isotope and since its diffusion mechanism is completely analogous to that of H in metals, we summarize here some relevant results. Technical details can be found in the book of Schenck (1985). Aluminium and copper are the metals in which the low-temperature diffusional behaviour of muons has been studied most. Because of the large nuclear moments of ^{27}Al, ^{63}Cu, and ^{65}Cu, large depolarization effects take place if the muons move sufficiently slowly. Futhermore, impurities can be removed from these fcc metals down to the ppm range whereas for the bcc metals such purity is not feasible. The first muon diffusion experiment was performed by Gurevich *et al.* (1972) with *copper* as host metal. The resulting depolarization rate exhibited textbook-like behaviour, i.e. at low temperatures the muon is frozen in

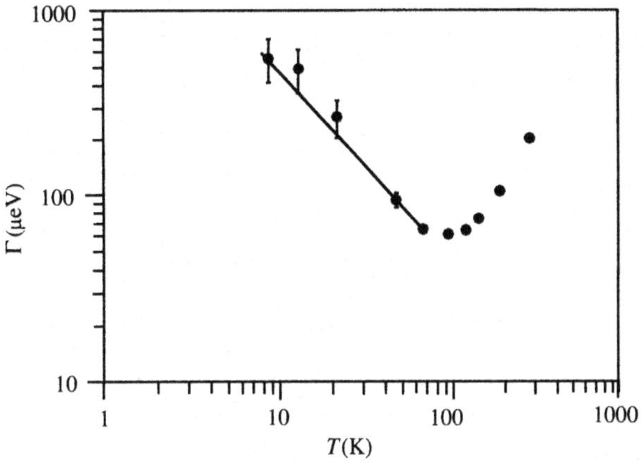

Fig. 8.26 Fitted Lorentzian linewidths (FWHM) for $ScH_{0.16}$. The solid line is the fit to the data below 100 K discussed in the text (from Anderson *et al.* 1990, copyright by the American Physical Society).

(static case, large Gaussian damping), and at high temperatures it moves quickly (motional narrowing, negligibly small exponential damping). So it was a surprise to observe (Hartmann *et al.* 1980) that the depolarization rate decreases again below 50 K in extremely pure copper. Later by careful line shape analysis of life time histograms taken in transverse geometry (Welter *et al.* 1983) as well as by zero-field and longitudinal field measurements (Kadono *et al.* 1984; 1985; 1989) it could be proved that the low-temperature drop in the depolarization rate also represents a motional narrowing; that means the muon diffusion coefficient, after passing through a minimum, increases again with decreasing temperature, like the H diffusion coefficient in Nb doped with O, N, and in Sc, see the previous section. As a measure of the jump rate and thus of the diffusion coefficient Fig. 8.27 displays the correlation rate τ_c^{-1} (Kadono *et al.* 1985) in the low temperature regime.

It soon became apparent that in pure *aluminium* muon diffusion is too fast to be directly measurable. Therefore Al has to be doped with well-defined amounts of dopants (in the ppm range) like Li, Mg, Ga, . . . ; then muon diffusion can be measured via diffusion-limited trapping. The resulting muon diffusion coefficient exhibits a minimum at about 3 K and, remarkably, increases again with decreasing temperature. Towards higher temperatures, it first passes through a regime with a linear T dependence and then goes over into the regime of phonon-assisted tunnelling with an Arrhenius-like temperature dependence.

Theoretically, the linear T dependence can be explained by processes in which only one phonon in the lattice is involved. Those processes are forbidden in ideal lattices because of energy conservation, but are possible in the presence of lattice distortions or for a transition between non-equivalent sites. Whether such a one-phonon regime also exists for Cu is still open to question.

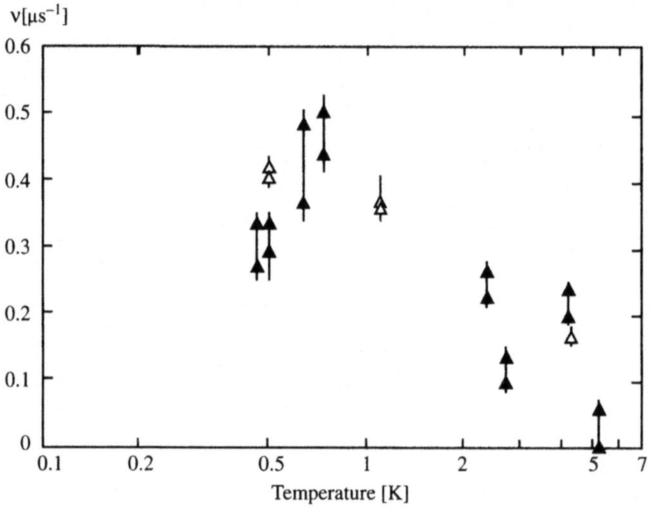

Fig. 8.27 Hopping rate for μ^+ diffusion in copper. The diffusion is sufficiently slow to be determined directly from measurements of the longitudinal μ^+ relaxation (from Kadono *et al.* 1985 and from Clawson *et al.* 1982, with permission from Elsevier Science).

The low-temperature slope of the correlation rate or of the diffusion coefficient, respectively, is proportional to $T^{-0.6}$ in Cu and to $T^{-0.7}$ in Al. These nearly coinciding temperature dependences are in agreement with Kondo's theory and confirm that electrons are involved in the low-temperature diffusion mechanism.

Primarily, scattering of electrons by small polarons takes place. This leads to a T^{-1} dependence of the diffusion coefficient what is approximately in agreement with experiment. But in this treatment the absolute value of the diffusion coefficient is overestimated by several orders of magnitude.

The second and much more important effect of the muon–electron interaction consists in an additional and temperature dependent reduction of the low–temperature tunnelling matrix element, see Eq. (4.127). The muon has a positive charge, and this is shielded by an electron cloud. The shielding cloud is assembled by positive interference of electrons in appropriate Bloch states. Since in the ground state, i.e. at very low temperatures, the charge density of the conduction electron gas is primarily homogeneous, electrons have to be excited into states above the Fermi energy in order to shield the muon charge. This electron–hole excitation increases the kinetic energy, but the shielding cloud decreases the Coulomb energy; for the total system the shielded state is energetically more favourable. Now when the muon, in a diffusional step, moves to an adjacent site, the shielding cloud has to be dismounted at the old site and reassembled at the new site. At very low temperatures with only very few electrons above the Fermi edge there is hardly any overlap of the wavefunctions forming the old and forming the new shielding cloud (*the orthogonality catastrophe*). Under these circumstances the electrons can no longer follow the movements of the muon adiabatically; the Born–Oppenheimer approximation, otherwise valid everywhere in solid state and molecular physics, breaks down in this case.

Both effects combined lead to the temperature dependence of Eq. (4.127), i.e. $D \propto T^{2K-1}$. For the muon–electron coupling parameter, K values were calculated which agree with the experimental results, and the absolute value of D was also in accordance. It is gratifying that theoretically for K^{Cu} a larger value is obtained than for K^{Al}, as is actually observed. This is due to the electron density which is larger in Al than in Cu. In the high-temperature range (phonon-assisted tunnelling, $T \geq 20\,\mathrm{K}$) the additional muon–electron coupling is only of minor importance.

The diffusional behaviour of muons in superconducting Al ($T_c = 1.2\,\mathrm{K}$) has been investigated using Al samples with 75 ppm, 20 ppm, 10 ppm, and 0 ppm Li doping (Karlsson *et al.* 1995). The depolarization functions at a temperature of 0.007 K taken in zero applied field indicate the existence of two fractions of muons, one of them corresponding to muons which are trapped and immobile over the whole observation period, the other for which there is no depolarization at all. The trapped fraction varies from almost 100% at 75 ppm Li doping to about 50% at 20 ppm, 25% at 10 ppm, and near zero for the purest Al. For pure Al muons are thus 100% mobile even below 0.1 K in the superconducting state as well as in the normal state.

When these experiments are repeated at higher temperatures, up to T_c, and again analysed in terms of two fractions of muons, then the *immobilised* fraction exhibits a T dependence of its depolarization functions as displayed in Fig. 8.28 for the lowest doping concentration. This behaviour can be understood in terms of Eq. (4.130) as a 'competition' between the energetic level broadening γ_s (Eq. 4.129) and the mean energy spread ξ due to the distortions near the doping atoms. Decreasing the temperature from T_c downwards, it is therefore expected that in a short T interval below T_c one would have $\gamma_s \gg \xi$, and Eq. (4.127) would still be valid leading to a strong increase of the mobility since $\gamma_s(T)$ decreases exponentially. In inhomogeneous samples, however, one will soon reach a condition where $\gamma_s \approx \xi$, which corresponds to a maximum in the mobility, before the condition $\gamma_s \ll \xi$ takes over and the diffusion rate is instead proportional to $\gamma_s(T)$, i.e. it decreases exponentially with decreasing temperature.

Thus the maximum of the zero-field depolarization rate in Fig. 8.28 at about 1 K is—in the framework of a trapping model—interpreted as an indication of particularly fast muon diffusion, whereas the low-temperature increase and the plateau are interpreted in terms of conventional motional narrowing, i.e. as an indication of unmeasurably slow or static muons. This argument is valid for muons that happen to be implanted in regions relatively far from the impurity centres. Muons which, on the other hand, are implanted close to the impurity centres will not be able to move from their initial positions and will show the polarization characteristic of static muons.

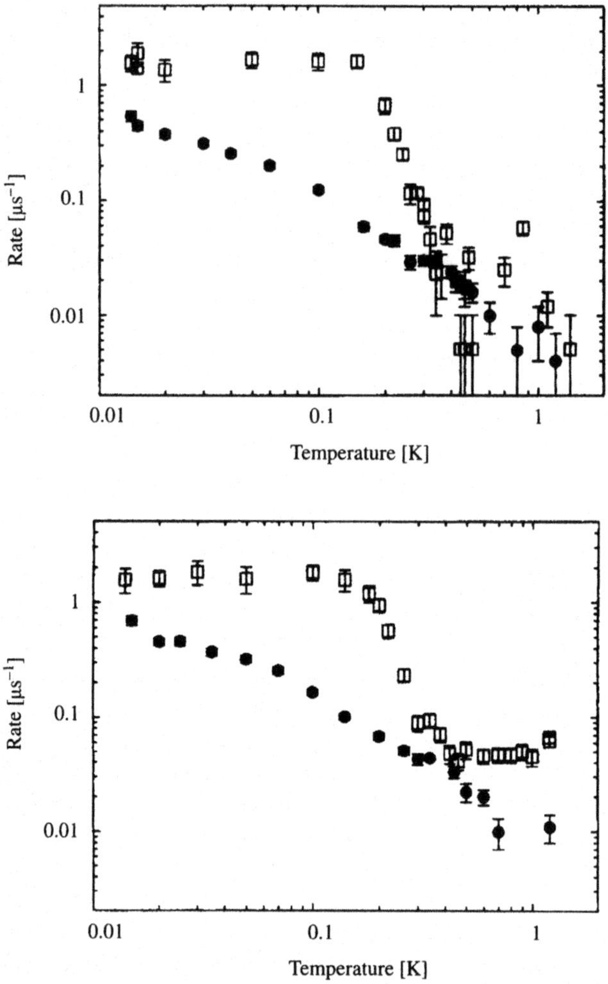

Fig. 8.28 Temperature dependence of the depolarization rate of the trapped fraction of muons in Al doped with 10 ppm Li (upper) and 20 ppm Li (lower) in the normal state (filled symbols) and in the super conducting state (open symbols). There is an indication of a peak just below $T_c = 1.2$ K, but the main trapping occurs below about $0.2\,T_c$ (from Karlsson *et al.* 1995, copyright by the American Physical Society).

The *non-trapped* fraction mentioned above represents muons which are implanted into a region of low energy level inhomogeneity ξ, far from impurities. They can tunnel coherently and develop into extended Bloch-like propagating states when the metal is superconducting. As soon as the propagating state has developed, it can lose coherence only by inelastic collisions. At the lowest temperatures the thermal energy is not sufficient for such an inelastic process, and the muon wave will be scattered elastically between different perturbed regions with coherence preserved.

In the purest Al samples, according to this interpretation, delocalization will be extensive and there is nothing to prevent the formation of a true band-like state for positive muons in the superconducting state. This long-range process of muon propagation corresponds to the coherent tunnelling of hydrogen in the double minimum potential in superconducting Nb doped with O or N. In the normal-conducting state, on the other hand, the muons in pure Al are also mobile, at a level of 10^8 jumps s^{-1}, but this is quantum diffusion with loss of coherence between each step, sometimes also called the hopping regime.

8.4 HYDROGEN DIFFUSION IN DISORDERED SYSTEMS

8.4.1 Hydrogen diffusion in the presence of trapping impurities

Interstitial atoms like nitrogen or oxygen, as well as certain substitutional metal atoms, randomly distributed in niobium or tantalum for example, induce *traps* for the dissolved hydrogen, either due to elastic distortions or to chemical affinities. This can be concluded from various experimental observations: the increase of the hydrogen solubility in the α phase, the appearance of an additional Snoek relaxation maximum apparently caused by jumps of the trapped hydrogen close to the impurity atom, the existence of slightly modified localized modes and of tunnelling states (Springer and Richter 1987).

The QENS spectra of hydrogen in niobium doped with nitrogen have been interpreted in terms of the two-state model, see Section 5.5, a modification by Richter and Springer (1978) of a model originally due to Singwi and Sjölander (1960) which has been developed in order to describe the self-diffusion of water. In nitrogen-doped niobium the hydrogen atoms alternately spend a time τ_0 on the average on a trapping site, and then, during a time τ_1, diffuse freely in undisturbed parts of the lattice. At small Q, a single Lorentzian is the dominating fraction of the quasielastic intensity. Its width is given by the effective diffusion coefficient

$$D_{\text{eff}} = \frac{\tau_1}{\tau_0 + \tau_1} D_{\text{free}} \tag{8.32}$$

which is reduced, compared to the free diffusion coefficient, by the fraction of time spend in the free state.

At larger Q, however, two quasielastic components appear; the narrow one has a width proportional to τ_0^{-1}, the broad one proportional to τ^{-1}. The weights correspond to the relative occupations of the free and trapped state. Figure 8.29 displays the temperature dependence of the trapping and escape rates resulting from a series of QENS experiments (Richter and Springer 1978) whereby a larger trap concentration yields a larger trapping rate, whereas the escape rate does not depend on the trap concentration. The resulting trapping radius is $r_t = 5$ Å and calculations of the elastic interaction energy between N and H as a function of

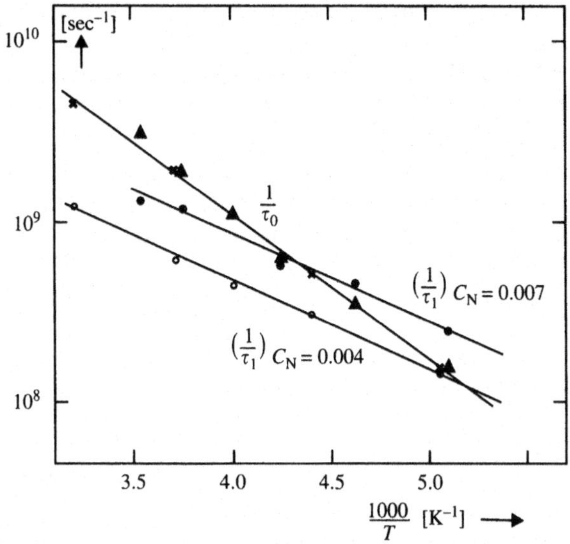

Fig. 8.29 Arrhenius representation of the hydrogen escape rate $1/\tau_0$ and trapping rate $1/\tau_1$ in $Nb(NH)_x$ (from Richter and Springer 1978, copyright by the American Physical Society).

distance and orientation in the lattice (Richter and Springer 1978) show that many sites around the impurity are energetically disturbed, leading to a trapping region with a spectrum of site energies. Theoretical calculations by Kehr *et al.* (1978) show that in this case the temperature effect of the impurities on the diffusion coefficient is much smoother than expected for a single trap.

The two-state trapping model has proved useful also for the interpretation of QENS spectra of intermetallic hydrides. Since the intermetallics are composed of different metal atoms, chemically different H sites exist as a consequence of the varying H affinities of the host atoms. In $Ti_{1.2}Mn_{1.8}$, for example, tetrahedra are formed from 4Mn atoms, 3Mn and Ti, and 2Mn and 2Ti, the latter containing the most attractive H site because of the large H affinity of Ti.

The QENS spectra of $Ti_{1.2}Mn_{1.2}H_3$ (Hempelmann *et al.* 1983) can be separated into components with Q and T dependences characteristic for diffusion in the presence of traps. As $Q \to 0$ only the linewidth of the narrow component tends to zero. The narrow component at large Q reveals a considerably higher activation energy.

As discussed above, in intermetallic compounds a whole spectrum of energetically different H sites is expected. In order to account for this feature, the available sites are divided into *trap sites*, comprising the energetically lowest interstitial positions, and *free sites* otherwise. Dissolved H preferentially occupies the trap sites and, according to the thermal occupation probability, saturates most of them. The remaining H is distributed over the free energetically less favourable sites and occasionally gets trapped in an empty trap site. Thus, in spite of the high density of traps, for a single H atom the diffusional process with respect to trapping is not very different from the situation in Nb with dilute N impurities, since the *empty* traps are dilute.

The existence of a third scattering component was concluded from the missing intensity at large Q. This was related to rapid motion confined to small regions in space. Correspondingly, the two-state model was extended to a three-state model. The experimental spectra obtained for eight temperatures $228 \leq T \leq 374$ K and for $0.17 \leq Q \leq 1.95$ Å were fitted simultaneously with this three-state model associating the local relaxation with the free state. The activation energies for τ^{-1}, τ_l^{-1}, and τ_0^{-1} are 210 meV, 209 meV, and 300 meV, respectively. While the trap molar fraction is evaluated to 0.24 ± 0.06, the concentration of free traps varies between 5% and 9% depending on temperature which is consistent with the application of the trapping model. The fraction of trapping sites resulting from the QENS evaluation is in accordance to the fraction of Ti_2M_n tetrahedral sites in the structure, whereas the resulting jump length of the local jump process, $l = 1.37$ Å, is in fair agreement with the nearest-neighbour distance of 1.3 Å. On the basis of the experimental findings it was suggested that the rapid local motion is connected with correlation effects due to the high H concentrations. Similar effects are discussed by Funke in the framework of his jump relaxation model for ionic conductors, see Chapter 12.

The QENS study on $Ti_{1.2}Mn_{1.8}H_3$ was the first of its kind and appears to have uncovered significant microscopic details of H diffusion in concentrated H storage materials. Similar phenomena have also been observed in $LaNi_5H_6$ (Richter *et al.* 1983) and in $Ti_{0.8}Zr_{0.2}CrMnH_3$ (Hempelmann *et al.* 1983). In particular, the reverse concentration dependence of the H diffusion coefficient as observed in $LaNi_5H_x$ can be explained in terms of a trapping model.

8.4.2 Hydrogen diffusion in metglasses

Hydrides of amorphous alloys have been investigated extensily in the last fifteen years mainly for their potential use in hydrogen storage technology; see for instance Maeland (1986) and references therein. Pressure-composition isotherms, for example, of Zr–Ni alloys (Aoki *et al.* 1984), deviate strongly from Sieverts's law at higher H concentration. These positive deviations indicate a distribution of site energies in the amorphous structure, and H atoms enter successively higher-energy states. Such a distribution of energy states can be naturally understood as a consequence of the varying local environments of the interstitial sites available for hydrogen: there may be different types of sites with different numbers and types of adjacent metal atoms, and with varying nearest neighbour distances. Kirchheim (1982) assumes that the site-energy distribution in $Pd_{1-y}Si_y$, $y = 0.15$–0.22, can be represented by a Gaussian function which allows one to describe both thermodynamic and diffusion data, measured electrochemically. For diffusion it is additionally assumed that the energy levels of the potential barrier (in the picture of classical over barrier jumps) have the same energy value throughout the sample, i.e. do not exhibit an energetic distribution (constant saddle-point energy). In the literature this model is known as the Gaussian model. The site-energy distributions of amorphous Ti–Ni- and Zr–Ni- alloys (Jaggy *et al.* 1989; Harris *et al.* 1987; Curtin and Harris 1988) exhibit several minima and maxima, which indicate the existence of different types of sites, each having a Gaussian distribution of site energies.

Conversely, a QENS study on H diffusion in a-$Pd_{85}Si_{15}H_{7.5}$ (Richter *et al.* 1986) gave experimental hints of the existence of energetically well-separated interstitial sites, whereas a QENS study on H diffusion in a-$Zr_{76}Ni_{24}$ (Schirmacher *et al.* 1990) has been interpreted in terms of a broad continuous distribution of activation energies.

Richter *et al.* (1986) have analysed their data on a-$Pd_{85}Si_{1.5}H_{7.5}$ with respect to different hypotheses which have been used to describe diffusive processes in amorphous materials. The Gaussian site energy distribution function was, at large Q, approximated by an appropriate superposition of Lorentzians, but this model turned out to be at variance with the experimental data. Later a model based on Gaussian distributed site and saddle-point energies was also compared to the data (Driesen and Kehr 1989); although an improvement was achieved over the model with Gaussian-distributed site energies only, the results still did not allow a proper description of H diffusion in amorphous $Pd_{1-y}Si_yH_x$. So-called anomalous diffusion—as pointed out by Schirmacher *et al.* (1990)—means that in the time scale of interest the mean square distance walked by the diffusing particle increases sublinearly instead of linearly with time. Also the conjecture that the network of diffusive paths exhibits fractal character leads to such a sublinear time dependence: $\langle r^2(t)\rangle \propto t^{2/(2+\Theta)}$, where Θ describes the range dependence of the diffusion coefficient on the fractal network ($D \propto r^{-\Theta}$) (Gefen *et al.* 1983). For the Q dependence of the quasielastic linewidth Λ this implies $\Lambda \propto Q^{2+\Theta}$ or Λ is expected to grow faster than with Q^2. The experimental observation of $\Lambda \propto Q^{1.54}$ does not support the assumption of anomalous diffusion or of a dominating fractal structure for the diffusive paths. The QENS data on amorphous $Pd_{85}Si_{15}H_{7.5}$ could successfully be evaluated in terms of a diffusion and trapping model. Figure 8.30 shows the Q dependences of the weight of the narrow component and of both linewidths. The temperature dependence of the weight shows that the protons are activated from energetically more stable sites, which are called traps, into a state of high mobility. From the Q dependence a mobility range of about 10 Å is obtained. For the mobile state the isotropic Chudley–Elliott model is assumed. The result of a fit with this two-state model to the observed linewidths and weights is shown by the solid lines in Fig. 8.30; obviously a reasonable agreement between model and experiment is achieved. In particular the diffusion process in between traps is considerably faster than in crystalline Pd, whereas, as a consequence of trapping, the long-range diffusion coefficient is of the same order. At small Q

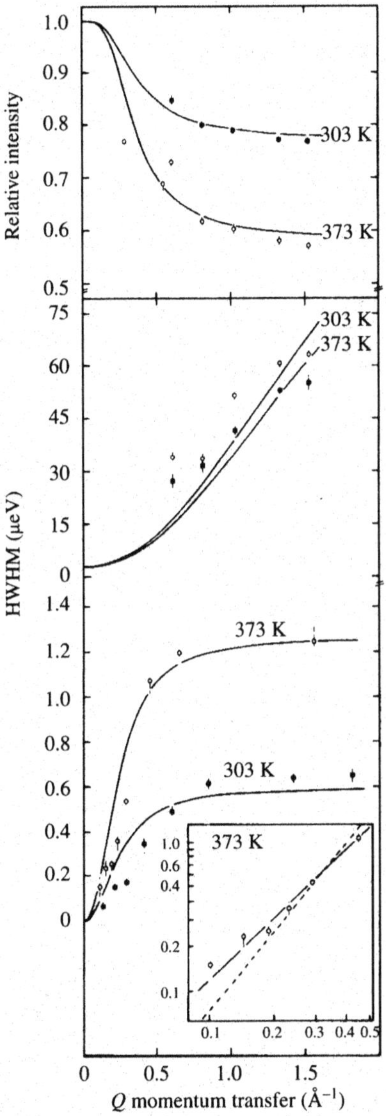

Fig. 8.30 (Upper) Weight of narrow QENS component of H in a-$Pd_{85}Si_{15}H_{7.5}$ as a function of temperature and momentum transfer. (Lower) T and Q dependence of the linewidths. The solid lines represent a fit of the two-state model to these data. Inset: magnification of the small Q behaviour of the linewidths at 373 K (dashed line: HWHM $\propto Q^2$) (from Richter *et al.* 1986, copyright by the American Physical Society).

the character of the narrow mode already crosses over from diffusive ($\Lambda \propto Q^2$) to localized behaviour ($\Lambda \propto 1/\tau_0$); this explains a Q exponent smaller than 2 as found experimentally. The observation of two well-separated regimes of jump rates is evidence for the existence of two different kind of interstices. Actually, by means of neutron vibrational spectroscopy (Rush *et al.* 1989), vibrational modes typical for octahedral and tetrahedral sites have been detected.

Schirmacher *et al.* (1990) claim that the above model does not appear to have a reasonable physical justification and is, as these authors say, in conflict with the current ideas of a continuous activation spectrum emerging essentially from all other investigations. They interpret their QENS data on a-$Zr_{76}Ni_{24}H_8$ in terms of an effective medium description of anomalous diffusion in disordered systems (Schirmacher 1991; Wagner and Schirmacher 1991) based on a model with a broad distribution of activation energies. Figure 8.31 shows combined spectra for $Q = 0.15\,Å^{-1}$ obtained with two different spectrometers. The high-frequency tail of the data seems to follow a $\omega^{-(1+\alpha)}$ law with $\alpha \approx 0.43$. This is in accordance with Schirmacher's theory which predicts a frequency-dependent diffusion coefficient $D(\omega) \propto t^{\alpha}$. Other experimental manifestations of anomalous diffusion are the a.c. conductivity of ionic glasses (see Wang and Angell 1973), anomalous transient photocurrents in disordered semiconductors (also called 'dispersive transport' by

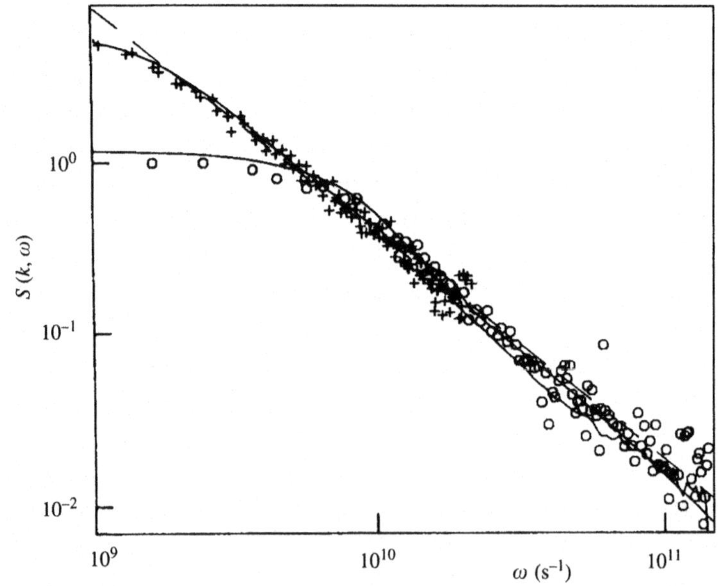

Fig. 8.31 Experimental incoherent scattering function of H in $Zr_{76}Ni_{24}H_8$ at $T \approx 445$ K and $Q = 1.5\,Å^{-1}$ compared to the theoretical expressions convoluted with the resolution functions of the corresponding neutron spectrometers (solid lines). The dashed line indicates a $\omega^{-1.43}$ law (from Schirmacher *et al.* 1990).

Scher and Montroll (1975)), NMR results on cation conducting glasses (Müller-Warmuth 1995) and β-NMR results on Li diffusion in $^7LiO \cdot B_2O_3$ glass (Schirmer *et al.* 1988). However, there are quite a number of theoretical approaches like the jump relaxation model (Funke 1991), relaxation dynamics in disordered ionic conductors (Dieterich *et al.* 1991), diffusion limited percolation (Bunde *et al.* 1991), and others which do not postulate a continuous activation spectrum.

Whether there is a 'universal' explanation for the 'universal' dynamic response in disordered solids (Funke 1991) is still an open question.

9

Metals and alloys

For the fcc noble metals and for bcc metals of group V and VI such as Mo, W, V, and Nb, atomic diffusion is slow compared to the case of bcc alkali metals Li, K, or Na. The group IV metals Ti, Zr, and Hf have still higher diffusivities. In the first section of this chapter, the investigation of sodium diffusion is described, in particular presenting the encounter model of vacancy diffusion in detail. The second part deals with similar experiments on bcc β-titanium where the relation between high diffusivity and soft modes as measured by triple axis spectrometry is explained.

9.1 VACANCY-INDUCED DIFFUSION IN BCC METALS

Self-diffusion in sodium was most extensively studied in particular by nuclear magnetic resonance and radioactive tracers (reviews of Mehrer (1978) and Peterson (1978)). Sodium nuclei have relatively large incoherent scattering cross-sections, so that the incoherent dynamical structure factor $S_i(Q, \omega)$ for the diffusing atoms can also be analysed in order to get information on the vacancy diffusion mechanism. QENS experiments on sodium were published by Ait Salem *et al.* (1979) and by Göltz *et al.* (1980).

In the bcc lattice of sodium, atomic diffusion occurs by a process where the atoms interact with diffusing vacancies. If a vacancy arrives at a site nearest to a selected Na atom of the lattice, this atom jumps into the vacancy. After a certain time, another vacancy may arrive at an adjacent site of the same atom, leading to a subsequent jump and so on. Assuming that the jumps occur to the eight nearest neighbour sites, with displacement vectors $\boldsymbol{s} = \sqrt{3}a/2\langle 111\rangle$, each sodium atom carries out a random walk which follows these vectors, where a is the cubic lattice constant. Consequently, the corresponding quasielastic width $\Gamma(\boldsymbol{Q})$ for sodium scattering is the same as that obtained in Chapter 5.3, namely

$$\Gamma = (1/8\tau_e) \sum_{n=1}^{8} (1 - \cos(\boldsymbol{s}_n \boldsymbol{Q})) \qquad (9.1)$$

where $\boldsymbol{s}_n$ are the jump vectors. This formula can easily be extended to also include jumps to next nearest neighbours. So far we have assumed that consecutive jumps are caused by independent vacancies arriving near the considered atom. However, it happens that the same vacancy arrives several times at the selected atom which leads to a sequence of consecutive and strongly correlated jumps. For the description of this process we refer to the general concept of the encounter (Wolf 1977a; 1978) which is the sequence of displacements caused by one and the same vacancy. This implies that during one encounter a selected atom rapidly moves by several jumps over a number of sites. Since the vacancy concentration is small ($c_v \leq 10^{-3}$) the overlap of encounters caused by different vacancy paths will be neglected.

The encounter formalism can be generalized for two (or several) kinds of defects, in view of the coexistence of single and divacancies in sodium, as will be described later. For two vacancy types we follow the procedure developed by Wolf (1977b; 1983). We first introduce the probability $q(S_1, S_2, t)$ that a lattice atom experiences S_1 encounters of type 1 and S_2 of type 2. Furthermore, we use the probability per space volume element $P(S_1, S_2, \boldsymbol{r})$ that, resulting from encounters of type 1 and 2, an atom is displaced from $\boldsymbol{r} = \boldsymbol{0}$ to a site $\boldsymbol{r}$. Then one gets for the self-correlation function of the atom (Wolf 1977b; Gissler and Rother 1970)

$$G_s(r, t) = \sum_{S_1, S_2=0}^{\infty} \sum q(S_1, S_2, t)\, P(S_1, S_2, \boldsymbol{r}). \tag{9.2}$$

Instead of the mixed and random sequence of encounters one can approximately introduce the probability P_e that in a 'representative encounter' (averaged over both types) an atom is displaced from $\boldsymbol{r} = \boldsymbol{0}$ to $\boldsymbol{r}_m$. This probability is given by

$$P_e(r) = \sum_{\boldsymbol{r}_m} [\beta_1 W_1(\boldsymbol{r}_m) + \beta_2 W_2(\boldsymbol{r}_m)]\, \delta(\boldsymbol{r} - \boldsymbol{r}_m) \tag{9.3}$$

where $W_{1,2}$ are the probabilities for an atomic displacement to a site $\boldsymbol{r}_m$ by type 1 and type 2 encounters, respectively. β_1 and β_2 denote the corresponding temperature-dependent probabilities ($\beta_1 + \beta_2 = 1$). If the two encounter types are independent one has

$$q(S_1, S_2, t) = w(S_1, t/\tau_{e1})\, w(S_2, t/\tau_{e2}) \tag{9.4}$$

where $1/\tau_{e1}$ and $1/\tau_{e2}$ are the average probabilities per time unit for the two processes. The functions w are the Poisson distributions for the occurrence of sequences with average length S_1 or S_2 during time t. We introduce the total number of encounters of both types per time unit which an atom experiences, namely

$$1/\tau_e = 1/\tau_{e1} + 1/\tau_{e2}. \tag{9.5}$$

Then one can express $q(S_1, S_2, t)$ by the Poisson distribution for a sequence of lengths $S = S_1 + S_2$, namely (Wolf 1977a)

$$\sum_{S_1} \sum_{S_2} q(S_1, S_2, t) = \sum_{S_1} w(S_1, t/\tau_{e1}) \sum_{S_2} w(S_2, t/\tau_{e2})$$

$$= \sum_{s=0}^{\infty} w(S, t/\tau_e). \tag{9.6}$$

We introduce $h_e(\boldsymbol{Q})$, the Fourier transform of $P_e(\boldsymbol{r})$, and the convolution in Q space

$$P_e(S, \boldsymbol{r}) = [h_e(\boldsymbol{Q})]^S. \tag{9.7}$$

With Eq. (9.2) we then obtain, in analogy to the random walk formalism in Section 5.3,

$$I_S(\boldsymbol{Q}, t) = e^{(-t/\tau_e)[1-h_e(\boldsymbol{Q})]} \tag{9.8}$$

where the resulting quasielastic width is then

$$\Gamma(\boldsymbol{Q}) = (1/\tau_e)[1 - h_e(\boldsymbol{Q})]. \tag{9.9}$$

From the formulation for $P_e(\boldsymbol{r})$ in Eq. (9.3) one finally obtains

$$h_e(\boldsymbol{Q}) = \sum_{\boldsymbol{r}_m} [\beta_1 W_1(\boldsymbol{r}_m) + \beta_2 W_2(\boldsymbol{r}_m)] \cos(\boldsymbol{Q}\boldsymbol{r}_m). \tag{9.10}$$

The sum goes over all sites adjacent to the origin which can be reached by atomic jumps during one representative encounter (including a term $\boldsymbol{r}_m = 0$ which refers to an atom which has returned to its origin by the jump sequence). Now the encounter time τ_e has to be related to the mean time between successive jumps of the atoms, $\bar{\tau}$. This quantitiy (and not τ_e) is directly accessible to experiments from the (mass transport) self-diffusion constant (see Section 5.3)

$$D_s = a^2/6\bar{\tau}. \tag{9.11}$$

During one representative encounter an atom performs $Z_1\beta_1 + Z_2\beta_2$ jumps on the average, where $Z_{1,2}$ is the number of jumps induced by a vacancy of type 1 or 2 during one encounter. The number of encounters per unit time, $1/\tau_e$, for an atom is then the number of atomic jumps per unit time, $1/\bar{\tau}$, divided by the average number of jumps per encounter, which means

$$\tau_e = \bar{\tau}(\beta_1 Z_1 + \beta_2 Z_2) \tag{9.12}$$

which has to be introduced into $\Gamma(\boldsymbol{Q})$ in Eq. (9.9).

The resulting quasielastic width is anisotropic due to the Fourier transform of $W(\boldsymbol{r}_m)$, and the anisotropy is temperature dependent because of the coefficients β_1 and β_2. Consequently, the experimental measurement of the quasielastic width yields $W(\boldsymbol{r}_m)$ combined with $\beta_{1,2}$, $Z_{1,2}$, and $\bar{\tau}$ which is obtained from the self-diffusion constant. Computer simulation allows one to calculate the quantities $Z_{1,2}$ and $W_{1,2}(\boldsymbol{r}_m)$ and the tracer correlation factor f_t. In the case of computer simulation for sodium, about 10^4 encounters had to be followed to obtain good statistics, and for each encounter virtually hundreds of jumps were followed, where about 50 jumps virtually completed one encounter (Wolf 1983).

The time between two vacancy jumps and, as well, the time for all atomic jumps during one encounter is small compared to τ_e. For the temperatures, we are dealing with, the time between successive jumps during one encounter is of the order of 10^{-12} s, compared to $\tau_e \cong 10^{-9}$ s or longer. Under these circumstances, the quasielastic width can be calculated as if the atomic displacement were a sudden process. Certainly, the rapid motion during the encounter leads to an additional quasielastic spectrum whose width, however, is 10^3–10^4 times larger than the width $\Gamma(\boldsymbol{Q})$. Therefore, this component remains unobserved in a high-resolution QENS experiment. In addition there is motion of the lattice atoms induced by the rapid displacement during a vacancy jump. The amplitude of these displacements is very small, and this contribution can also be neglected. The motion of the atoms also induces rapid displacements of its neighbours. These displacements, δ, are very small (δ is of the order of a few tenths of an Å, $Q^2\delta^2 \leq 1$) and their contribution is negligible.

Seeger and Mehrer (1970) have argued that mono- and divacancies exist and that their motion contributes to atomic diffusion in sodium. A divacancy 'tumbles' through the lattice by jumps where the vacancy pair changes between different orientations (Ho 1972). As sketched in Fig. 9.1, there are three orientations, 1n, 2n, and 4n, and a vacancy moves by a sequence of such orientations of the 1n-2n type, or of the 2n-4n reorientation. Each of these processes induces a $(a/2)\langle 111\rangle$ jump. The model assumes that the lifetime of a divacancy is sufficiently large such that it maintains its identity over several encounters.

The first quasielastic scattering by Ait-Salem *et al.* (1979) on poly- and single-crystals of sodium were carried out by the backscattering method with an energy window of a few μeV. The observed Q dependence agrees with a mono-vacancy process caused by nearest neighbour jumps of the $(a/2)\langle 111\rangle$ type. Results in the symmetry direction are shown in Fig. 9.2 (upper part) together with results of Göltz *et al.* (1980). There are small discrepancies which suggest the admixture of other processes. The figure shows that $\langle 111\rangle$ jumps of the atoms can be clearly excluded. Fig. 9.2 (lower part) also presents selected results of Göltz *et al.* (1980)

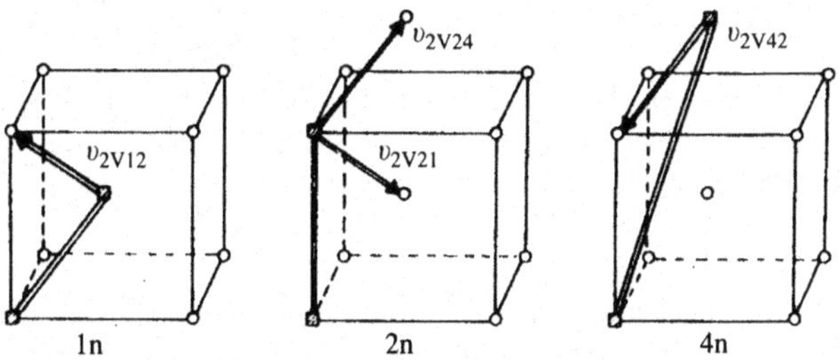

Fig. 9.1 Divacancies in the bcc lattice. □: vacancy, o: atom. First illustration: 1n configuration; second illustration: 2n configuration; third illustration: 4n configuration. The migration of the divacancy can occur by $(a/2)\langle 111\rangle$ jumps of atoms which lead to reorientations between 1n and 2n, or between 2n and 4n configuration (from Göltz *et al.* 1980).

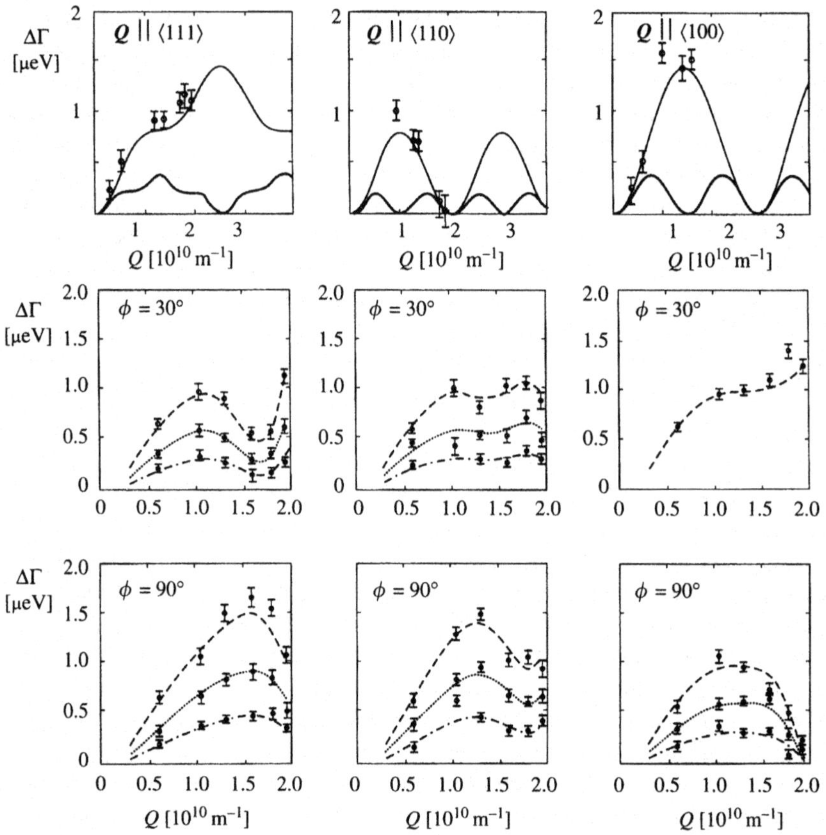

Fig. 9.2 Quasielastic width $\Delta\Gamma = 2\Gamma(Q)$ as a function of scattering vector $\boldsymbol{Q}$. First row $\boldsymbol{Q}$ parallel to symmetry directions, —o— experiments of Göltz *et al.* (1980) at 369 K, —x— experiments of Ait-Salem *et al.* (1979). Curves calculated from the encounter model, introducing the tracer diffusion constants with a mono-vacancy mechanism and $(a/2)\langle 111\rangle$ jumps (solid lines), and a mono-vacancy mechanism with a $\langle 111\rangle$ jumps (dotted lines). Second and third row: Measurements on single-crystals: $T = 343$ K (o), $T = 358$ K (Δ), $T = 369$ K (Lower curves). Neutron wavevector in the (110) plane, where ϕ is the angle between neutron wavevector $\boldsymbol{k}_1$ and $\langle 111\rangle$. Dashed upper curve: pure divacancy mechanism with 2n-4n reorientations at 369 K. The other (lower) curves were calculated with different mixtures of mono-vacancy and divacancy mechanism (from Göltz *et al.* 1980).

where $\boldsymbol{k}_0$, $\boldsymbol{k}_1$ is in a (110) plane of the sample, and ϕ is the angle between $\boldsymbol{k}_1$ and the $\langle 111\rangle$ direction, such that the orientation of $\boldsymbol{Q}$ changes along the abscissa of the diagrams. The agreement with a 2n-4n divacancy mechanism is very good, but the experiment cannot clearly identify the degree of admixture of divancancies to the monovacancy mechanism. The theoretical curves in the figure were obtained by using Eq. (9.9), where, by computer simulation of about 10^4 encounters, the parameters entering Eq. (9.9), namely f_t, $W_{1,2}$, and $\beta_{1,2}$, were obtained. The time $\bar{\tau}$ was calculated from tracer diffusion experiments yielding $D_t(T)$ (Mundy 1971). A combined analysis of $\Gamma(\boldsymbol{Q})$, together with the other parameters mentioned above, as a function of temperature, leads to best consistency for the admixture of mono- and divacancy process, where the 2n-4n process is probably dominant.

9.2 DIFFUSION IN β-TITANIUM AND PHONONS

Quasielastic scattering experiments on bcc β-titanium were performed by Vogl *et al.* (1989) on single crystals. The results for the quasielastic width as a function of sample orientation are shown in Fig. 9.3. These experiments had to be performed at very high temperatures. The experiments are compared with calculations by means of the encounter model as described in the previous paragraph for sodium, considering different jump geometries. Obviously, jumps of the $(a/2)\langle 111\rangle$ type are the dominating process and other processes considering a $\langle 100\rangle$ jump admixture

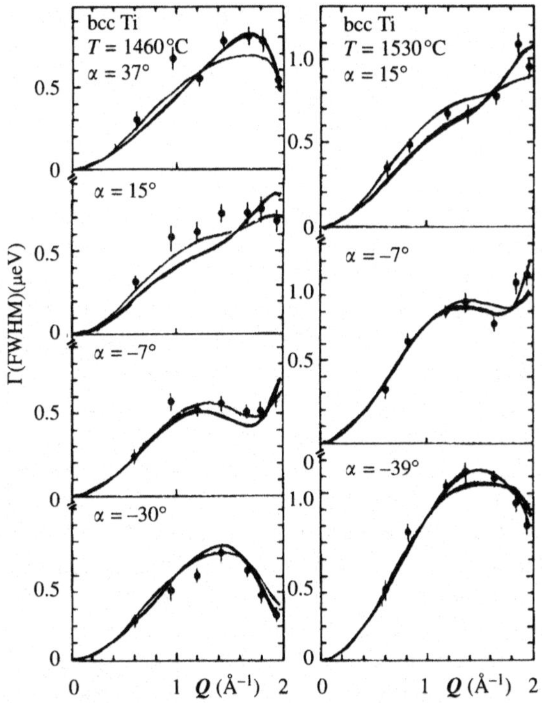

Fig. 9.3 Quasielastic linewidth $\Gamma(\boldsymbol{Q})$ as a function of scattering vector $\boldsymbol{Q}$ for diffusion in bcc Ti at 1460 °C for different crystal orientations: Wavevectors $\boldsymbol{k}_0$, $\boldsymbol{k}_1$ in (001) direction; α is the angle between $\boldsymbol{k}_0$ and a $\langle 110\rangle$ orientation. Experimental data (•), calculations: Solid line assuming $(a/2)\langle 111\rangle$ atomic jumps. Thin line: 75% $(a/2)\langle 111\rangle$ and 25% a $\langle 100\rangle$ jumps (from Vogl *et al.* 1989, copyright by the American Physical Society).

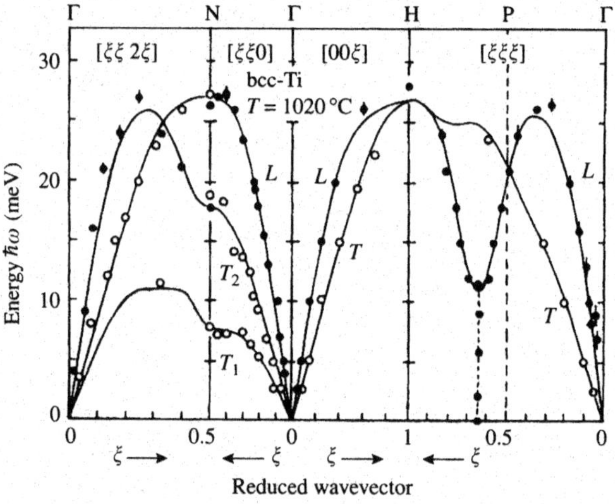

Fig. 9.4 Phonon dispersion curves $\omega(\boldsymbol{q})$ in symmetry directions for bcc Ti ($\boldsymbol{q}$ = wavevector). A low energy T_1 branch is identified along $\langle\xi\xi 0\rangle$ and $\langle\xi\xi\, 2\xi\rangle$. The mode at $\boldsymbol{q} = (2/3a)\langle 111\rangle$ completely degenerates to a quasielastic line. The soft modes, in particular those surrounding $\boldsymbol{q} = (2/3a)\langle 111\rangle$, contribute strongly to the high Ti diffusion constants in this metal (from Petry *et al.* 1991b, copyright by the American Physical Society).

can be practically excluded in the temperature range of the experiments. It was tempting to relate the diffusion process in titanium to the striking phonon anomalies appearing in the metals which were observed by Petry *et al.* (1991a, b) by phonon spectroscopy. As can be seen from Fig. 9.4, a soft longitudinal mode appears at $(2a/3)\langle 111\rangle$ which virtually 'collapses' at this point. It obviously degenerates into a Lorentzian-shaped quasielastic coherent peak centred at energy transfer $\hbar\omega = 0$. This is shown in Fig. 9.5 where the superimposed narrow line is identified as purely elastic due to incoherent scattering on the sample.

In Fig. 9.6 the $(2/3a)\langle 111\rangle$ mode is sketched indicating the motion of the corresponding lattice planes. If planes 1 and 3 are moving in opposite directions, with plane 2 at rest, at sufficiently large amplitude these two planes would collapse with neighbouring planes leading to a new structure, called the ω phase (de Fontaine and Buck 1973). However, at normal pressure this phase transition is not accessible and there is no indication of precursors of the ω phase, i.e. no diffuse scattering was identified which could develop into a peak due to a new structure. However, it is tempting to relate the longitudinal $(2/3a)\langle 111\rangle$ mode and the modes in its vicinity to the vacancy jumps described above. A possible scenario is included in Fig. 9.6 (Petry *et al.* 1991a) where both the jump vector and the phonon propagation are parallel to the $\langle 111\rangle$ direction. The corresponding titanium atom and the adjacent barrier move towards each other. The barrier is supposed to be low as concluded from the soft modes. Penetrating the second barrier (in the immobile plane) leads

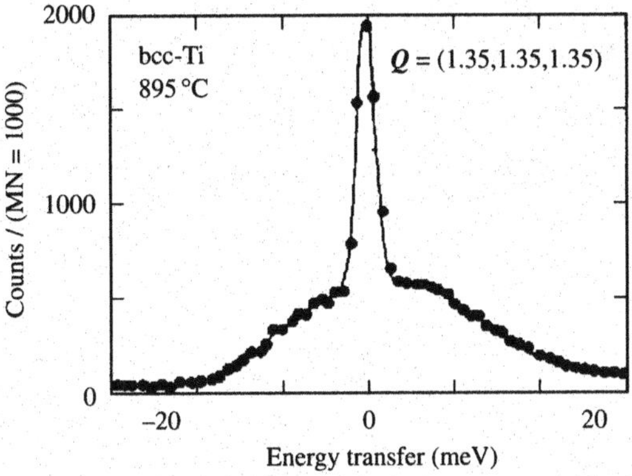

Fig. 9.5 Spectral distribution at $\boldsymbol{q} = (2/3a)\langle 111\rangle$ for bcc Ti (see Fig. 9.4). The coherent quasielastic line centred at $\hbar\omega = 0$ can be considered as a 'degenerate phonon'. The superimposed narrow peak is due to elastic incoherent scattering (its width is the spectrometer resolution). The quasielastic line could be fitted by an overattenuated harmonic oscillator with an attenuation constant of 15 meV and a phonon energy of 11 meV (from Petry 1991, copyright by OPA N.V., with permission from Gordon and Breach Publ.).

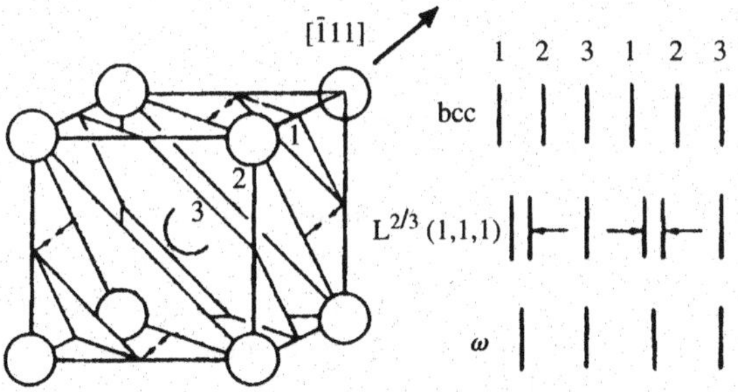

Fig. 9.6 Motion of (111) planes for the longitudinal phonon $(2/3a)\langle 111\rangle$ in Fig. 9.4. Planes 1 and 3 move in opposite sense, plane 2 is at rest. An atomic jump can occur with a jump vector $(a/2)\langle 111\rangle$ parallel to $\boldsymbol{q}$ which leads into the vacancy (hollow square in the cube corner). The insert shows qualitatively the potential which the jumping atom passes, penetrating two neighbouring planes. If plane 1 and 3 would collaps, this would lead to the ω phase (not existent at normal pressure) (from Petry *et al.* 1991b, copyright by the American Physical Society).

to the vacancy sitting in the corner of the cube. The authors point out that other modes also exist which could lead to phonon enhanced diffusion. A whole branch, namely $\langle 2\xi\ \xi\xi\rangle$ along the zone boundary, is expected to come into play.

The relation between phonons and vacancy diffusion has led to attempted calculations of activation energies and diffusion constants for this process (Schober *et al.* 1992). Four parameters are required to obtain the diffusion constants

(Section 4.4.1), namely the entropies of formation and of migration, S_V^f, S_V^m, and the corresponding enthalpies of formation and migration, H_V^f, H_V^m of the vacancies, respectively. The entropies can directly be calculated from phonon energies using Eq. (4.115). A Born–von Karman model helps to get all phonons, knowing the high-symmetry branches from the measurements. The activation energy E_a or the enthalpy of migration $H_V^m \simeq E_a$ is estimated under the assumption that the energy $E(x)$ for an atom or vacancy to move by a distance x between two equilibrium sites is sinusoidal, namely

$$E(x) = (E_a/2)\left[1 - \cos\left(\frac{\pi}{d}x\right)\right] \tag{9.13}$$

where d is the distance between the equilibrium site and the saddle point. Expanding for small x, this yields the relation between the activation energy E_a and the curvature of the potential at $x = 0$, namely $(\partial^2 E/\partial x^2)_{x=0} = E_a\pi^2/2d^2$. Assuming harmonic phonons, the activation energy can be calculated from the vibrational frequency, which follows from the Green's function G of the lattice (Leibfried and Breuer 1978).

For the case of a cubic crystal E_a or H_V^m is then inversely proportional to the Green's function and one gets a simple relation with the density of states $Z(\omega)$, namely

$$G_{\text{cubic}} = M^{-1}\int_0^\infty Z(\omega)\,\mathrm{d}\omega/\omega^2 \tag{9.14}$$

where $Z(\omega)$ can be obtained by extrapolation of the measured phonons. During the calculation of the diffusion constant in this procedure, soft phonons in the vicinity of the $(2/3a)\langle 111\rangle$ modes or along $\langle 2\xi\xi\xi\rangle$ are responsible for the largest contributions in lowering the entropies S_V^m, S_V^f and the enthalpy H_V^m. The 'mode' exactly at $(2a/3)\langle 111\rangle$ cannot be treated as a harmonic phonon at all, which creates an inherent uncertainty in this procedure. The enthalpy H_V^f is not known experimentally in the case of β-Ti. However, this quantity can be calculated from electron theory (Willaime and Massobrio 1991) as was successfully demonstrated in the case of β-Zr. In this case the consistency between the temperature dependence of the diffusion constant with these calculations was satisfactory. The described procedure demonstrates that an important step in the understanding of diffusivity in this class of metals was achieved in terms of the lattice dynamics.

9.3 ALLOYS AND INTERMETALLIC COMPOUNDS

In numerous *metallic alloys* the atoms of the minority compound diffuse several orders of magnitude faster than the host atoms; for a review see, e.g., Herzig (1989). A model system which allows us to study this phenomenon by means of QENS in Co is β-Zr (bcc high-temperature phase) because (i) Co has a comparatively large incoherent neutron scattering cross-section, whereas the host atoms have a

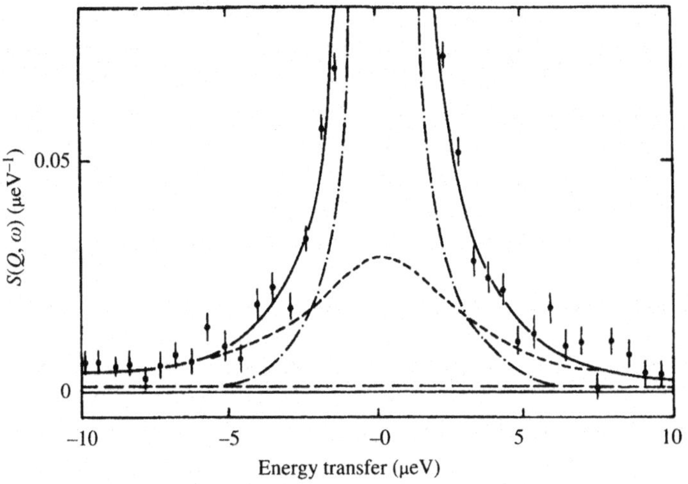

Fig. 9.7 Normalized intensity $S(Q, \omega)$ of neutrons scattered at Zr-2 at.% Co for $Q = 1.5\,\text{Å}^{-1}$. The dashed curves represent the quasielastic intensity, the continuous curve the sum of the elastic and quasielastic intensity with fractions of 0.28 and 0.72, respectively (from Petry *et al.* 1987).

small one, and (ii) Co exhibits a solubility of several at.% in β-Zr. Figure 9.7 shows a set of QENS spectra measured at the IN10 backscattering spectrometer by Petry *et al.* (1987). They consist of an elastic component due to Zr and a quasielastic component due to Co. The Q dependence of the quasielastic widths can be well described by the isotropic Chudley–Elliott model and is compatible with a jump length of 3.1 Å, the distance of nearest neighbours in β-Zr. The resulting diffusion coefficient, amazingly, is about eight times smaller than that determined from tracer experiments on the same alloy, Zr-2 at.% Co (Kidson and Kirkaldy 1969; Kidson and Young 1969). The authors (Petry *et al.* 1987) suspect the existence of an additional, very broad quasielastic component and point to some hints in their QENS spectra which may support this suggestion. Combining the information available from QENS and from tracer experiments, the authors suggest two alternative diffusion models between which no definite decision can be made due to experimental uncertainties:

(i) In a 'vacancy-interstitial' model 72% of the Co atoms dissolve on substitutional sites and diffuse via tightly bound vaccancies. The remaining 28% of Co atoms dissolve in the interstitial lattice and diffuse very fast.
(ii) In a 'two-interstitial model' Co atoms dissolve on two different interstitial sites with different site energies and perform, in other words, a diffusion in the presence of traps. However, no-one has tried to apply a two-state model in order to describe the data quantitatively.

Only for very few *intermetallic compounds* has metal atom diffusion been studied by means of QENS using single crystalline samples. An example is NiSb (Vogl *et al.* 1993) which crystallizes in the NiAs or B8 structure; in this structure the Sb atoms form a hexagonal close packed lattice, and the majority of the Ni atoms occupy the octahedral interstices, but some Ni atoms occupy the 'double tetrahedral interstices' (DTI), sometimes also called 'trigonal bipyramidal interstices'. In this way vacancies are created on the octahedral sublattice which are probably the reason for the fast Ni diffusion which is about 500 times faster than the Sb diffusion.

The QENS measurements were performed using single-crystals of the exact composition $Ni_{53}Sb_{47}$ welded into vacuum-tight thin-walled Nb sample holders. A backscattering spectrometer was used at different sample temperatures close to 1100 °C. A jump model with Ni diffusional jumps only on the octahedral sublattice turned out to be in contradiction to the data.

Therefore the QENS data were (successfully, as it turns out) compared with a complex jump model which comprises four sublattices, i.e. four sites of different local symmetry in the unit cell: two regular sites of octahedral geometry (O_1, O_2) and two of the DTI type (D_1, D_2): a Ni atom leaves its regular lattice site, jumps to a DTI, but then cannot continue jumping in the DTI sublattice but rather uses the next opportunity to jump back into an octahedral site. The eigenvalues of the jump matrix of this model and the corresponding weights were calculated for all

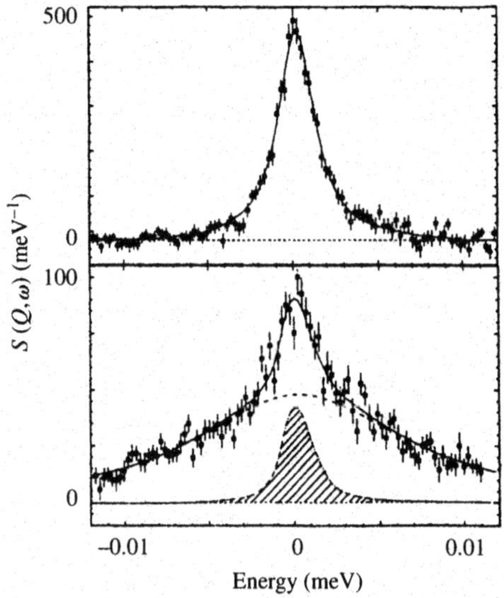

Fig. 9.8 Quasielastic neutron scattering spectra for a $Ni_{72.5}Sb_{27.5}$ single crystal in a particular orientation to the neutron beam; the upper and lower panels correspond to $Q = 0.41\,Å^{-1}$ and $Q = 1.82\,Å^{-1}$, respectively, at $T = 800\,°C$ (from Sepiol *et al.* 1994, with permission of IOP Publishing Ltd).

crystal orientations of the experiment. From the resulting ratio of the jump rates, $\tau_{D\to O}^{-1}/\tau_{O\to D}^{-1} = 6\pm2$, the authors obtain an occupation ratio $c_D/c_O = 0.14\pm0.03$, which corresponds to an occupation of 16% of the DTI sites in $Ni_{53}Sb_{47}$. This result indicates an appreciable vaccancy concentration on the octahedral sites, even for a Ni surplus.

Also Ni diffusion in the high-temperature phase of the intermetallic compound Ni_3Sb has been investigated by the same group (Sepiol *et al.* 1994). The difficulty in this case is that the DO_3 phase of Ni_3Sb is stable only above 530 °C. Therefore the necessary single crystals had to be grown *in situ* in a combined growth-and-measuring furnace (Flottmann and Vogel 1987; Vogl *et al.* 1989). Figure 9.8 shows QENS spectra for two different values of the momentum transfer Q. At small Q, fits with one broadened line are sufficient, but this does not hold true for $Q > 1\,\text{Å}^{-1}$.

The DO_3 structure is a cubic superstructure consisting of four sublattices labelled $\alpha_1, \alpha_2, \beta, \gamma$. In the fully ordered structure the Ni atoms occupy the α_1, α_2, and γ sublattices. This corresponds to a 3×3 jump matrix that takes into account the different jump possibilities between the three different Ni sites in the DO_3 unit cell. From a comparison between theory and experiment it is concluded that jumps between α and γ sublattices are the principal mechanism for the surprisingly fast diffusion of Ni atoms in the DO_3 structure, which corresponds to $Ni_{75}Sb_{25}$. In $Ni_{72.5}Sb_{27.5}$, the actual composition of the sample of Sepiol *et al.* (1994), Ni diffusion in particularly fast, certainly due to the very high vacancy concentration on the Ni sublattices. Recently high concentrations of ordered vacancies have been discovered in Ni_3Sb. Again QENS has been used to determine the elementary diffusion jump (Vogl *et al.* 1996); the authors found that the elementary diffusion jump is unambiguously a jump into a nearest neighbour site.

10

Intercalation compounds

10.1 GRAPHITE INTERCALATION

Graphite has a layer structure of condensed carbon hexagons with covalent bonds in the layers and van der Waals forces in between. The lattice is capable of dissolving metal atoms or molecules which are 'intercalated' between the layers, forming superstructures with sequences of full and empty layers in the hexagonal c-direction. Within the layers, the atoms or molecules can arrange themselves in commensurate or incommensurate structures at low temperatures, and with liquid-like short-range order at higher temperatures.

First, we discuss intercalation compounds with alkali metals which form two classes, namely the stage 1 compounds C_6Li, C_8K, and C_8Rb, and the stage 2 compounds $C_{24}K$, $C_{24}Rb$, and $C_{24}Cs$. The stage 1 group shows commensurate superstructures that melt at higher temperatures where the metal atoms diffuse with jumps between the potential minima in the graphite layers. On the other hand, the stage 2 compounds lead to an incommensurate distribution of the metal atoms over the planes at low temperatures with domain formation. At higher temperatures the arrangement is a nearly free two-dimensional liquid, possibly with a weak modulation caused by the underlaying graphite planes. The corresponding structures are explained in Figs. 10.1 and 10.2. The activation energies for diffusion in the second group are smaller and the mobility is higher than in the stage 1 intercalates. The melting transition is not sharp but extends over more than 100 K. These compounds are of physical interest since they are two-dimensional liquid-like systems, and we refer to the corresponding theoretical work of Coppersmith *et al.* (1982), Zippelius *et al.* (1980), and Fan *et al.* (1990). Also molecular intercalates were studied, for instance graphite with HNO_3, which are very complicated as concerns structure dynamics. Reviews of the whole field are found, for example, in the monograph of Zabel and Solin (1990). In particular, neutron diffraction work is reported by Sinha (1987) and neutron phonon work by Zabel (1990).

The diffusional dynamics in intercalation compounds has been intensively studied by means of NMR (Müller-Warmuth 1995). But neutron scattering has also been applied in a few cases to investigate the diffusion and also the vibrational states in intercalates. In many cases the diffusion constants are high so that the quasielastic spectra from the intercalated atoms or molecules can be resolved. A difficulty occurs if the scattering has a coherent contribution. In this case, the experiments were interpreted with the line narrowing correction as described in

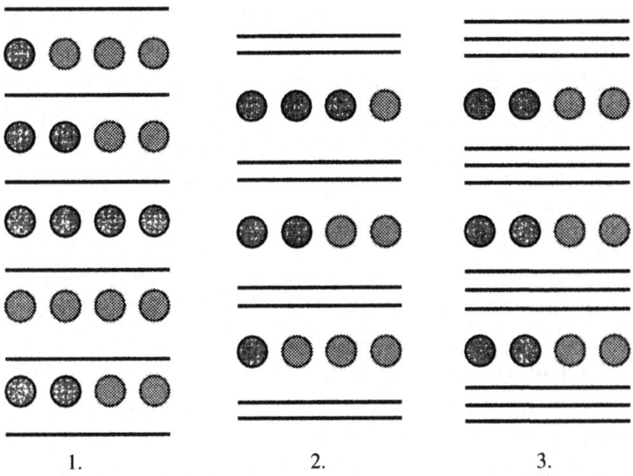

Fig. 10.1 Stacking sequence of carbon layers (solid lines) and alkali layers (circles) for different stages of loading.

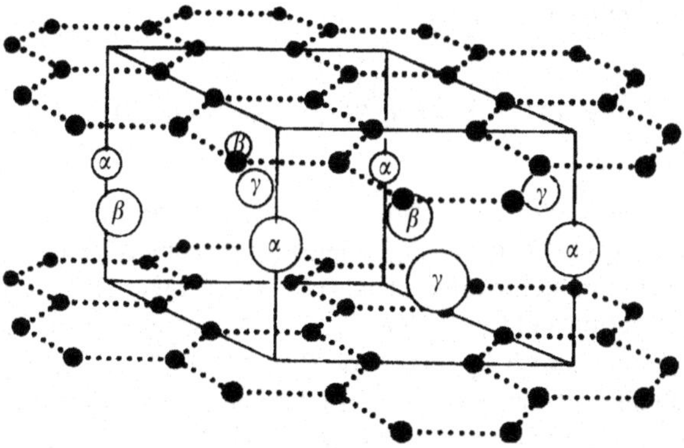

Fig. 10.2 Typical intercalate structure in the graphite plane: C_8K. The sites α, β, γ, and δ are the positions of the K atoms occupied in subsequent layers (from Robinson and Salamon 1982, copyright by the American Physical Society).

Section 6.2. Another problem in the interpretation of the experiments occurs if the diffusing particles are molecules. In addition to the motion of the centre of gravity, simultaneous rotations occur around the molecular axes. The resulting spectrum then has a diffusive component whose width is $\Gamma_d = Q^2 D$ for smaller Q. This is superimposed on a quasielastic line whose width is of the order of $1/\tau_r$ where τ_r is the characteristic time of rotation (see Section 5.1). The intensity of the rotational component increases with increasing Q. Both lines have comparable intensity if $QR \approx 1$ where R is the path radius of the rotating atoms. Consequently, the translational part can be separated if Q is not too large. A reliable separation of both components requires that $DQ^2 < 1/\tau_r$. Otherwise the interpretation of the data is uncertain.

Zabel *et al.* (1989) have carried out quasielastic scattering experiments on several alkali intercalates. C_6Li has an ordered 2d Li lattice which 'melts' at $T_m = 715\,K$ (Rossat-Mignod *et al.* 1982). The intercalate 2d lattice has three equivalent sites where only one is occupied in the ordered state. The occupied sites form a stacking sequence of Li atoms in the c-direction. The measurements were carried out below T_m where the width Γ is in the μeV region, and the IN10 backscattering spectrometer was used. Measurements were performed at 5 K above T_m with the TOF spectrometer IN5 where Γ is in the 100 μeV range. Pyrolytic graphite was used as a sample in the form of disks of about 20 mm diameter with the c-axis perpendicular to the disks, and stacked together to a height of 40–50 mm. Loading occurs from the vapour phase or by immersion in the liquid for the case of Li. ^7Li is a dominating incoherent scatterer which simplifies the evaluation of the quasielastic scattering results. $\boldsymbol{Q}$ was oriented in the basal graphite planes which are randomly oriented around the c-axis. Out-of-plane misorientation is negligible. Experiments with $\boldsymbol{Q}$ perpendicular to these planes reveal no out-of-plane diffusion below 675 K, with $D < 10^{-8}\,cm^2\,s^{-1}$ as an upper limit. The expected width $\Gamma(\boldsymbol{Q})$ for $\boldsymbol{Q}$ in the basal plane is obtained as

$$\Gamma = (4/3\tau)\left[\sin^2(\boldsymbol{Q}\boldsymbol{S}_1/2) + \sin^2(\boldsymbol{Q}\boldsymbol{S}_2/2)\right] \qquad (10.1)$$

where $\boldsymbol{S}_1$ is the jump vector connecting nearest neighbour sites, called α and β in Fig. 10.2, whereas $\boldsymbol{S}_2$ goes to next nearest neighbours (namely δ and β in the figure). Orientational averaging in the planes leads to $\Gamma = (2/\tau)[1 - J_0(Ql)]$ where J_0 is the zero Bessel function and $l = |\boldsymbol{S}_1|$ or $|\boldsymbol{S}_2|$. τ is the mean residence time on a site. The data are consistent with $\boldsymbol{S}_2$ jumps for $T < T_c$ and $\boldsymbol{S}_1$ jumps above T_c. This indicates that all sites of the hexagonal graphite lattice become accessible above the transition. Because of the restricted $\boldsymbol{Q}$ range and the error bars, strictly conclusive results on the jump model were not obtained. From temperature-dependent measurements the activation energy below T_m was $E^* = 1.0 \pm 0.3\,eV$ which agrees roughly with β-NMR data (Freilander *et al.* 1986; Schirmer and Heitjans 1995), but disagrees with the NMR results of Estrade *et al.* (1980). Also for the stage 2 systems $C_{24}Rb$, where one expects an incommensurate liquid behaviour above the transition temperature of 165 K, quasielastic scattering experiments were performed by Zabel *et al.* (1989). The interesting observation from these measurements is that the spectrum has to be described by the superposition of two lines centred at $\hbar\omega = 0$. One of them has the shape and width of the resolution curve of the spectrometer; the second one shows measurable broadening. Its width and intensity increases with increasing temperature. As a function of Q, the coherent structure factor, i.e. the intensity of the two superimposed lines, has a pronounced maximum at $Q = 1.2\,Å^{-1}$ like the structure factor of a liquid, shown in Fig. 10.3. Figure 10.4 presents both intensities as functions of temperature. The sharp line

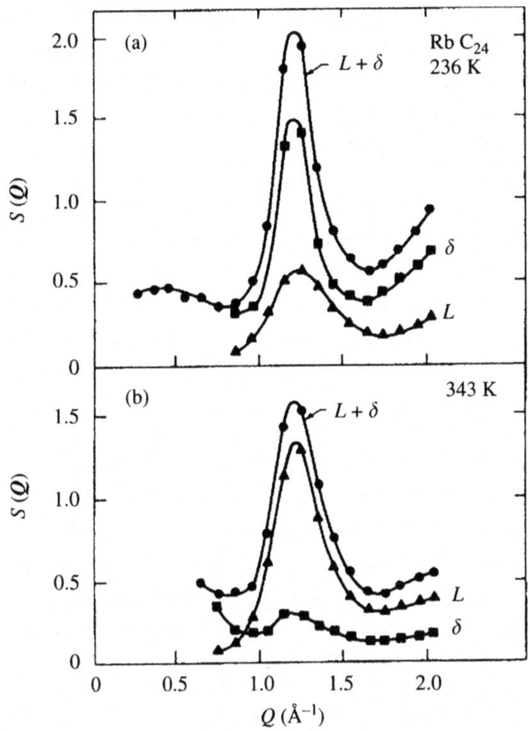

Fig. 10.3 The structure factor for Rb in the intercalate C_{24}Rb: S(Q) is the elastic intensity as a function of scattering vector $\boldsymbol{Q}$, and $S_L(\boldsymbol{Q})$ the quasielastic line intensity. $L + \delta$ denotes the total integrated intensity which is the structure factor $S(\boldsymbol{Q})$ of the liquid-like distribution of Rb atoms in the 2d graphite layer, as also measured by X-ray diffraction (from Zabel *et al.* 1989, copyright by the American Physical Society).

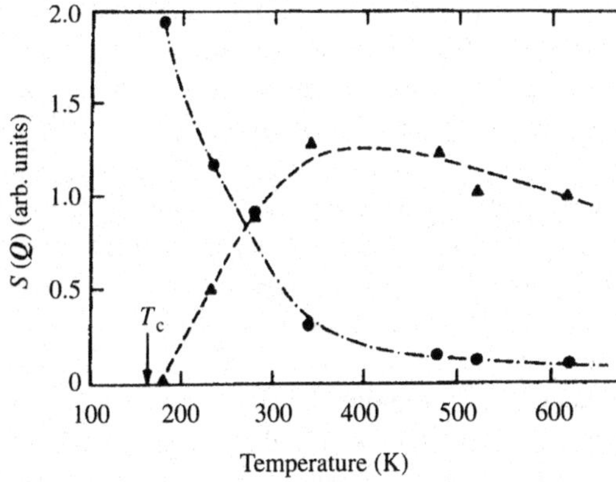

Fig. 10.4 Temperature dependence of the elastic contribution in Fig. 10.3 (solid points) and the integrated quasielastic intensity (triangles) for C_{24}Rb. At a 'critical temperature' T_c the Rb atoms become mobile (from Zabel *et al.* 1989, copyright by the American Physical Society).

as measured at the structure factor peak is attributed to atoms which are relatively immobile. Their intensity contribution decreases with increasing temperature. At the same time, the quasielastic contribution at $1.2\,\text{Å}^{-1}$ becomes measurable at a temperature $T_m \approx 165$ K and then increases steeply. It is noteworthy that an anomaly appears in the $\langle 100 \rangle$ transverse acoustic modes (Zabel *et al.* 1986) near T_m, namely a softening of the mode, and also of c_{44} by more than 30%. The result of these experiments suggests the coexistence of a 2d liquid and a 2d solid like component above T_m where in these measurements 'solid' means a diffusion constant $< 10^{-8}\,\text{cm}^2\,\text{s}^{-1}$ or an atomic rest time $> 10^{-8}$ s. One can speculate that there are regions or domains which are partly commensurate and more or less immobile, surrounded by floating domain walls which are incommensurate with a higher mobility. From the quasielastic width the self-diffusion constant D_s was also determined. The influence of coherence on the quasielastic width was taken into account by multiplying the width by the coherent structure factor, in order to obtain $Q^2 D_s$ (see Eq. (6.34)). Since the structure factor does not strongly change with temperature, this line-narrowing correction (at fixed Q) may lead to reliable results, at least as concerns the temperature dependence of D_s.

Further investigations were carried out on HNO_3 intercalated in graphite. This system has been investigated in great detail with X-ray diffraction by Samuelsen *et al.* (1985). Quasielastic neutron scattering experiments by Batallan *et al.* (1985) on this system reveal the dynamical aspects of the intercalated molecules. Compared to the Rb and Li experiments, in this case the statistical quality of the data is fairly good since incoherent scattering on the proton is very strong. At low temperatures, for $T < 210$ K, the intercalation is commensurate and no measurable quasielastic broadening was observed. At higher temperatures the quoted X-ray experiments show a semicommensurate phase: the molecules are incommensurate with one basis vector of the 2d graphite unit cell ($\boldsymbol{A} = 1.43\,\boldsymbol{a}$), whereas they are commensurate in the other direction $\boldsymbol{B} = 2\boldsymbol{a} + 9\boldsymbol{b}$ where $\boldsymbol{a}$ and $\boldsymbol{b}$ are the graphite basis vectors. This phase has a very complex behaviour; in particular it reveals sliding of the graphite planes, a change of the hexagonal stacking order, the formation of domains, and hysteresis near the transition which occurs in the region of 250 K. This transition leads into a disordered and completely incommensurate phase with a liquid-like X-ray pattern. The quasielastic spectra in this phase were measured with about 0.1 μeV resolution with a backscattering spectrometer and with a TOF spectrometer in the 100 μeV resolution region. The results show clearly a composite spectrum (Fig. 10.5). In the semicommensurate phase there is an elastic line (not broader than the resolution width) superimposed on a relatively broad component whose width is Q independent and increases with temperature as expected for a molecule which rotates rapidly, whereas its centre of gravity is practically fixed (Section 5.1). Above the transition, however, the elastic line begins to broaden and the width follows roughly the prediction of the Chudley–Elliott model with a jump distance of about 2.5 Å. The Q dependence is in favour of jumps over discrete sites, and does not resemble the case of a continuous liquid. Figure 10.5 shows these results. From the measurements one also finds that

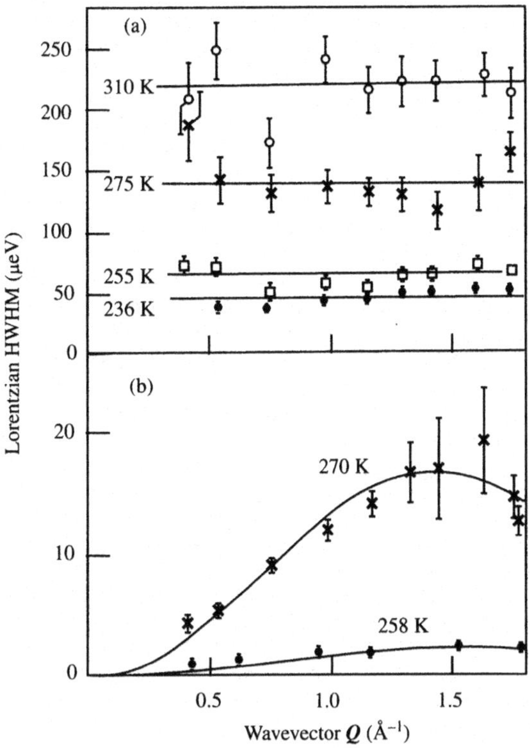

Fig. 10.5 The Q dependence of the quasielastic linewidth for HNO_3 intercalated in graphite. (a) Quasielastic component due to a rotational motion of the H atom in the molecule which is typically Q independent. (b) Component from a diffusive motion of the whole molecule. Obviously, the diffusion starts at a temperature above ≈ 250 K (from Battalan *et al.* 1985, copyright by the American Physical Society).

the energy-integrated quasielastic intensity for fixed Q in the semicommensurate phase decreases with decreasing temperature with a discontinuity at the transition temperature near 250 K. Since the radius of the rotation (i.e. the corresponding Q-dependence of the line) does not change with temperature, this observation implies that a certain fraction of the molecules is at rest whereas the others are rotating. This fraction obviously decreases with increasing temperature. Similarly to the alkali intercalates described above it seems that neutron scattering experiments are consistent with the picture of domains with mobile stripe walls where also the rotation is more or less unlocked. The fraction of the molecules in these stripe regions grows with temperature. The temperature dependence is not Arrhenius-like, but indicates that there is an anomaly near the transition as predicted (Zippelius *et al.* 1980). The X-ray results from Touzain *et al.* (1979) have led to the assumption that the NO_3 groups stand with one NO-axis perpendicular to the graphite basal planes as depicted in Fig. 10.6. Accurate measurements of the EISF,

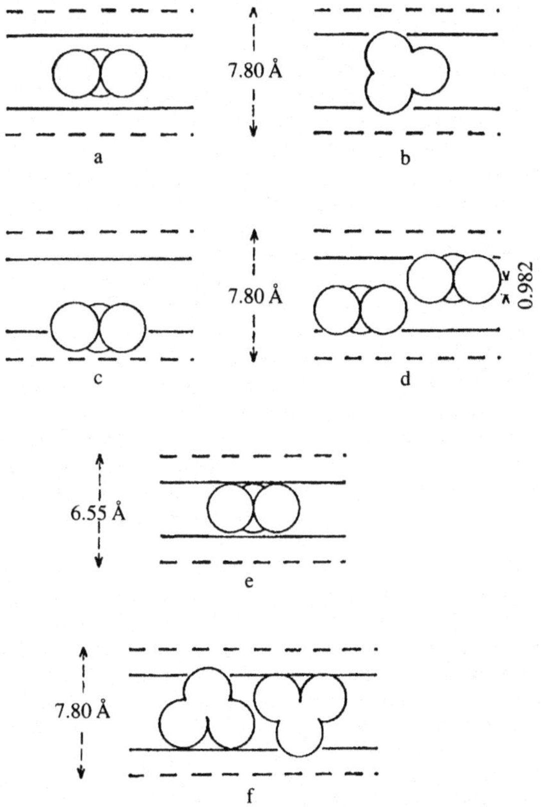

Fig. 10.6 Cross-section through the graphite lattice with intercalated HNO_3 molecules. The picture is derived from X-ray diffraction (from Touzain 1979/80, with permission from Elsevier Science).

i.e. the elastic line intensity, as a function of Q, could confirm this assumption. The dynamics of this system were investigated by NMR experiments by Avogadro and Villa (1979).

Measurements of the molecular mobility by quasielastic neutron scattering by Bindl (1986) on intercalated H_2SO_4 have also revealed an anomaly of the mobile fraction similar to the transition described above in the case of HNO_3. It turned out that the mobility strongly decreases with decreasing water content of the compound. Experiments on stage 1 $K(NH_3)_{4.3}C_{24}$ by Qian *et al.* (1985; 1986) revealed translational diffusion and molecular rotation. The latter has an extraordinarily low activation energy of 5 meV. Consequently, the molecule is nearly a free rotator which may reveal the behaviour of a quantum rotator at lower temperatures.

The experiments in this section show that the evaluation of translational diffusion by neutron scattering exhibits features of the phase transition of a two-dimensional liquid which experiences a weak influence of the graphite substrate and, in certain cases, the behaviour is related to mobile domains.

10.2 IONIC COMPOUNDS WITH CHANNELS OR LAYERS

Binary and ternary molybdenum cluster chalcogenides (Chevrel phases) with the composition Mo_6X_8 and $A_xMo_6X_8$ (X = S, Se, Te; A = main group or transition metal) have been of great interest due to their superconducting properties and due to the large variety of chemical and physical properties which depend strongly upon the nature of the chalcogen anion X, the nature of the ternary metal A, and the stoichiometry of the latter (Schöllhorn 1980). They are able to undergo reversible *topotactic* redox reactions at ambient temperature via electron/ion transfer according to

$$xA^+ + xe^- + \square_n[Mo_6X_8] \Leftrightarrow (A^+)_x\square_{n-x}[Mo_6X_8]^{x-} \qquad (10.2)$$

$\square$ = vacant site.

The electrons are accepted by appropriate levels of the host lattice conduction band, and the A^+ or A^{2+} ions occupy empty sites $\square$ in the lattice channels.

Reaction (10.2) describes the conversion of an electronic conductor to an electronic/ionic mixed conductor by an intercalation process. Since in such a process the host lattice is not changed (except for a small lattice expansion), topotatic chemical reactions exhibit high reaction rates even at room temperature, quite unusual for a solid state reaction, which are connected with high diffusivities. Actually, electrochemical work gave evidence of fast long-range diffusion for, e.g., $Li_3Mo_6S_8$ with $D_{300\,K} = 1.4\cdot10^{-9}\,cm^2\,s^{-1}$ or $Ni_2Mo_6S_8$ with $D_{300\,K} = 1.7\cdot10^{-9}\,cm^2\,s^{-1}$ (Gocke *et al.* 1987). Since Ni exhibits an appreciable incoherent neutron scattering cross-section (see Table 5.1), Ritter *et al.* (1993) studied the Ni^{2+} mobility in the Chevrel phase $Ni_2Mo_6S_8$ by high-resolution QENS and observed a pronounced line broadening even at room temperature. The Q-dependence of the linewidths establishes fast long-range motion of the intercalated Ni^{2+} ions, with a diffusion coefficient of $3\cdot10^{-9}\,cm^2\cdot s^{-1}$ at room temperature.

Applying the isotropic Chudley–Elliott model, a jump length of 2.1 Å and an activation energy of $24\,kJ\,mol^{-1}$ were derived. The observation of fast mobility of transition metal ions like Ni^{2+} at ambient temperature is a highly unusual phenomenon (Schöllhorn 1980), since bivalent ions with rather high charge/radius ratios should exhibit high activation energies for site changes. No other solid host matrix seems to be known so far with these high diffusion coefficients of these ions.

Systems of great practical importance are swelling clays (Poinsignon 1997). They are lamellar minerals made by stacks of two tetrahedral silica sheets with one octahedral layer inserted. The structure and microdynamics of adsorbed water has also been studied by high-resolution QENS measurements (Lechner 1993;

Conrad *et al.* 1984). The proton diffusional motion ($D = 6 \cdot 10^{-6}\,\mathrm{cm^2\,s^{-1}}$) fits the Einstein relation well with the residence time $\tau_0 = 10^{-10}$ s and the jump length 2.8 Å.

Even more applied are neutron scattering studies of hydrating cement pastes (Berliner *et al.* 1998). The progress of the hydration reactions of tricalcium silicate has been followed using QENS at an energy resolution of 60 μeV. The degree of reaction in the hydrating cement paste is inferred from the fraction of water that is chemically bound to the cement reaction products and thus becomes immobile.

11

Solid state proton conductors

Solid state protons conductors nowadays form a very active field of science. Reviews are given, for example, by Chandra (1989), in the book of Kudo and Fueki (1990), in the monograph of Colomban (1992), and by Kreuer (1996). The fundamental scientific interest in this class of materials has increased strongly in recent years mainly because of the prospects of applications in certain types of fuel cells. Solid oxide fuel cells (SOFC) operate at temperatures of 1000 °C or slightly less; according to present technology they use oxygen-conducting yttrium-stabilized zirconia as solid electrolyte; but as is indicated in Fig. 11.1, on decreasing the operating temperature—a main aim of development—proton-conducting oxides can become competitive.

In general, bare H^+ ions (protons) are not found in solids under equilibrium conditions ($\tau > 10^{-11}$ s). In metals, the proton is shielded by a charge cloud formed from the conduction electrons and, due to the undirectional metallic bonding, exhibits a high coordination number, e.g. on a tetrahedral or octahedral site. In

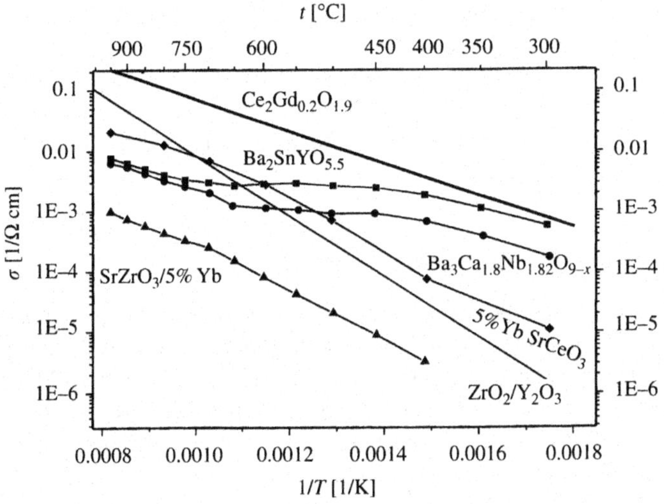

Fig. 11.1 Comparison of the conductivity of high-temperature oxygen conductors (straight lines) and high-temperature proton conductors. Note that the activation energy of oxygen conductors is considerably larger, with the consequence that at lower temperatures proton conductors are competitive.

non-metallic solids, however, H^+ is always covalently bonded to an electronegative atom or ion in the structure, e.g. O–H or N–H. Such a covalent bond is directional and thus, for large oxygen–oxygen distances, the proton's coordination number is just one. But for shorter interatomic distances protons are shared between two electronegative atoms, e.g. O–H---O forming the so-called hydrogen bonds. Oscillations of H from one side to the other side in a hydrogen bond,

$$\text{O–H---O} \quad \leftrightarrow \quad \text{O---H–O}, \tag{11.1}$$

represent an essential and necessary step in the proton conduction mechanism of hydrogen-bonded systems. If this charge shift happens collectively, it can lead to a macroscopic polarization of the sample (ferroelectricity). But for long-range charge transport a second elementary step is necessary, namely the (jump) rotation or reorientation of the (dipolar) proton carrier, e.g. of the O–H group:

$$\cdots\text{O---H–O}\cdots\text{O}\cdots \quad \leftrightarrow \quad \cdots\text{O}\cdots\text{O–H---O}\cdots \tag{11.2}$$

This mechanism is called the *Grotthuss mechanism* and was suggested long ago (von Grotthuss 1806) in order to explain the anomalously high molar ionic conductivity of H^+ in aqueous electrolytes. Because of the reorientation step the host lattice (the 'solvent') also contributes to the proton diffusion. The relevant rates for this mechanism are those of proton transfer Γ_{trans} and reorganization of its environment Γ_{reor}. The slower one, which is different for different systems, would then be rate determining for long-range diffusion. Whether this reorientation has some cooperative or collective character in concentrated systems has not yet been established experimentally (Kreuer *et al.* 1992).

The term 'proton conductor' also includes materials with complex proton transporting groups like NH_4^+, H_3O^+, or OH^-. This is called the *vehicle mechanism* of proton conductivity and involves, of course, also the back-diffusion of the empty 'vehicles' NH_3, H_2O, or O^{2-}.

11.1 ALIOVALENTLY DOPED PEROVSKITES

Yb-doped $SrCeO_3$ can be considered as a prototype system for non-hydrogen-bonded solid oxide proton conductors which contain protons as defects and thus in non-stoichiometric (low) quantities (Iwahara *et al.* 1981, 1988). Macroscopic proton conductivity studies are numerous (Norby 1990); for a microscopic investigation of the proton conductivity/diffusivity mechanism this system is particularly suited for several reasons:

(i) Yb-doped $SrCeO_3$ in moist atmosphere is proton conducting in the temperature range 400–1000 °C with increasing electronic contributions at the more elevated temperatures but without any measurable contribution of oxygen ions to the total conductivity under these conditions (Uchida *et al.* 1989).

This is favourable because without oxygen conductivity a vehicle mechanism of proton conductivity can be excluded *a priori*.

(ii) The proton content of several mol% of Yb-doped $SrCeO_3$, although large in comparison to other proton conducting oxides, is still to be considered as small for a quasielastic neutron scattering experiment. Therefore it is essential that the main constituents Sr, Ce, and O exhibit hardly any incoherent neutron scattering cross-section; however, due to the random distribution of Yb and Ce ions on the Ce sublattice there is significant (elastic) Laue scattering.

Figure 11.2 shows some high-resolution QENS spectra of $SrCe_{0.95}Yb_{0.05}H_{0.02}O_{2.985}$ (Hempelmann *et al.* 1995; Karmonik *et al.* 1995) which consist of an elastic

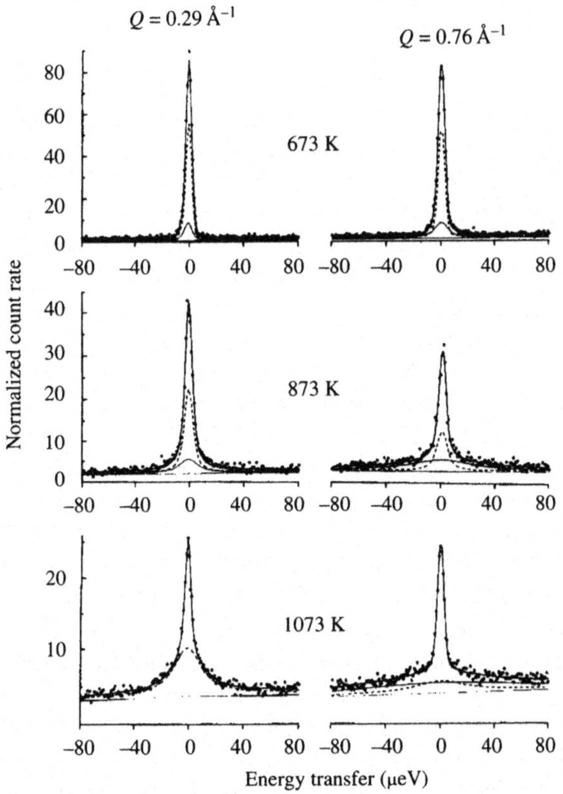

Fig. 11.2 Neutron scattering spectra for three temperatures at the momentum transfers $Q = 0.29\,\text{Å}^{-1}$ and $Q = 0.76\,\text{Å}^{-1}$. The solid lines show the total scattering function resulting from the final fit with the two state model (including a sloped background), the dotted and dashed lines represent the two Lorentzians of the two state model (from Karmonik *et al.* 1995).

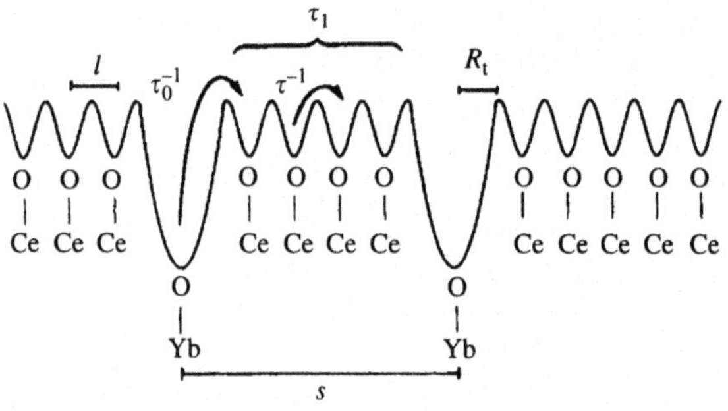

Fig. 11.3 Schematic view of the two state model for proton diffusion in Yb-doped $SrCeO_3$. The temporal parameters are the jump rate τ^{-1} in the free state, the escape rate from the trap τ_0^{-1} and the trapping rate τ_1^{-1} (or the time between two trapping events τ_1). The jump length l in the free state, the distance s between two traps and the radius R_t of the trap represent the spatial parameters of this model.

term (due to the host lattice) and two quasielastic terms of different widths. From the Arrhenius-like temperature dependences of the two quasielastic linewidths two different activation energies were derived, which is clear evidence for the existence of energetically different proton sites. Therefore the authors modelled the energy profile of the proton sites and the proton diffusion mechanism as is schematically shown in Fig. 11.3: Yb^{3+} ions substituting Ce^{4+} carry an effective negative charge and, additionally, create a local elastic distortion of the lattice. Thus they exert an attractive interaction on protons. Proton sites on the oxygen ions adjacent to a Yb^{3+} dopant are therefore energetically lowered and act as *proton traps*. Protons perform a random walk over the '*regular proton sites*', adjacent to Ce^{4+} ions, with a jump rate τ^{-1}; the distance between two regular sites represents the jump length l. This random walk is called the *free state* of the protons, with a corresponding *free diffusion coefficient* $D_{\text{free}} = l^2/6\tau$. In the course of its random walk, after a mean time τ_1 in the free state, a proton hits a trap and stays there for a certain mean time τ_0 (the *trapped state*) before it manages to escape again due to thermal fluctuations. Thus τ_1^{-1} is called the trapping rate and τ_0^{-1} the escape rate. The square of the distance between two traps, denoted by s^2, corresponds to the mean square displacement of the protons in the free state, i.e. within the time τ_1. Hence $l^2/6\tau = s^2/6\tau_1$. Thus proton diffusion in $SrCe_{0.95}Yb_{0.05}H_{0.02}O_{2.985}$ was treated in the framework of the two-state model superimposed on the isotropic Chudley–Elliott model; see Chapter 5. The resulting microscopic spatial parameters are the jump length $l = 4$ Å and $s = 17$ Å; the former is in accordance with the mean O–O distance, the latter with the Yb concentration (reciprocal cubic root of the particle number density). From these primary parameters the effective proton self-diffusion coefficient was calculated according to $D_{\text{eff}} = \tau_1 D_{\text{free}}/(\tau_1 + \tau_0)$; it is compared in Fig. 11.4 with the conductivity diffusion coefficient obtained on the same sample batch by means of impedance spectroscopy.

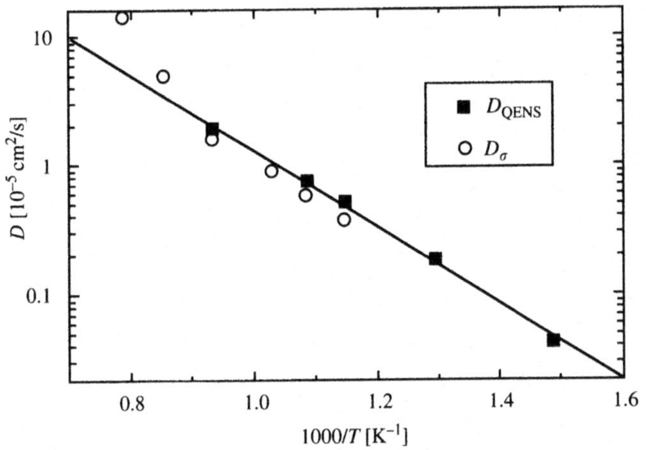

Fig. 11.4 Self-diffusion coefficient D_{QENS} obtained from QENS compared to the conductivity diffusion coefficient D_σ from impedance spectroscopy. Generally, both diffusion coefficients are different, but for the low H concentration involved they should coincide. In the evaluation of the proton conductivity the (temperature dependent) proton transport numbers, determined in a separate experiment, were taken into account. The solid line in the figure represents the Arrhenius fit of the D_{QENS} values (from Matzke et al. 1996, with permission from Elsevier Science).

In the common temperature interval the agreement is striking. Further support of the model and confidence in the data evaluation procedure is given by two observations:

(i) The activation energies of the jump rate and of the trapping rate agree within the limits of error as expected according to the relation $\tau_1^{-1} = 4\pi R_t N_t D_{free}$ if the trap radius, R_t, and the number density of traps, N_t, are considered to be temperature independent.
(ii) From the ratio of the quasielastic scattering intensity (due to H) and the elastic scattering intensity (due to the host lattice comprising incoherent and Laue scattering) a hydrogen concentration of 1.95 mol% is deduced which is in agreement with the value of 2 mol% from the chemical sample preparation.

In a subsequent QENS experiment with a broad energy window (and correspondingly broad energy resolution) (Matzke *et al.* 1996) very fast localized motion of the protons was discovered. Figure 11.5 displays the resulting EISF. The data were fitted with a model corresponding to rotational diffusion on the surface of a sphere. The resulting radius of this motion was 1.1 Å which is in agreement with the OH-distance in the water molecule. The rotational diffusion coefficient, $D_{rot} = 3 \cdot 10^{12}\ s^{-1}$, is three orders of magnitude larger than the transport jump rate. Thus, although this fast motion is not at all rate-determining for the proton

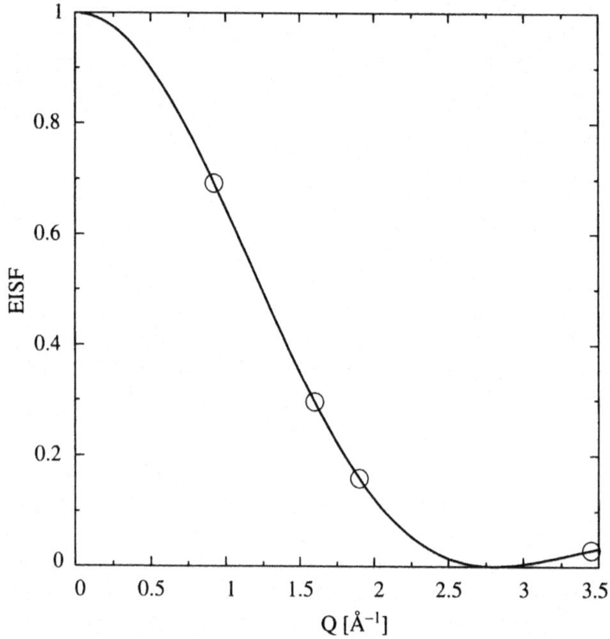

Fig. 11.5 Elastic incoherent structure factor of the fast localized H motion in Yb-doped $SrCeO_3$ at $T = 1023$ K; the solid line is a simultaneous fit with the model of rotational diffusion on the surface of a sphere. The symbols indicate at which Q values data were taken (from Matzhe *et al.* 1996, with permission from Elsevier Science).

diffusion process in $SrCeO_3$, it is nevertheless essential because the proton on the oxygen ion, in spite of the covalent and thus directional OH bonding, is effectively distributed over all directions.

The structure of the traps could be elucidated by means of μSR measurements on $SrZr_{1-x}Sc_xO_{3-x/2}$ (Hempelmann *et al.* 1998). This system has been chosen because it exhibits representive physicochemical properties and favourable nuclear properties: Sc has a large nuclear magnetic moment whereas all the other constituents have essentially none, and Zr does not tend to valence fluctuations like Ce and thus does not give rise to paramagnetic 'impurities' which strongly disturb. It was shown that in $SrZr_{1-x}Sc_xO_{3-x/2}$ the implanted muons form muoxid ions, $(OMu)^-$, which are diamagnetic entities. The temperature dependence of the μ^+ depolarization rates exhibits a trapping peak at 200 K; from the static dipolar linewidth a μ^+-Sc distance of 2.49 Å was calculated. Assuming for the O–Mu distance the well-known O–H distance (Lutz 1995), this means that the μ^+ (serving as a radioactive tracer for H^+) is located on the bisection of oxygen–oxygen connection lines. The resulting structure is schematically displayed in Fig. 11.6.

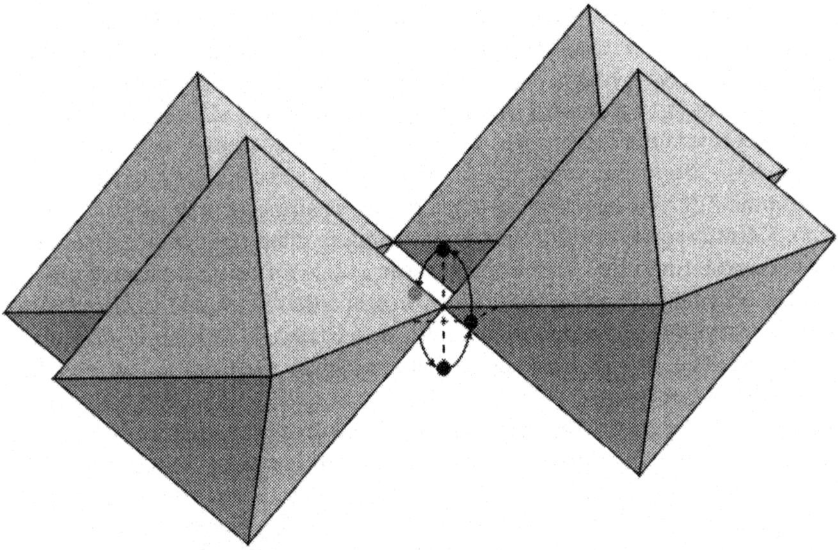

Fig. 11.6 Schematic representation of the proton or muon sites and the reorientational motion in aliovalently doped perovskites.

A second type of proton conducting oxide is formed by off-stoichiometric mixed perovskites like $Ba_3Ca_{1.18}Nb_{1.82}O_{9-\delta}$ (Liang *et al.* 1994; Yang Du and Nowick 1996); in this compound Ca^{2+} ions partly substitute Nb^{5+} ions; for charge neutrality reasons vacancies are formed in the oxygen sublattice, which can be filled by dissociative absorption of water vapour. Karmonik *et al.* (1996) have investigated the proton diffusion in this compound by means of QENS; they again observed trapping effects, and they found, like Matzke *et al.* (1996) for $SrCe_{1-x}Yb_xO_{3-\alpha}$, a rapid rotation of the OH groups. These results were later confirmed by Pionke *et al.* (1997).

11.2 HYDROGEN BONDED SYSTEMS

In spite of their high hydrogen content, systems with hydrogen bonds in stoichiometric quantity, like solid hydroxides (Lutz 1995), often exhibit only very low proton conductivity (which is mostly due not to the compound itself but to defects). This missing dynamics can be understood by site-blocking: each hydrogen bond contains one proton, and transport would only be possible by a cooperative displacement of all protons, which is very improbable. The situation changes drastically if, for instance by a phase transition to a high-temperature high-conductivity ('superprotonic') phase, a situation is created where several energetically equivalent sites are available for each proton.

In 1982 measurements of the temperature dependence of the ionic conductivity of crystalline $CsHSO_4$ revealed that the high-temperature improper ferroelastic phase transition at $T = 414\,K$ is accompanied by a drastic (four orders of magnitude) increase of the conductivity (Baranov *et al.* 1984). Later it was shown that the charge carriers are highly mobile protons. Phases with high protonic mobility were also found in a number of related systems such as $CsHSeO_4$, $RbHSeO_4$, NH_4HSeO_4 and their deuterated counterparts (Baranov *et al.* 1984; Moskvich *et al.* 1984). Some other crystals such as $RbHSO_4$ and NH_4HSO_4 have shown superprotonic properties at elevated temperatures and high pressure (Sinitsyn *et al.* 1988). X-ray and neutron diffraction studies of the structural aspects of the superprotonic phase transitions have revealed some general features characteristic of this class of proton conductors with hydrogen bonds (Belushkin *et al.* 1991a). Here we consider only the QENS result on $CsHSO_4$ (Colomban *et al.* 1987; Belushkin *et al.* 1992); for the other compounds we refer to the summary by Belushkin *et al.* (1994).

In the low-temperature (below 414 K) low-conductivity phase of $CsHSO_4$ the number of protons equals the number of positions for them. Hydrogen bonds link SO_4 tetrahedra so as to form zigzag chains. Hydrogen atoms are localized on the bond, and their mobility is low. The superionic high-temperature phase has a higher symmetry (tetragonal) with four formula units per unit cell as shown in Fig. 11.7. From the structural data it turns out that the number of protons in a unit cell is six times lower than the number of available positions. QENS measurements were performed over a wide range of momentum transfers using different spectrometer resolutions.

The translational diffusion was studied with high resolution (15 μeV). Figure 11.8 shows the Q^2 dependence of the linewidth which was described with the isotropic approximation of the Chudley–Elliott model. At $T = 433\,K$ the resulting diffusion coefficient agrees well with that from a previous NMR experiment (Blinc *et al.* 1984), whereas the resulting jump length is in accordance with crystallographic distances. The activation energy of 100 ± 30 meV, however, disagrees with the value of 300 meV from conductivity measurements.

The reorientation of the HSO_4 groups in the superionic phase of $CsHSO_4$ also involves the protons; this process turned out to be faster by about two orders of magnitude. Belushkin *et al.* (1992) describe their scattering data by a superposition of a single Lorentzian of Q-independent width and an elastic scattering contribution, i.e. they consider the reorientation of the HSO_4^- groups as a back and forth jumping between two sites ('dumb-bell'). These are two crystallographically and thus energetically non-equivalent sites; consequently the residence time on these two sites is different. The EISF is then given by Eq. (5.25).

Figure 11.9 displays the EISF; the experimental data do not seem to extrapolate to EISF $= 1$ for $Q = 0$. This is indicative of multiple scattering. The authors did not correct their data for multiple scattering. The result for the residence times ratio is $\rho = 0.38$ in agreement with the occupancy ratio obtained from diffraction work (Belushkin *et al.* 1991b). These measurements were performed only at one temperature, $T = 433\,K$; therefore activation energies for the jump rates are not given.

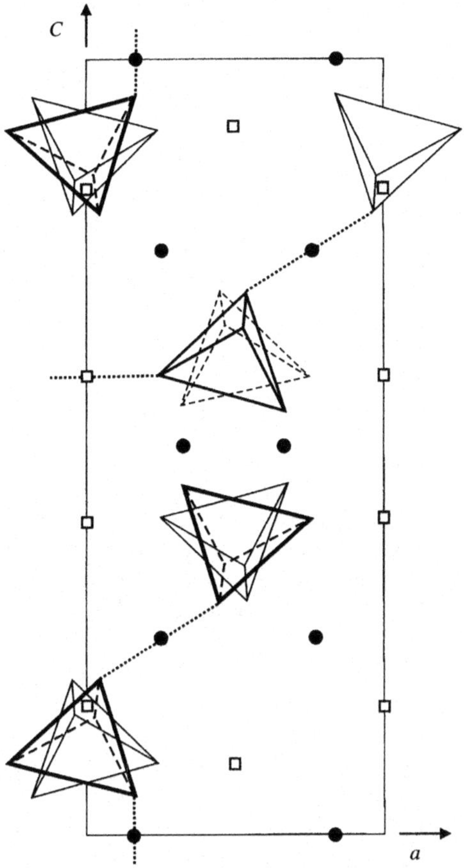

Fig. 11.7 The crystal structure of the $CsHSO_4$ superionic phase as projected on the *a–c* crystal plane according to Belushkin *et al.* (1991). Only oxygen and hydrogen atoms are marked for clarity. One of the possible hydrogen binding schemes is shown. The bold tetrahedron represents one of the four SO_4 group orientations, and the thin tetrahedron shows another possible orientation. The closed circles and open squares represent crystallographically different hydrogen sites (from Belushkin *et al.* 1992, with permission from IOP Publishing Ltd).

In summary, proton transport in $CsHSO_4$ (and analogously in the related compounds mentioned above) comprises a two-stage process. The first step is the proton motion along the hydrogen bond and the formation of a HSO_4^- defect. The second step is the breaking of the longest (weakest) half of the hydrogen bond and reorientation (or high-amplitude libration) of the tetrahedral defects. The proton is then transferred to the new site and a new hydrogen bond is formed if another SO_4 group is found in the appropriate orientation. If not, the reorientation can be considered as not successful, and a fast (correlated) back-reorientation

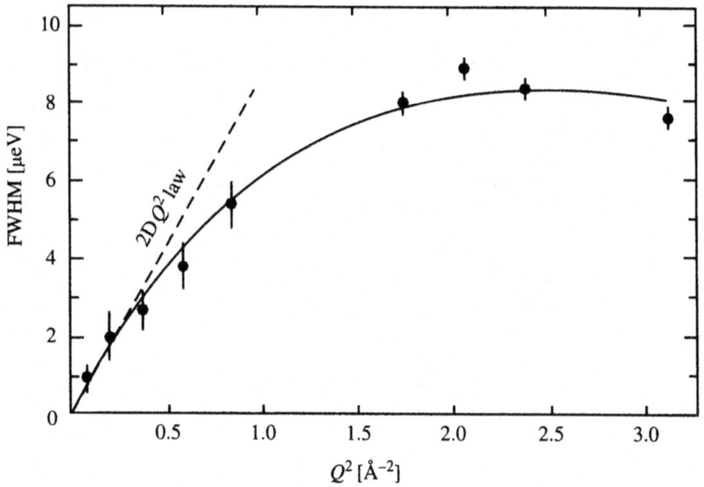

Fig. 11.8 The dependence of the Lorentzian FWHM versus Q^2 at $T = 433$ K for $CsHSO_4$ (from Belushkin *et al.* 1994) •, experimental data; —, best fit to the data using the isotropic Chudley–Elliott model; - - - Q^2 law with a diffusion coefficient obtained from previous NMR results (Blinc *et al.* 1984) (from Belushkin et al. 1992, with permission from IOP Publishing Ltd).

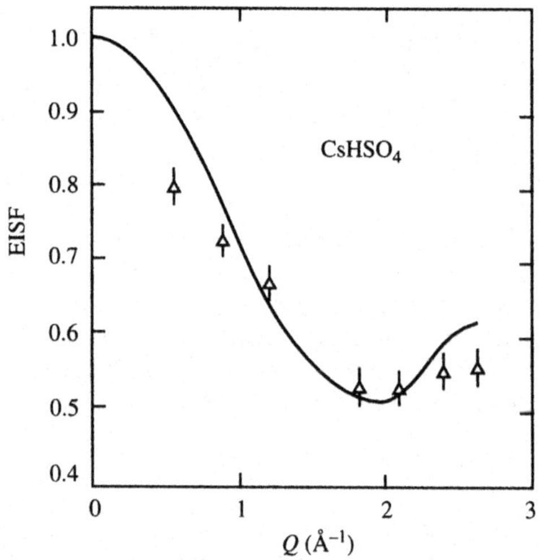

Fig. 11.9 The EISF for jump reorientation of the HSO_4 groups in $CsHSO_4$; △, experimental values at 433 K; —, best fit using Eq. (5.25) (from Belushkin *et al.* 1992, with permission from IOP Publishing Ltd).

is anticipated by the authors; discrepancies in the activation energy between the QENS and the conductivity results are attributed to these back-jumps.

Another well-investigated hydrogen-bonded proton conducting system is cesium hydroxide monohydrate, $CsOH \cdot H_2O$ (or CsH_3O_2); detailed QENS studies of this system have been performed by Lechner and coworkers (Lechner 1993; Lechner *et al*; 1991, 1993). This system exhibits several solid phases with layered structures, where Cs^+ ion layers are alternating with layers which are formed by two-dimensional networks of oxygen atoms connected by hydrogen bridges. Whilst below 229 K $CsOH \cdot H_2O$ presents a monoclinic structure with chemically inequivalent hydrogens, above this temperature it has been shown to crystallize in three different hexagonal or pseudohexagonal structures (Jacobs *et al.* 1982; Marx *et al* 1990). Above 340 K the space group is P6/mmm with Cs atoms at the origin and thus in the basal plane and O atoms at $\left(\frac{1}{3}, \frac{2}{3}, \frac{1}{2}\right)$ and $\left(\frac{2}{3}, \frac{1}{3}, \frac{1}{2}\right)$, i.e. in the plane $z = \frac{1}{2}$. In the high-temperature phases, there are no distinct OH^- and H_2O groups in the layers, i.e. all the H atoms appear to be equivalent, as displayed in Fig. 11.10. A Grotthuss-type mechanism for H diffusion is expected such that any particular proton will—after some residence time τ—change over from one minimum of the H bond double-well potential to the other; the second step would then be a proton exchange between neighbouring H bridges due to a reorientation of a H_2O group.

The authors observe a fast process with a jump rate of nearly 10^{11} s^{-1} at 402 K in the framework of a two-site jump model, with a jump distance of 1.03 Å (Lechner *et al.* 1991). It was difficult to allocate this to one of the two diffusion steps, and therefore a careful comparison of the experimental EISF with different models

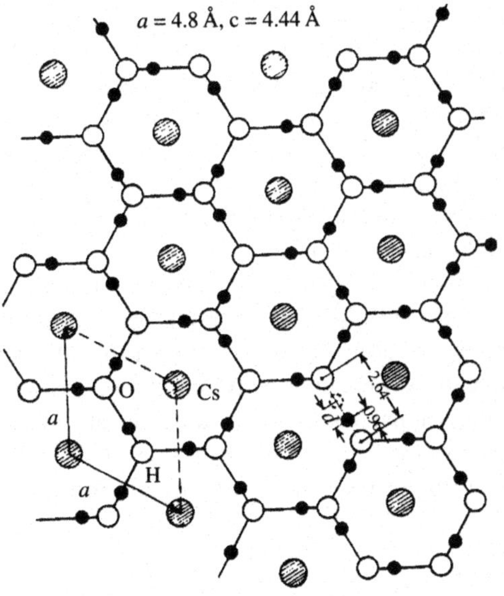

Fig. 11.10 Simplified projection of the hexagonal structure (T > 340 K) of $CsOH \cdot H_2O$ on to the *a–b* plane (from Stahn *et al.* 1983, with permission from IOP Publishing Ltd).

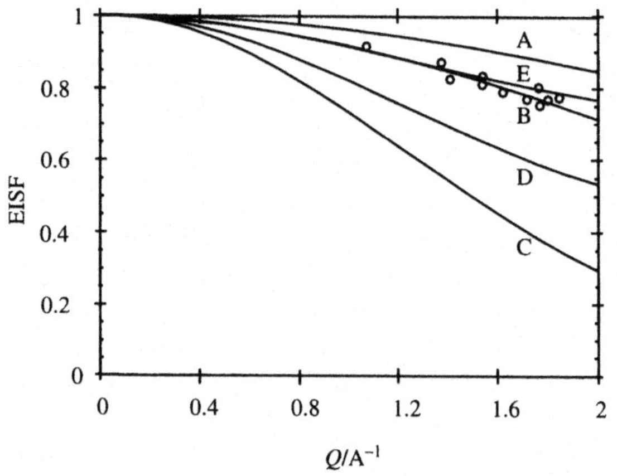

Fig. 11.11 EISF of the localized proton motion in $CsOH \cdot H_2O$; experimental data are compared to 5 different models: A: H-bond double-well jump model, linear version; B: H-bond double-well jump model, non-linear version; C: uniaxial 120°-reorientations of H_2O and OH^- groups; D: uniaxial 120°-reorientations of H_2O molecules alone; E: uniaxial 120 °-reorientations of OH^- groups alone (from Lechner *et al.* 1993, with permission from Elsevier Science).

was performed (Lechner *et al.* 1993); this comparison is shown in Fig. 11.11. The data are compatible both with the non-linear version of the H-bond double-well jump model (curve B) and with uniaxial 120°-reorientations of OH-groups (curve E); in the former case the proton jumps within the hydrogen bond are fast, i.e. the reorientation is rate-determining for diffusion, whereas in the latter case the reorientations of the OH^- groups are fast and jumps between the two minima of the H-bond double-well potential are slow and rate-determining. Assuming that, with respect to the respective slow motion, the proton is effectively located in the centre of the sites visited in the framework of the respective fast motion, the authors decide between the two models from the Q^2 dependence of the HWHM of the Lorentzian observed at a high-resolution backscattering spectrometer. These linewidths are displayed in Fig. 11.12.

According to the authors the experimental data are more in favour of model B, i.e. jumps within the hydrogen bonds are the slow process; this is mainly concluded from tentative multiple scattering corrections which shift the data towards curve B. All these data evaluations were done by means of models for three-dimensional diffusion; the two-dimensional features of the diffusion are apparently not obvious from powder data. This emphasizes the importance of QENS measurements on single crystals, which should yield direct and straightforward results for layered structures.

11.3 SYSTEMS WITH PROTONIC VEHICLES

Protonic charge transport by means of carrier molecules or ions occurs mainly in systems with channels in the structure like hydrogen tungsten bronzes, H_xWO_3,

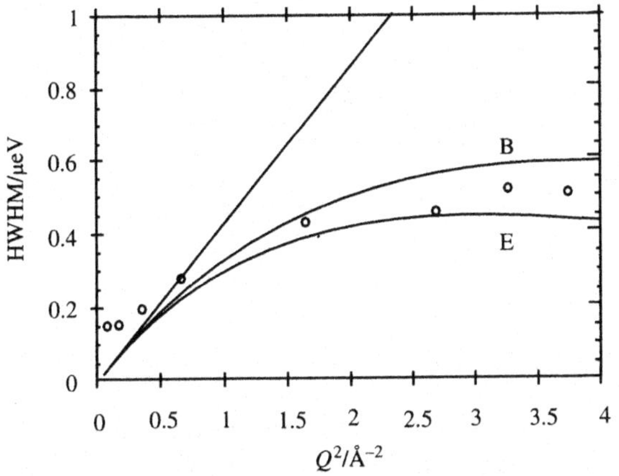

Fig. 11.12 Quasielastic linewidths due to long-range translational diffusion, shown as a function of Q^2: Open circles: $CsOH \cdot H_2O$ data at 402 K. Straight line: $D_s Q^2$-behaviour of the linewidth in the low Q limit, calculated using the D_s-value from PFG-NMR measurements at 402 K (from Lechner *et al.* 1993, with permission from Elsevier Science).

and hydrogen molybdenum bronzes (Slade *et al.* 1989) or in layered structures like clays or β alumina containing either NH_4^+ or more or less hydrated H_3O^+ cations. Proton conduction in these systems is often restricted to low dimensionality and some systems actually show features typical of low dimensional diffusion. The vehicles do not belong to the original structure but are introduced a posteriori by means of ion exchange reactions, which are sometimes also called topochemical reactions because the topology of the host is not changed during the insertion. Zeolites also belong to this class of materials; under certain conditions they can exhibit proton conductivity.

In solid state inorganic acids (hetero- and isopolyacids) the vehicles occur in stoichiometric quantity and belong to the structure. A well-known example is $HUO_2PO_4 \cdot 4H_2O$ (hydrated uranyl phosphoric acid: HUP) with a H^+ conductivity of $5 \cdot 10^{-3}\ \Omega^{-1}\ cm^{-1}$ at room temperature.

Other types of systems with protonic vehicles are organic and inorganic polymers. Nafion®, the proton exchange membrane of room-temperature fuel cells, contains acid $\text{-}SO_3^- H^+$ groups and exhibits protonic conduction only if it is swollen with water. Other organic examples are polyamides, polysulphinimides $(SONH_2)_n$, and mylar. Analogous inorganic systems are gels which in some sense can be considered as the non-crystalline precursors of the above-mentioned hetero- and isopolyacids. A QENS study exists for $HV_2O_5 \cdot 1.6H_2O$ gels (Poinsignon 1989). Protonic transport in these polymer systems is brought about by protonic vehicles (H_3O^+) in the liquid; the mechanism is essentially equal to that in strong aqueous acids.

12

Solid ionic conductors

Solid ionic conductors and mixed conductors (with appreciable fractions of both ionic and electronic conductivity) form the materials basis of solid state ionics; this is a new and interdisciplinary science comprising physics, chemistry, and materials science of ion transport phenomena in solids, and properties resulting from ion mobility. Activity in this field is growing rapidly, to a certain extent concerning fundamental science, but mainly with respect to the many applications in batteries, chemical sensors, fuel cells, electrochemical displays, and other electrochemical and photoelectrochemical devices.

The most fundamental property of solid ionic conductors is of course their ionic conductivity, and this is—via the Nernst–Einstein equation—connected to their diffusivity. Therefore appropriate QENS studies can considerably contribute to an understanding of the atomistic mechanism of ionic transport in solids; however, due to the limited energy resolution of neutron spectrometers only ions with exceptionally high ionic mobilities are suited for QENS investigations. This implies that the ions possess rather small ionic radii and carry only a single charge (an exception is O^{2-}) because in general the large Coulomb interaction of double or triply charged ions inhibits mobility. For these reasons only H^+, Li^+, Na^+, Cu^+, and Ag^+ (i.e. group IA and IB in the periodic table) are cation candidates for fast-ion conductivity and only O^{2-}, OH^-, H^-, F^-, and Cl^- are anion candidates. The second requirement for QENS investigations is favourable neutron scattering and absorption cross-sections, of course. In Table 12.1 for the ions in question (with their isotops) the ionic radii as well as σ_c, σ_i, and σ_a are listed.

Incoherent QENS studies have been performed only on $^7Li^+$, Na^+, H^-, and Cl^- ions conductors. For these investigations $\boldsymbol{Q}$ regimes were selected where the coherent intensity was negligibly small. Purely coherent QENS concerns F^- and O^{2-} conductors, whereas for Ag^+ a mixture of coherent and incoherent QENS prohibits quantitative data evaluation. Last but not least the structure of the solid ion conductor has to fulfil the requirement that the mobile ion only partially occupies the lattice sites on its sublattice. We distinguish three cases:

(i) The large fraction of vacancies is a structural pecularity. An example is α-AgI where the I-ions from a bcc sublattice and the Ag^+ ions occupy the tetrahedral 'interstices'. There are six energetically equivalent sites for each Ag^+ ion, i.e. ample space for diffusion via a vacancy mechanism. Similar arguments hold for oxygen diffusion in Bi_2O_3.

Table 12.1 Ionic radii, coherent and incoherent neutron scattering cross-sections, and neutron absorption cross-sections (in barns) for possible fast-conducting ions and appropriate isotopes (without H^+, D^+).

ion	r(Å)	σ_c	σ_i	σ_a
Li^+	0.60	0.45	0.83	70.5
$^6Li^+$	0.60			94.0
$^7Li^+$	0.60	0.62	0.68	0.045
Na^+	0.95	1.66	1.62	0.530
Cu^+	0.96	7.49	0.52	3.78
Ag^+	1.26	4.41	0.58	63.3
H^-	2.08	1.76	79.90	0.333
O^{2-}	1.40	4.23	0.00	0.000
OH^-	0.96		79.9	0.333
F^-	1.36	4.02	0.00	0.0096
Cl^-	1.81	11.53	5.2	33.5

(ii) The large fraction of vacancies on the regular sites is due to 'excitations' of the mobile ions to anti-structure sites at high temperatures, close to the melting point. This kind of Frenkel disorder occurs in fluorite fast-ion conductors like $SrCl_2$ or CaF_2 and anti-fluorite fast-ion conductors like Li_2S at temperatures within a few 100 K of the melting point and involves a kind of diffuse phase transition.

(iii) Appreciable vacancy concentrations can be artificially created by aliovalent doping: when, for example, in zirconia the four-valent Zr^{4+} ions are partially substituted by three-valent Y^{3+}, the charge deficit is compensated by vacancies on the oxygen sublattice.

Layer and tunnel-structured compounds exhibit the characteristic features of intercalation compounds and are treated in Chapter 10.

A large amount of work has been performed on ion dynamics in structurally disordered materials. In particular we mention the theoretical work by Knödler *et al.* (1996) and by Bunde *et al.* (1998). We are, however, not aware of QENS work concerning diffusion in those systems.

12.1 SILVER ION CONDUCTORS

Silver iodide has a phase transition at 149 °C, and the high-temperature form, α-AgI, has been known for a long time as a 'superionic' conductor with an ionic conductivity of 200 $\Omega^{-1}\,m^{-1}$ at 200 °C. This high value is a direct consequence of the crystal structure. The I-sublattice is body centred cubic, as shown in Fig. 12.1. There are three inequivalent sets of sites for Ag^+ in the interstices of the anion lattice, denoted by 6b, 12d, and 24h. The octahedral 6b sites are located at the centre of each face and edge of the cubic unit cell. The tetrahedral 12d sites are located on the faces (four on each face) and correspond to the sites which

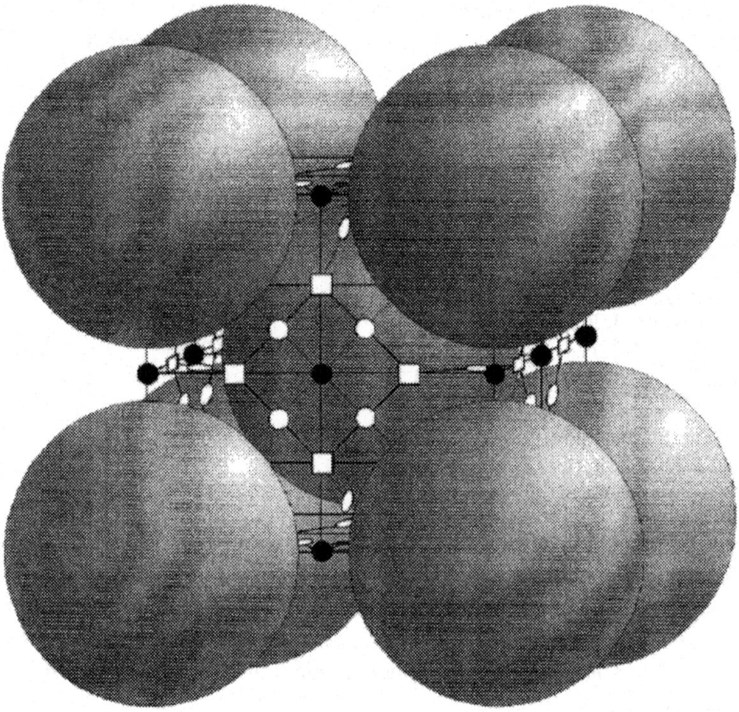

Fig. 12.1 The stucture of 'superionic' α-AgI.

hydrogen atoms occupy in bcc metals. The trigonal bipyramidal 24h sites are located at the centre of a triangle of anions but can more properly be considered as being coordinated with two anion triangles. Thus there are 42 possible sites for Ag^+ per unit cell as shown in Fig. 12.1, whereas there are only two Ag^+ ions, thus leaving 40 sites unoccupied. Although the site energies are different and therefore also the occupancies differ, this structural disorder (to distinguish it from point defects) is sometimes even called a 'half-fused' state with a molten cation sublattice. The jump lengths to the adjacent sites are very short, and silver ion transport cannot properly be interpreted in terms of hopping between distinct lattice sites. Whereas the transport mechanism in AgI is considered as a blend of hopping and liquid-like characteristics (Catlow 1990), it is nearly completely liquid-like in Ag_2S and Ag_2Se. Their high-temperature phase also exhibits a bcc anion sublattice; well-defined maxima of the Ag^+ probability density have so far not been detected. Instead, the silver ions appear to be rather homogeneously distributed over a three-dimensional system of intersecting channels.

A number of QENS experiments on the Ag^+ diffusivity in α-AgI, α-Ag_2Se, and $RbAg_4I_5$ have been performed by Funke and coworkers (Funke 1987; 1989; 1993; 1996). Funke takes into account that many-body interactions play an important role in producing slow relaxation processes in defective solids; he has developed

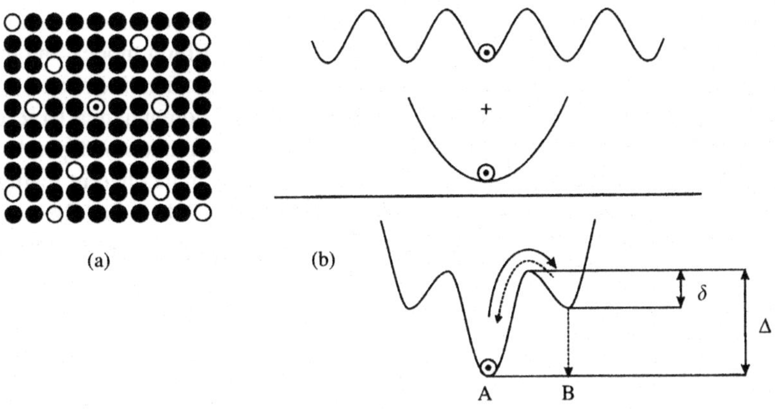

Fig. 12.2 Jump relaxation model of Funke (a) Ions (o) on a sublattice, (b) the single particle potential (from Funke 1991).

a particularly successful phenomenological approach relying on Debye–Hückel interactions between moving defects. His jump-relaxation model yields dispersive transport via backward correlations among successive particle hops and accounts for a remarkable set of experimental data. The *leitmotif* of Funke's jump relaxation model is illustrated in Fig. 12.2. The mobile ions are represented by open circles. Due to the repulsive Coulomb interaction between them, they tend to stay at some distance from each other. Each of them feels a 'Coulomb-cage potential' and is 'expected' by it neighbours to be at the position of its cage-effect minimum. In Fig. 12.2b, the neighbourhood is relaxed with respect to the 'central' ion. Superposition of the periodic lattice potential yields the single-particle potential actually experienced by the ion. From the shape of the potential it can immediately be seen that this interaction results in preference for backward jumps. Suppose the ion at site A is thermally excited and hops to an adjacent site B, surmounting an energy barrier Δ as in Fig. 12.2b. After this initial forward jump two competing relaxation processes can occur:

(i) The ion may hop back to A, passing an energy barrier smaller than Δ. In this case a correlated forward–backward hopping sequence has been performed.
(ii) The surrounding ion cloud ('the neighbourhood') relaxes with respect to the newly occupied site B. As a result, a new absolute potential minimum has been formed at site B and the initial forward jump has eventually proved successful.

The shifting of the cage-effect potential has the consequence that the backward barrier height, $\delta_{B\to A}(t)$, increases as the ion stays at B. Therefore, the 'time constant' of the back-hop process increases as time progresses, and the resulting spectra hence give the impression that there were a 'distribution of relaxation times'. Funke introduces a time-dependent correlation factor $W(t)$, which decays from unity at $t = 0$ to a constant value $W(\infty)$, displaying a t^{-p} power-law behaviour at intermediate times. $W(t)$ is introduced into the Chudley–Elliott model, and the resulting incoherent scattering function no longer has a Lorentzian line shape, but looks like a narrow Lorentzian at small Q and like a broad Lorentzian at large Q with a Q-independent width. With parameters derived from frequency-dependent conductivity measurements Funke and coworkers calculate an incoherent scattering function, which seems to be compatible with their QENS data. For $RbAg_4I_5$ this is shown in Fig. 12.3. A quantitative statement about the agreement between

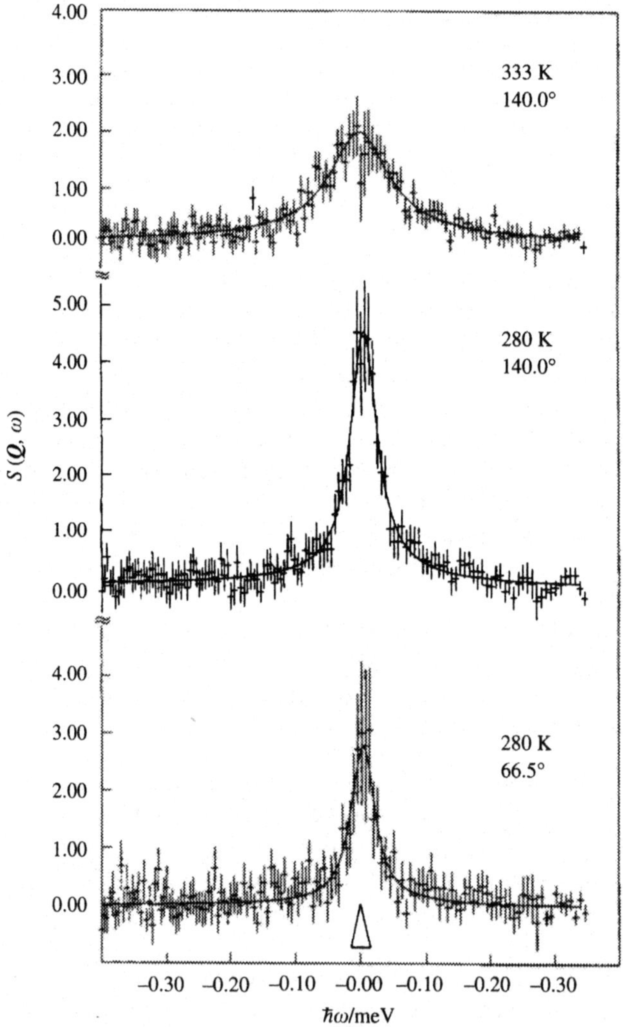

Fig. 12.3 QENS data of $RbAg_4I_5$ at certain temperatures and scattering angles (from Funke 1993, with permission from Elsevier Science).

theory and experiment is, however, difficult, because:

(i) Funke *et al.* do not evaluate their data in such a way that they give the Q and T dependence of some fitting parameters and compare these in detail with their model calculation; and
(ii) Ag is a mixed scatterer with a considerably larger coherent than incoherent neutron scattering cross-section; the superposition of coherent and incoherent scattering in the experimental spectra is only mentioned, but does not seem to be taken into account in the data evaluation. For a 'molten' liquid-like Ag^+ sublattice (sometimes also called a lattice-liquid) due to the pronounced Ag^+–Ag^+ repulsion one would expect Ag^+ short-range order and therefore a liquid-like coherent structure factor.

12.2 SODIUM ION CONDUCTORS

An example of QENS applied to sodium diffusion in a solid sodium conductor is the work by Lucazeau *et al.* (1987) on the Na^+ motions in β-Al_2O_3. β-alumina is a very important solid electrolyte not only because it is practically useful in Na/S batteries, but also because it is a typical two-dimensional ionic conductor. The ideal composition of β-Al_2O_3 is $Na_2O{\cdot}11\ Al_2O_3(= Na\ Al_{11}\ O_{17})$, which is however, not obtainable. The compounds take a non-stoichiometric composition, and Lucazeau *et al.* have investigated $Na_{1+x}\ Al_{11}\ O_{17+x/2}$ with $x = 0.25$, in the powder state. Figure 12.4 shows the crystal structure of β-alumina and the sodium sites in the conduction plane. The distribution of the cations in the conducting planes has been extensively studied by X-ray and neutron diffraction (Collin *et al.* 1980; Roth *et al.* 1976). Three kinds of sites can be occupied, called BR (Beevers–Ross), aBR (anti-Beevers–Ross), and mO (mid-oxygen) sites, as shown in Fig. 12.4. In stoichiometric β-alumina, due to the excess Na^+, double occupancy of the planes occurs with Na^+ also on mO sites. Other authors consider the situation as a 'half-fused' state of the Na^+ sublattice, analogously to α-AgI. In their QENS study Lucazeau *et al.* observe a large fraction of immobile Na ions and attribute this to slightly occupied planes. In addition, they find a clear indication of a fast localized Na^+ motion and compare their QENS spectra to seven different possible localized jump (rotation) models. These models give similar results, i.e. with the available counting statistics it is hard to decide between them. The authors conclude that below 300 °C the local motions of pairs of Na^+ ions jumping between mO–mO and BR–aBR positions give rise to the dominant feature in the quasielastic region. Above 300 °C a narrow feature in the quasielastic region appears which is attributed to a long-range diffusion process. In summary, as in AgI or $RbAg_4I_5$ a fast localized ion motion is experimentally observed, but in the case of β-alumina this localized motion is explained as being due to structural pecularities. Like for the Ag^+ ion conductors coherent QENS is disregarded, but for Na^+ coherent scattering is much less than for Ag^+.

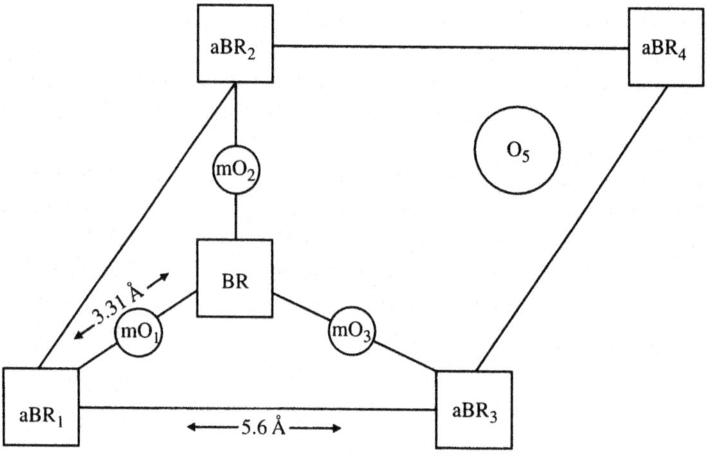

Fig. 12.4 Site of the conduction plane available for Na^+ ions in β-alumina. BR: Beever and Ross site; aBR: anti-Beever and Ross site; mO: middle oxygen site (at mid-distance between two O_5 (oxygen) atoms, these oxygen atoms ensure the spinel block bondings) (from Lucazeau *et al.* 1987, with permission from Elsevier Science).

One possibility of avoiding the coherent part of neutron scattering of mixed scatterers is to choose a Q regime where the coherent scattering is negligibly small. This was done by Hempelmann *et al.* (1994) in their QENS study of Na^+ diffusion in sodium silicate glass, $Na_2O{\cdot}2\,SiO_2$. They used the long neutron wavelength of $\lambda = 19.8$ Å to analyse the scattered neutrons and thus limited the Q range of their experiment to $Q < 0.65$ Å^{-1}. Since the Si and O atoms in this glass do not diffuse on the time scale of QENS, all Si–Si, O–O, and Si–O scattering contributions are elastic. The Si–Na and O–Na scattering terms are quasielastic, since they involve correlations to a mobile ion. Integrated over ω each of these coherent quasielastic terms yields a partial structure factor. For $Q < 0.65$ Å^{-1} the static X-ray structure factor $S(Q)$ (Weigelt 1991) and thus also the partial neutron structure factors do not deviate from their value at $Q = $ O, i.e. from their value in the thermodynamic limit. S(O) is determined by fluctuations in number density and, for an ionic solid, also by fluctuations in charge density. For a solid, density fluctuations or, what is equivalent because of the fluctuation–dissipation theorem, compressibility are vanishingly small. For an ionic compound, the requirement of microscopic charge neutrality in combination with the long-range Coulomb potential severely inhibits charge fluctuations. For these two reasons all coherent scattering terms can be neglected in the whole Q range of that experiment. Simultaneously, even multiple coherent scattering is excluded: the most probable double scattering process is to scatter twice elastically with a Q value corresponding to the maximum of the structure factor, which could heavily contaminate the forward scattering intensity. However, with $\lambda = 19.8$ Å, scattering processes involving values of $Q \geq 0.65$ Å^{-1} are impossible. The Na self-diffusion coefficient resulting from the quasielastic linewidths displayed in Fig. 12.5 is in full agreement with ^{23}Na

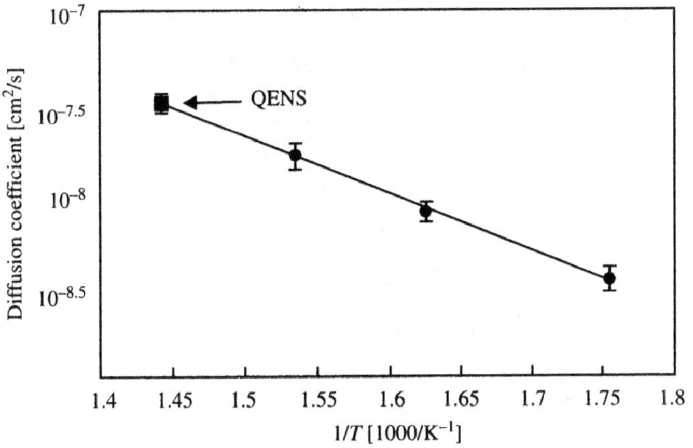

Fig. 12.5 Na self-diffusion coefficient from QENS compared to the Na tracer diffusion coefficient from a ^{23}Na tracer measurement, in an Arrhenius representation (from Hempelmann *et al.* 1994, copyright by Springer-Verlag).

tracer measurements performed on the same sample batch. The linewidth data in Fig. 12.5 seem to show a slight upward curvature, and in a log-log plot a straight line with a slope of 2.46 results, i.e. it seems that $\Gamma \propto Q^{2.46}$. Growth of the linewidth faster than Q^2 would indicate some kind of anomalous diffusion, e.g. diffusion on a network of diffusive paths with fractal character or diffusion-limited percolation or perhaps jump relaxation extending over more than a single forward–backward jump sequence.

12.3 LITHIUM ION CONDUCTORS

Since Li^+ is a very small ion, see Table 12.1, it exhibits high ionic conductivity in many compounds. A particularly detailed investigation has been performed on lithium sulfide, Li_2S (Altorfer *et al.* 1994a), using single crystalline samples and polarization analysis. Li_2S crystallizes in the anti-fluorite structure (space group Fm3m) which, like the fluorite structure for anion conductors, provides one of the simplest systems in which fast ion conduction can be observed. The crystal structure is shown in Fig. 12.6. The cation sublattice consists of the superposition of two fcc sublattices, hence two energetically equivalent but crystallographically different Li sites. The structure can also be considered as a simple cubic array of Li^+ ions (with half of the fcc lattice constant) where every second cube is centred by an S^{2-} ion, whereas the empty cube centres are possible interstitial or anti-structure Li^+sites. Above a diffuse phase transition at about 800 K these interstitial or anti-structure sites are appreciably occupied (at $T = 1300$ K approximately 15% of the Li^+ ions are on these sites) and Li_2S exhibits fast (or super-) ionic conductivity.

For their QENS experiment Altorfer *et al.* used ^{7}Li isotope pure samples; by means of polarization analysis they proved that the quasielastic intensity was mainly ($\geq$ 90%) incoherent and that the elastic diffuse intensity showed hardly

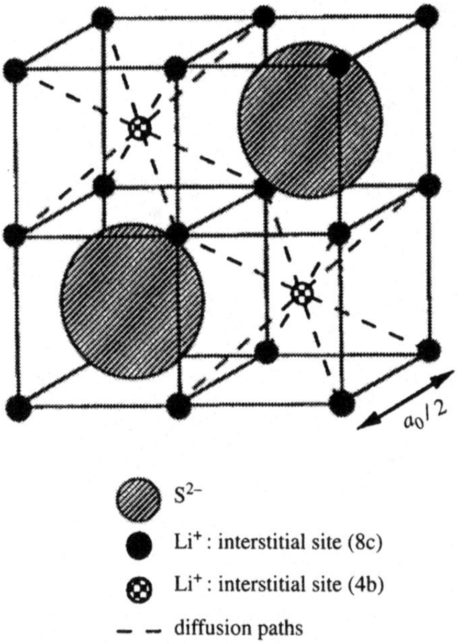

Fig. 12.6 The crystal structure of Li_2S. For clarity only half of the unit cell is shown. Possible Li^+ diffusion paths (jump vectors) between tetrahedral (8c) and octahedral (4b) sites are indicated by dashed lines (from Altorfer *et al.* 1994a, with permission from IOP Publishing Ltd).

any Q dependence as would be expected in the case of no correlation between interstitials (Laue scattering). Therefore QENS was subsequently done without polarization analysis. The data were analysed in terms of an extended Chudley–Elliott model which allows for jumps from regular sites to interstitial sites and vice versa, but without any direct jumps between regular sites; the possible jump vectors are shown in Fig. 12.6. The formalism corresponds to that one described in Section 5.4.2 and yields, due to the three Li^+ sublattices, a superposition of three Lorentzians as scattering function:

$$S_i(\boldsymbol{Q}, \omega) = \frac{1}{\pi} \sum_{\delta=1}^{3} \frac{w_\delta |\lambda_\delta|}{|\lambda_\delta|^2 + \omega^2} \tag{12.1}$$

where the $\boldsymbol{Q}$-dependent linewidths λ_δ are the eigenvalues of the jump matrix and where the $\boldsymbol{Q}$ dependent weights w_δ of the Lorentzians are determined by the orthonormal eigenvectors v_δ^l and the site occupation factors ρ_l

$$w_\delta(\boldsymbol{Q}) = \left| \sum_{l=1}^{3} \sqrt{\rho_l} v_\delta^l \right|^2 \quad \text{and} \quad \sum_\delta w_\delta(\boldsymbol{Q}) = 1. \tag{12.2}$$

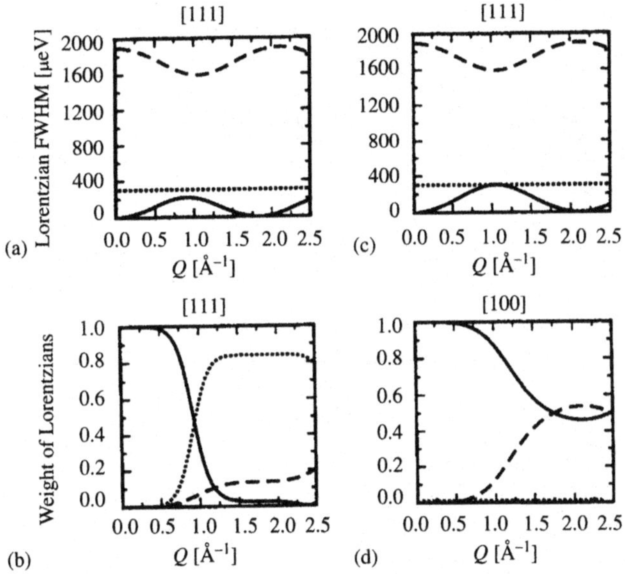

Fig. 12.7 Model calculations of Lorentzian FWHM's and Lorentzian weights as a function of Q with residence time $\tau_{(8c)\rightarrow(4b)} = 4.3$ ps: —, model 1; - - - mode 2; — — —, mode 3. (a) Halfwidths along [111]; (b) weights along [111]; (c) halfwidths along [100]; (d) weights along [100] (from Altorfer *et al.* 1994a, with permission from IOP Publishing Ltd).

The site occupancies for the three sublattices are known from neutron diffraction, the jump vectors are known from the structure, and the jump rates for the jumps into the interstitial site, $\tau_{13}^{-1} = \tau_{23}^{-1}$, and out of the interstitial sites, $\tau_{31}^{-1} = \tau_{32}^{-1}$, are related to the occupation factors by the detailed balance condition:

$$\tau_{ij} = \tau_{ji}\frac{\rho_i}{\rho_j}. \tag{12.3}$$

Therefore the model contains only one free parameter, i.e. the jump rate from a regular Li site to an interstitial site, τ_{13}^{-1}. The results of the model calculation with $\tau_{13} = 4.3$ ps are presented in Fig. 12.7, where FWHM and weights along [111] and [100] are plotted. The scattering law is split into three Lorentzian components: a narrow mode 1 which dominates at small Q with a Q^2 dependence of the width, a dispersionless mode 2 which does not exist in the [100] direction, and a very broad mode 3 with an oscillating width and appreciable weight only at large Q.

Figures 12.8 and 12.9 compare the experimental results to the model calculations. The Q-independent width of the dominant mode 2 (largest weight) along [111] for $Q > 1.0\,\text{Å}^{-1}$ is well reproduced, and also the mode 1 behaviour along [100] with the Q dependence both of the quasielastic broadening and of the observed intensity. The fit of the model to the broad component, measured in a different spectrometer configuration, leads to a residence time, which is in satisfactory agreement with the value derived from the high-resolution experiments.

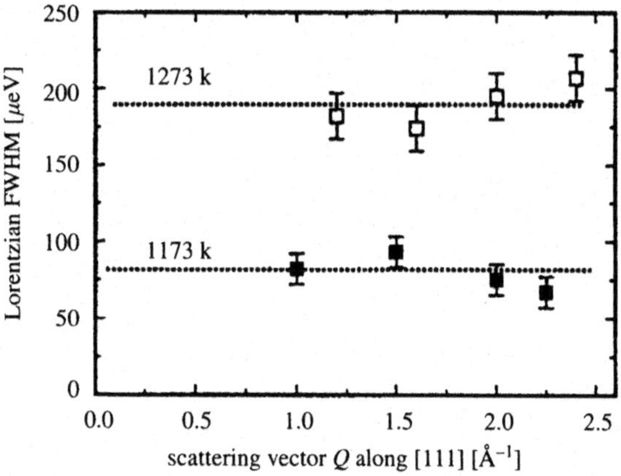

Fig. 12.8 The linewidth of Lorentzian mode 2 along [111] in Li_2S. Experimental points: □, 1273 K, ■, 1173 K. As predicted by the model (Fig. 12.7a) the width is not Q dependent (from Altorfer *et al.* 1994a, with permission from IOP Publishing Ltd).

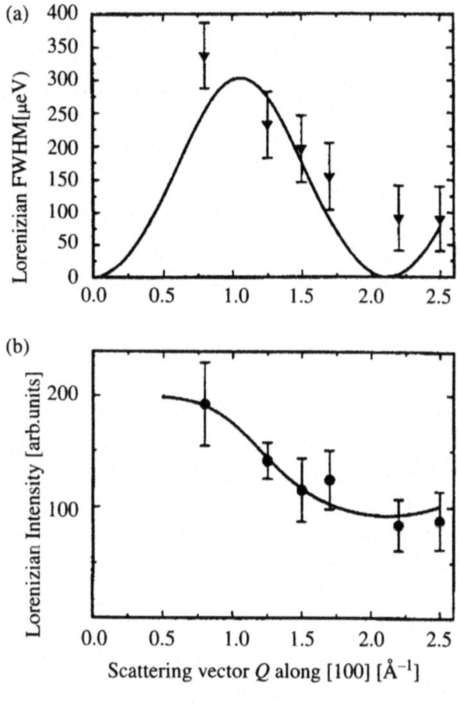

Fig. 12.9 (a) The linewidth of Lorentzian mode 1 along [100]: ▼, experimental points at 1363 K; — model calculation. (b) The intensity of Lorenztian mode 1 along [100]:•, experimental points at 1363 K; —, model calculation (from Altorfer *et al.* 1994a, with permission from IOP Publishing Ltd).

The experimental determination of three quasielastic scattering components clearly demonstrates that Li diffusion in anti-fluorite Li_2S takes place via interstitial hopping. Apparently the Li^+ cations, which are much smaller than the immobile S^{2-} anions, can easily form Frenkel defects although this mechanism has not been reported for Li_2O (Farley *et al.* 1989). In contrast, in fluorites with anionic conductivity the large mobile anions do not have the possibility of forming Frenkel defects, as will be discussed below.

12.4 ANIONIC CONDUCTORS

As listed in Table 12.1, among the mobile anions only H^- and Cl^- exhibit sufficient incoherent neutron scattering cross-sections that their self-diffusion can be measured by QENS. The compound Ba_2NH is one of the few documented H^- ionic conductors (Wegener *et al.* 1992). The electronic part of its conductivity is extremely low (about 10^{-4} of the ionic conductivity). The structure of Ba_2NH consists of (distorted) Ba^{2+} octahedra which are connected over edges. The H^- ions and the N^{3-} ions occupy the centres of the octahedra in such a way that H^- layers and N^{3-} layers are formed. The hydrogen conductivity increases rapidly at temperatures higher than 525 K which resembles the diffuse phase transition of the fluorites and anti-fluorites and reaches a value of 1 Ω^{-1} cm^{-1} at 693 K. In addition to the regular octahedral H^- sites interstitial tetrahedral H^- sites are also occupied with an occupancy between 10 and 15% depending on temperature. Therefore the regular H^- plane contains many vacancies. Altorfer *et al.* (1994b) have performed a QENS experiment on Ba_2NH and evaluate their data with the assumption that the H^- diffusion is caused by in-plane hopping of H^- ions between regular sites and that out-of-plane jumps are unlikely due to the N^{3-} layers. Correspondingly, a Chudley–Elliott model for a two-dimensional triangular lattice was applied. Since, for lack of single crystals, the measurements were performed on a powder sample, an orientational average had to be performed. The isotropic Chudley–Elliott model in this case is not a useful approximation because for a two-dimensional diffusional system for Q perpendicular to the diffusion plane the scattering is elastic, i.e. the linewidth is zero. Instead Altorfer *et al.* numerically calculated a density of states of Lorentzian halfwidths, $g(\Gamma, Q)$, and fitted their data with the scattering function

$$S_i(Q, \omega) = \frac{1}{\pi} \int_{\Gamma=0}^{\Gamma_{max}} g(\Gamma, Q) \frac{\Gamma}{\Gamma^2 + \omega^2} \, d\Gamma. \tag{12.4}$$

The experimental data, unfortunately measured with comparatively low counting statistics, are compatible with the model, and from the low Q data the diffusion coefficient $D = 2.1 \cdot 10^{-5}$ cm^2 s^{-1} is derived. How this self-diffusion coefficient compares to the conductivity diffusion coefficient to be derived from the ionic conductivity is not indicated. Furthermore it is not clear what the scattering contribution of the 10–15% interstitial H^- ions is and how they are involved in the diffusion process.

$SrCl_2$ is probably the best investigated Cl^- ionic conductor. It crystallizes in the fluorite structure which may most easily be viewed as a simple cubic array of anions with cations occupying alternate cube centres. Like many other fluorites $SrCl_2$ exhibits a diffuse transition to a state of relatively high, dynamic anion disorder, in this case at $T_c = 1000\,K$. Dickens *et al.* (1983) and Schnabel *et al.* (1983) report on an incoherent QENS study. Since from other experiments these authors know that the characteristic coherent diffuse quasielastic scattering peaks over an anisotropic shell in reciprocal space with radius $2.4 \cdot 2\pi/a_0$ (see the next section) they can easily avoid this Q regime and thus measure predominantly incoherent Cl^- scattering. The sample was a single crystal, and the data were evaluated in terms of three models:

(i) A basic Chudley–Elliott model comprising only nearest neighbour (nn) jumps along [100] yields a single Lorentzian as scattering function (the anion sites in fluorites, space group Fm3m, originate from two crystallographically inequivalent sublattices and should thus give rise to a superposition of two Lorentzians but this was not taken into account).

(ii) A simple extension of the Chudley–Elliott model also allows next nearest neighbour (nnn) jumps along [110] to regular anion sites with the jump probability p_2 as free parameter:

$$\Gamma(Q) = \frac{1}{2\pi\tau} \sum_{i=1}^{2} \frac{p_i}{n_i} \left(\sum_j [1 - \exp(-\mathrm{i}\boldsymbol{Q}\boldsymbol{S}_{ij})] \right) \tag{12.5}$$

where the index j runs over the n_i equivalent neighbours of type i of a given site, situated at Bravais lattice vector $\boldsymbol{S}_{ij}$, to which the ion can hop with probability p_i. The macroscopic self-diffusion coefficient D may be determined using

$$D = \frac{\sum p_i\, S_{ij}^2}{6\tau \sum p_i}. \tag{12.6}$$

(iii) In the encounter model of Wolf (Wolf 1977a; Göltz *et al.* 1980), the ionic motion is viewed as resulting from encounters with a defect, in the simplest case a single vacancy, which undergoes truly stochastic motion. As a result of each encounter the anion may make a number of essentially instantaneous jumps to nearest neighbour sites $\boldsymbol{S}_{ij}$ or back to its original site $i = 0$. The probabilities, $W(i)$, of reaching each type of site have to be calculated by Monte Carlo simulations. The mean time between encounters is denoted as τ_{enc}. The scattering function is a single Lorentzian with a linewidth

$$\Gamma(Q) = \frac{1}{2\pi\,\tau_{enc}} \sum_i \frac{\mathrm{W}(i)}{n_i} \left(\sum_j [1 - \exp(-\mathrm{i}\boldsymbol{Q}\boldsymbol{S}_{ij})] \right) \tag{12.7}$$

This model has the possibility of taking account of correlated motion and thus of the possibility that $W(0)$ is non-zero, i.e. the particle returns to its original site.

In the limit $W(0) = 0$, $W(1) = 1$, all other $W(i) = 0$, the basic Chudley–Elliott model is recovered, with $W(1) = p_1$ and $W(2) = p_2$ the simple extension. The main difference is that the probabilities in the encounter model are predetermined and not adjustable parameters. In Fig. 12.10 the experimental linewidths along the three main crystallographic directions are compared to the three models. All of them give reasonable fits. The best agreement is obtained if both [100] nn and [110] nnn hops are included. Thus a simple model of a random jump-diffusion mechanism involving only regular anion sites is found to account well for data in the complex fast-ion phase of $SrCl_2$. The anion diffusion mechanism in fluorites appears to be simpler than the cation diffusion mechanism in anti-fluorites, see the previous section.

Essentially the same results were obtained by Goff *et al.* (1992) for chlorine diffusion in single-crystal $(Sr,Y)Cl_{2.03}$ with a concentration of the dopant in the range 2.5–3.2 mol%. Figure 12.11 shows the FWHM resulting from a Lorentzian fit for three different sample orientations; the solid line corresponds to a fit of Eq. (12.5), i.e. uncorrelated nn and nnn jumps account very well for the data. Within the accuracy of the experiment correlation effects obviously are not relevant. The diffusion coefficient derived by means of Eq. (12.6) is found to be higher at a given temperature for $(Sr,Y)Cl_{2.03}$ than for the pure compound, possibly due to the higher concentration of defects in the anion-excess compound, in accordance with analogous conductivity measurements.

The most mobile anions, apart from H^-, are F^- and O^{2-}, both purely coherent neutron scatterers with zero incoherent scattering cross-section. Cl^- as a mixed scatterer also gives rise to intense coherent scattering. These anions show high mobility in compounds with the fluorite structure. This structure tolerates high levels of disorder which may be generated both intrinsically (i.e. at high temperatures) or extrinsically (by doping or deviation from stoichiometry).

The relevant fluorides, like CaF_2 or PbF_2, and the only relevant chloride, $SrCl_2$, with melting points between 1100 and 1700 K and 'superionic transitions' a few hundred degrees lower, are much more accessible to neutron scattering studies than the oxides for which temperatures well above 2000 K are necessary. Extensive investigations of diffuse coherent QENS from PbF_2, CaF_2, and $SrCl_2$ have been performed by Hutchings, Hages, and coworkers and have been reviewed by Hutchings *et al.* (1984) in detail and by Catlow (1989). Hutchings *et al.* observe pronounced coherent diffuse neutron scattering with a maximum in intensity along [100], just beyond (200). There is an anisotropic shell of intensity with peaks in certain directions, and little intensity at low Q. This implies correlation between disordered anions. The data account well for a defect model involving independent clusters of one or two anion Frenkel pairs with lattice relaxation around the interstitials. The Frenkel interstitials and the relaxed anions are located at positions within the empty cube; the former are displaced in $\langle 110 \rangle$ directions from mid-regular anion sites and the latter in $\langle 111 \rangle$ directions from regular anion sites. Thus the interstitial is stabilized by the relaxation of the neighbouring anions;

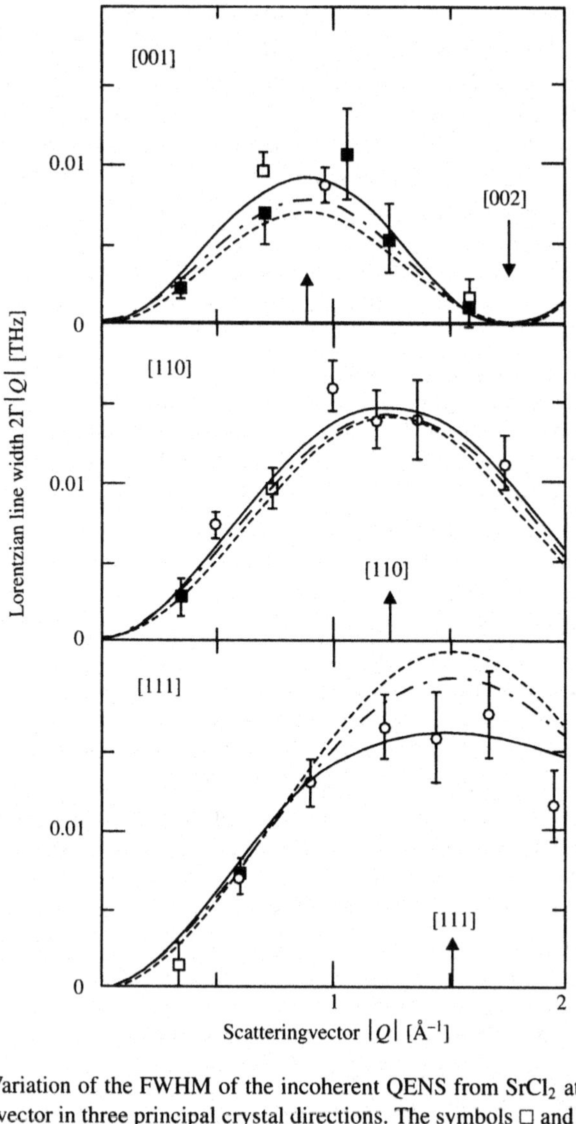

Fig. 12.10 Variation of the FWHM of the incoherent QENS from $SrCl_2$ at 1053 K with the scattering vector in three principal crystal directions. The symbols □ and ○ denote data taken with $k_i = 1.047$ Å, respectively. The broken (nn hops) and full (nn and nnn hops) curves are the best fits of the Chuley–Elliott expression, and the chain curve that of Wolf's encounter model, as described in the text (from Dickens *et al.* 1983, with permission from IOP Publishing Ltd).

the compensating vacancy is located in the vicinity of the cluster. The cluster are short lived with life times of about 10^{-12} s, which are derived from the energetic broadening of the diffuse scattering. In this simplest model the clusters are assumed to decay exponentially with time, with life time τ_{coh}. In this case the Q-independent

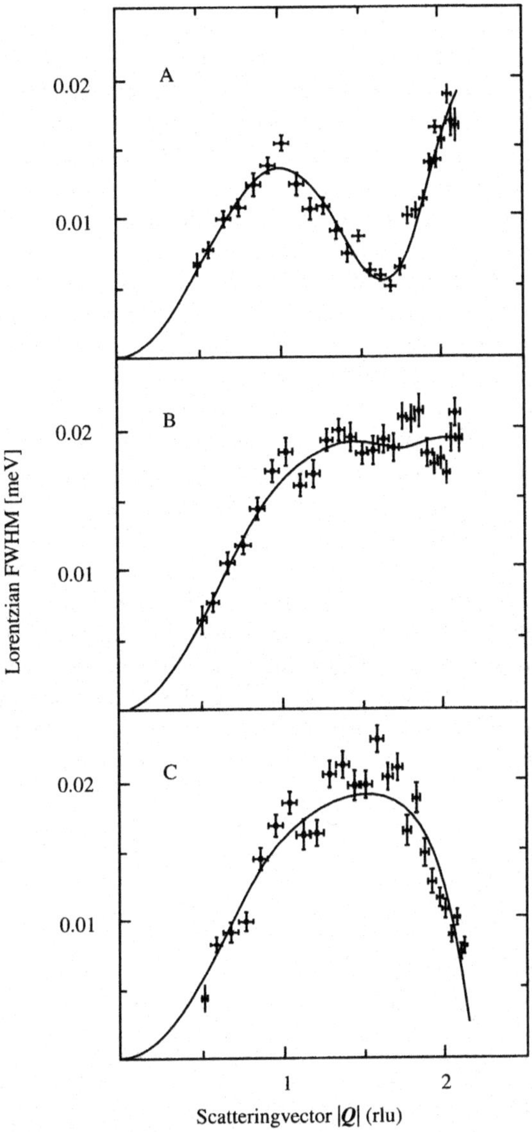

Fig. 12.11 The FWHM of the incoherent spectra of (Sr, Y) $Cl_{2.03}$, measured at 973 K, plotted against $\boldsymbol{Q}$ for three different orientations A, B, C of the sample (from Goff *et al.* 1992, with permission from Elsevier Science).

Lorentzian FWHM, $2\Gamma(Q)$ in meV is related to τ_{coh} in ps by

$$\tau_{coh} = 4.14/\pi \left[2\Gamma(\boldsymbol{Q})\right]. \tag{12.8}$$

For all three materials the clusters have a finite life time which decreases as the temperature increases. In the case of $SrCl_2$ and CaF_2 the energy width of the coherent diffuse scattering is dependent on the wavevector q (where $Q = (2+q, 0, 0)$), varying linearly in $SrCl_2$ and CaF_2 at lower temperatures, but as q^2 in CaF_2 at higher temperatures, whereas that for PbF_2 is largely q-independent. For $SrCl_2$ and CaF_2 this q-dependence suggests that an effective translational motion of the cluster as a whole takes place. This follows from the fact that the widths tend to zero as $q = 0$, and may be associated with correlated decay and reformation, either with the same or different constituent particles. The relatively q-independent energy widths in PbF_2, on the other hand, suggest that in this case clusters simply form and disappear with no effective translational movement. In connection with the coherent scattering energy widths it is interesting to recall the result of incoherent QENS from $SrCl_2$ presented in the previous section, which is an uncorrelated jump diffusion of the Cl^- ions over the regular sites, i.e. a vacancy diffusion mechanism, with a mean residence time τ_{incoh}. Obviously the random thermal creation and annihilation of Frenkel defects causes a stochastic generation of vacancies which randomize the diffusion.

For $SrCl_2$ the coherent linewidths are larger than the corresponding incoherent ones, i.e. the residence times τ_{incoh} are generally longer than the characteristic times of coherent QENS. A possible explanation by Hutchings *et al.* is that the clusters are intimately involved in the single-particle diffusion mechanism, in that the anion forms a Frenkel pair and associated cluster for a short time whilst in transit. But it is also possible that the cluster is merely the means by which a vacancy is created in the lattice into which another regular site anion can then hop directly.

13

Bibliography

BOOKS AND GENERAL REVIEWS

Chapter 1: Introduction

Bacon, G.E. (ed.) (1986). *Fifty Years of Neutron Diffraction: The Advent of Neutron Scattering*. Adam Hilger, Bristol.

Willis, B.T. (ed.) (1973). *Chemical Application of Thermal Neutron Scattering*. Oxford University Press.

Lechner, R.E. and Riekel, C. (1982). *Anwendungen der Neutronenstreuung in der Chemie*. Akademische Verlagsgesellschaft, Wiesbaden; Application of Neutron Scattering in Chemistry, in *Neutron Scattering and Muon Spin Rotation*, Springer Tracts in Modern Physics, Springer, Berlin.

Sköld, K. and Price, D.L. (eds) (1986). *Methods of Experimental Physics* Vol. 23 Part A, B, C: Neutron Scattering. Academic Press, New York.

Chapter 2: The neutron

Bacon, G.E. (1975). *Neutron Diffraction*. Clarendon Press, Oxford.

Mayer-Kuckuk, T. (1984). *Kernphysik*. Teubner Studienbücher; or any other textbook on nuclear physics.

Lieser, K.H. (1980). *Einführung in die Kernchemie*. Verlag Chemie, Weinheim; or any other textbook on nuclear chemistry.

Chapter 3: The basic theory of neutron scattering

Squires, G.L. (1978). *Introduction to the Theory of Thermal Neutron Scattering*. Cambridge University Press.

Lovesey, S.W. (1987). *Theory of Neutron Scattering from Condensed Matter, Vol. 1: Nuclear Scattering; Vol. 2: Polarization Effects and Magnetic Scattering*. Clarendon Press, Oxford.

Chapter 4: Diffusion in solids

Brenig, W. (1989). *Statistical Theory of Heat: Nonequilibrium Phenomena*. Springer, Berlin.

Ghez, R. (1988). *A Primer of Diffusion Problems*. Wiley, New York.

Haken, H. (1982). *Synergetik*. Springer, Berlin.

Haus, J.W. and Kehr, K.W. (1987). Diffusion in Regular and Disordered Lattices. *Physics Reports*, **150**, 263.

Kärger, J., Heitjans, P., and Haberlandt, R. (eds) (1998). *Diffusion in Condensed Matter*. Vieweg, Wiesbaden.

Laskar, A.L., Bocquet, J.L., Brebec, G., and Monty, C. (eds) (1990). *Diffusion in Materials*. NATO ASI Series E: Applied Sciences Vol. 179, Kluwer, Dordrecht.

Murch, G.E. and Nowick, A.S. (eds) (1984). *Diffusion in Cristalline Solids*. Academic Press, New York.

Nowick, A.S. and Burton, J.J. (eds) (1975). *Diffusion in Solids*. Academic Press, New York.

Philibert, J. (1985). *Diffusion et Transport de Matière dans les Solides*. Monographies de Physique, Les Éditions de Physique, Paris.

Philibert, J. (1991). *Atom Movements: Diffusion and Mass Transport in Solids*. Les Éditions de Physique, Paris.

Shockley, W., Hollomon, J.H., Maurer, R., and Seitz, F. (1952). *Imperfections in Nearly Perfect Crystals*, Wiley, New York.

van Kampen, N.G. (1981). *Stochastic Processes in Physics and Chemistry*. North-Holland, Amsterdam, new edition 1992.

Chapter 5: Incoherent quasielastic neutron scattering

Springer, T. (1972). *Quasielastic Neutron Scattering for the Investigation of Diffusive Motions in Solids and Liquids*. Springer Tracts in Modern Physics Vol. 64, Springer, Berlin.

Bée, M. (1988). *Quasielastic Neutron Scattering. Principles and Applications in Solid State Chemistry, Biology and Materials Science*. Adam Hilger, Bristol.

Chapter 6: Coherent quasielastic neutron scattering

Mezei, F. (1991). Neutron Scattering and Collective Dynamics in Liquids and Glass, in J.P. Hansen, D. Levesque, and J. Zinn-Justin (eds), *Liquids, Freezing and Glass Transition*. NATO-ASI in Les Houches 1989, North Holland, Amsterdam.

Chapter 7: Experimental techniques

Windsor, C.G. (1981). *Pulsed Neutron Scattering*. Taylor & Francis, London.

Sköld, K. and Price, D.L. (eds) (1986). *Methods of Experimental Physics* Vol. 23 Part A, B, C: Neutron Scattering. Academic Press, New York.

Williams, W.G. (1988). *Polarized Neutrons*. Clarendon Press, Oxford.

REFERENCES

Ait-Salem, M., Springer, T., Heidemann, A., and Alefeld, B. (1979). *Philosophical Magazine A*, **39**, 797.

Aldebert, P., Dianoux, A.J., and Traverse, J.P. (1979). *Le Journal de Physique*, **40**, 1005.

Alefeld, B. and Springer, T. (1990). *Nuclear Instruments and Methods in Physics Research*, **A295**, 268.

Alefeld, G. and Völkl, J. (eds) (1978). *Hydrogen in Metals I, II*. Topics in Applied Physics, Vol. 28, 29. Springer, Berlin.

Alefeld, B., Birr, M., and Heidemann, A. (1968). *Neutron Inelastic Scattering, 1967*, International Atomic Energy Agency, Vienna, p. 381.

Alefeld, B., Birr, M., and Heidemann, A. (1969). *Naturwissenschaften*, **56**, 410.

Alefeld, B., Springer, T., and Heidemann, A. (1992). *Nuclear Science and Engineering*, **110**, 84.

Alefeld, G., Völkl, J., and Schaumann, G. (1970). *Physica Status Solidi*, **37**, 337.

Altorfer, F., Bührer, W., Anderson, I.S., Schärpf, O., Bill, H., and Carron, P.L. (1994a). *Journal of Physics: Condensed Matter*, **6**, 9937.

Altorfer, F., Bührer, W., Winkler, B., Coddens, G., Essmann, R., and Jacobs, H. (1994b). *Solid State Ionics*, **70/71**, 272.

Anderson, I.S., Ross, D.K., and Carlile, C.J. (1978). *Neutron Inelastic Scattering 1977*, Vol. II, International Atomic Energy Agency, Vienna, p. 421.

Anderson, I.S., Carlile, C.J., Ross, D.K., and Vinhas, L.A. (1982). *Journal of the Less Common Metals*, **88**, 343.

Anderson, I.S., Heidemann, A., Bonnet, J.E., Ross, D.K., Wilson, K.P., and McKergow, M.W. (1984). *Journal of the Less Common Metals*, **101**, 405.

Anderson, I.S., Berk, N.F., Rush, J.J., Udovic, T.J., Barnes, R.G., Magerl, A., and Richter, D. (1990). *Physical Review Letters*, **65**, 1439.

Anderson, N.H., Clausen, K.N., and Kjems, J.K. (1987). Fast Ionic Conductors, in K. Sköld and D. L. Price (eds) *Methods of Experimental Physics—Neutron Scattering*, Vol. 23B. Academic Press, New York.

Anderson, P.W. (1958). *Physical Review*, **109**, 1492.

Anderson, P.W. (1967). *Physical Review*, **164**, 352.

Anderson, P.W., Happerin, B.I., and Varma, C.M. (1972). *Philosophical Magazine*, **25**, 1.

Aoki, K., Kamachi, M., and Sasumoto, T. (1984). *Journal of Non-Crystalline Solids*, **61/62**, 679.

Avogadro, A. and Villa, M. (1979). *Journal of Chemical Physics*, **70**, 109.

Bacon, G.E. (1975). *Neutron Diffraction*. Clarendon Press, Oxford.

Bacon, G.E. (ed.) (1986). *Fifty Years of Neutron Diffraction: The Advent of Neutron Scattering*. Adam Hilger, Bristol.

Baranov, A.I., Shuvalov, L.A., and Shchagina, N.M. (1982). *JETP Letters*, **36**, 459.

Baranov, A.I., Fedosyuk, R.M., Shchagina, N.M., and Shuvalov, L.A. (1984). *Ferroelectrics Letters*, **2**, 25.

Baranov, A.I., Ponsyatovskii, E.G., Sinitsyn, V.V., Fedosyuk, R.M., and Shuvalov L.A. (1985). *Soviet Physics Crystallography*, **30**, 1121.

Baranov, A.I., Merinov, B.V., Tregubchenko, A.B., Shuvalov, L.A., and Shagina, N.M. (1988). *Ferroelectrics*, **127**, 257.

Bardeen, J. and Herring, C. (1952). In W. Shockley (ed.) *Imperfections in Nearly Perfect Crystals*, Wiley, New York, p. 261.

Barnes, J.D. (1973). *Journal of Chemical Physics*, **58**, 5193.

Batallan, F., Rosenman, I., Magerl, A., and Fuzellier, H. (1985). *Physical Review B*, **32**, 4810.

Bauer, G.S. (1979). Treatise on Materials Science and Technology. In G. Kostorz (ed.) *Neutron Scattering*, Vol. 15. Academic, New York.

Bauer, G.S. (1991). *SINQ-Status Report*, in Proceedings of ICANS-XI, KEK-Report **90-25**, 41, Tsukuba, Japan.

Bauer, G.S. (1992). *Physica B*, **180 & 181**, 883.

Bauer, G.S. and Thamm, G. (1991). *Physica B*, **174**, 476; and other reviews in the same volume.

Bauer, G.S., Seitz, E., Horner, H., and Schmatz, W. (1975). *Solid State Communications*, **17**, 161.

Bauer, G.S., Schmatz, W., and Just, W. (1977). *Proceedings of the Second International Congress on H in Metals*, Pergamon Press, Oxford.

Bauer, H.C., Völkl, J., Tretkowski, J., and Alefeld, G. (1978). *Zeitschrift für Physik B*, **29**,17.

Bauer *et al.* (1985). SINQ Project Study.

Beckurts, K.H. and Wirtz, K. (1964). *Neutron Physics*. Springer, Berlin.

Bée , M. (1988). *Quasielastic Neutron Scattering: Principles and Applications in Solid State Chemistry, Biology and Materials Science*. Adam Hilger, Bristol.

Belissent-Funel, M.-C., Lal, J., and Bosio, L. (1993). *Journal of Chemical Physics* **98**, 4246.

Belushkin, A.V., David, W.I.F., Ibberson, R.M., and Shuvalov, L.A. (1991a). *Acta Crystallographica B*, **47**, 161.

Belushkin, A.V., Carlile, C.J., David, W.I.F., Ibberson, R.M., Shuvalov, L.A., and Zajac, W. (1991b). *Physica B*, **174**, 268.

Belushkin, A.V., Carlile, C.J., and Shuvalov, L.A. (1992). *Journal of Physics: Condensed Matter*, **4**, 389.

Belushkin, A.V., Carlile, C.J., and Shuvalov, L.A. (1994). The diffusion of protons in some hydrogen bonded crystals by quasielastic neutron scattering, in J. Colmenero (ed.) *Quasielastic Neutron Scattering*. World Scientific, Singapore.

Berliner, R., Popovici, M., Herwig, K., Jennings, H.M., and Thomas, J. (1998). *Physica B*, **241–243**, 1237.

Bermejo, F.J., Chahid, A., García-Hernández, M., Martínez, J.L., Mompean, F.J., Howells, W.S., and Enciso, E. (1992). *Physica B*, **182**, 289.

Bindl, M. (1986). *Thesis, Technical University München*.

Birr, M., Heidemann A., and Alefeld, B. (1971). *Nuclear Instruments and Methods*, **95**, 435.

Blaesser, G., Perretti, J., and Toth, G. (1968). *Physical Review*, **171**, 665.

Blanc, Y. (1983). ILL Report 83BL21G.

Blaschko, O., Krexner, G., Baon, J.N., and Vajda, P. (1985). *Physical Review Letters*, **55**, 2876.

Blinc, R., Dolinsek, J., Lahainar, G., Zupancic, I., Shuvalov, L.A., and Baranov, A.I. (1984). *Physica Status Solidi*, **B123**, K83.

Böni, P. (1997). *Physica B*, **234–236**, 1038.

Bonnet, J.E. and Daou, J.N. (1974). *Journal of Physics and Chemistry of Solids*, **40**, 421.

Boureau, G. (1984). *The Journal of the Physics and Chemistry of Solids*, **45**, 873.

Bragg, W.L. and Williams, E.J. (1934). *Proceeding of the Royal Society* (London) A, **145**, 699.

Bragg, W.L. and Williams, E.J. (1935). *Proceeding of the Royal Society* (London) A, **151**, 540.

Brenig, W. (1989). *Statistical Theory of Heat: Nonequilibrium Phenomena*. Springer, Berlin.

Brockhouse, B.N. (1958). *Nuovo Cimento*, **9**, Suppl. 1, 45.

Brockhouse, B.N. (1959). *Physical Review Letters*, **2**, 287.

Buchholz, J., Völkl, J., and Alefeld, G. (1973). *Physical Review Letters*, **30**, 318.

Bührer, W. (1994). *Nuclear Instruments and Methods in Physics Research A*, **338**, 44.

Bührer, W., Bührer, R., Isacson, A., Koch, M., and Thut, R. (1981). *Nuclear Instruments and Methods*, **179**, 259.

Bunde, A. and Havlin, S. (eds) (1996). *Fractals and Disordered Systems*, 2nd edn. Springer Verlag, Heidelberg.

Bunde, A. and Kantelhardt, J.W. (1998). *Percolation*, in J. Kärger, P. Heitjans, and R. Haberlandt (eds) *Diffusion in Condensed Matter*. Vieweg, Braunschweig.

Bunde, A., Maas, P., and Ingram, M.D. (1991). *Berichte der Bunsengesellschft Physikalische Chemie*, **95**, 977 and 1002.

Bunde, A., Maass, Ph., and Meyer, M. (1998). *Ionic Transport in Disordered Materials*, in J. Kärger, P. Heitjans, and R. Haberlandt (eds) *Diffusion in Condensed Matter*. Vieweg, Braunschweig.

Cahn, J.W. (1961). *Acta Metallurgica*, **9**, 795.

Cahn, J.W. (1962). *Acta Metallurgica*, **10**, 179; 907.

Campbell, S.I., Kemali, M., and Ross, D.K. (1998). *Physica B*, **241–243**, 326.

Cantù. L., Cavatorta, F., Corti, M., Del Favero, E., and Derin, A. (1997). *Physica B*, **234–236**, 281–282.

Carlile, C.J. and Adams, M.A. (1992). *Physica B*, **182**, 431.

Carlile, C.J., Kearley, G.J., Lindsell, G., and White, J.W. (1998). *Physica B*, **241–243**, 491.

Carpenter, J.M. and Yelon W.B. (1986). Neutron Sources, in K. Sköld and D.L. Price (eds) *Neutron Scattering, Methods in Experimental Physics*, Vol. 23A, Academic Press, New York.

Catlow, C.R.A. (1989). Suprionic Fluorites, in A.L.A.L. Laskar and S. Chandra (eds) *Superionic Solides and Solid Electrolytes*, Academic Press, Boston, p. 339.

Catlow, C.R.A. (1990). *Journal of the Chemical Society Faraday Transactions*, **86**, 1167.

Catlow, C.R.A. (1992). *Solid State Ionics*, **53–56**, 955.

Chadwick, J., (1932). *Nature* (London), **129**, 312.

Chandra, S. (1989). Proton Conductors, in A.L. Laskar and S. Chandra (eds) *Superionic Solids and Solid Electrolytes*, Academic Press, San Diego 1989, p. 185.

Chudley, C.T. and Elliott, R.J. (1961). *Proceedings of the Physical Society* (London), **77**, 353.

Clapp, P.C. and Moss, S.C. (1968). *Physical Review*, **171**, 754 and 764.

Clawson, C.W., Crowe, K.M., Kohn, S.E., Rosenblum S.S., Hung, C.Y., Smith, J.L., and Brewer J.H. (1982). *Physica B*, **109–110b**, 2164.

Clemens, D., Vananti, A., Terrier, C., Böni, P., Schnyder, B., Tixier, S., and Horisberger, M. (1997). *Physica B*, **234–236**, 500.

Cocking, S.J. (1992). *J. Phys. C: Solid State Physics*, **11**, 2047.

Collin, G., Comes R., Boilot J.P., and Colomban, P.H. (1980). *Solid State Ionics*, **1**, 59.

Colomban, Ph. (ed.) (1992). *Proton Conductors*. Cambridge University Press.

Colomban, Ph., Lassegues, J.C., Novak, A., Pham-Thi., M., and Poinsignon, C. (1987). NASICON: an intermediate structure between glass and crystal, in J. Lascombe (ed.) *Dynamics of Molecular Crystals*, Elsevier, Amsterdam, pp. 269–275.

Compaan, K. and Haven, Y. (1958). *Transaction of the Faraday Society*, **54**, 1498.

Conrad, J., Estrade, H., Poinsignon, C., and Dianoux, A. J. (1984). *Journal de Physique*, **45**, 1361.

Cook, C.J., Richter, D., Schärpf, O., Benham, M.J., Ross, D.K., Hempelmann, R., Anderson, I.S., and Sinha, S.K. (1990). *Journal of Physics: Condensed Matter*, **2**, 79.

Cook, C.J., Richter, D., Hempelmann, R., Ross, D.K., and Züchner, H. (1991). *Journal of the Less Common Metals*, **172–174**, 5585.

Cook, J.C., Petry, W., Heidemann, A., and Barthélemy, J.-F. (1992). *Nuclear Instruments and Methods in Physics Research A*, **312**, 553.

Copley, J.R.D. (1992). *Physica B*, **180 & 181**, 914.

Coppersmith, S.N., Fisher, D.S., Halperin, B.I., Lee, P.A., and Brinkman, W.F. (1982). *Physical Review B*, **25**, 349.

Cowley, J.M. (1950). *Journal of Applied Physics*, **21**, 24.
Crank, J. (1975). *The Mathematics of Diffusion*. Oxford University Press.
Culvarhouse, J.W. and Richards, P.M. (1988). *Physical Review B*, **38**, 10 020.
Curtin, W.A. and Harris, J.H. (1988). *Materials Science and Engineering*, **99**, 463.
Dawber, P.G. and Elliott, R.J. (1963). *Proceedings of the Royal Society of London* A, **273**, 222; see also Kagan, Y. and Iosilevskii, Y. (1963). *Soviet Physics JETP*, **17**, 925.
de Fontaine, D. and Buck, O. (1973). *Philosophical Magazine*, **27**, 967.
de Gennes, P.G. (1959). *Physica*, **25**, 825.
Dederichs, P.H. (1973). *Journal of Physics F: Metal Physics*, **3**, 471.
Dianoux, A.J. (1992). *Physica B*, **182**, 389.
Dianoux, A.J., Volino, F., and Hervet, H. (1975). *Molecular Physics*, **30**, 1181.
Dickens, M.H., Hayes, W., Schnabel, P., Hutchings, M.T., Lechner, R.E., and Renker, B. (1983). *Journal of Physics C: Solid State Physics*, **16**, L1.
DiCola, D., Deriu, A., Sampoli, M., and Torcini, A. (1996). *Journal of Chemical Physics*, **104**, 4223.
Dietrich, S. and Fenzl, W. (1989). *Physical Review B*, **39**, 8873.
Dieterich, W., Knödler, D., and Petersen, J. (1991). *Berichte der Bunsengesellschft Physikalische Chemie*, **95**, 964.
Dietze, H.-D. and Nowak, E. (1981). *Zeitschrift für Physik B—Condensed Matter*, **44**, 245.
Dosch, H., Peisl, J., and Dorner, B. (1987). *Physical Review B*, **35**, 3069.
Dosch, H., Schmid, F., Wiethoff, P., and Peisl, J. (1992). *Physical Review B*, **46**, 55.
Douchin, F., Lechner R.E., and Blanc, Y. (1973). ILL Report 73D26T.
Driesen, G. and Kehr, K.W. (1989). *Physical Review B*, **39**, 8132.
Edwards, S.F. and Anderson, P.W. (1975). *Journal of Physics F: Metal Physics*, **5**, 965.
Egelstaff, P.A. (1951). *Nature* (London), **168**, 290.
Egelstaff, P.A. (1992). *An Introduction to the Liquid State*, Oxford Series on Neutron Scattering in Condensed Matter, Clarendon Press, Oxford.
Egelstaff, P.A. and Schofield, P. (1962a). *Nuclear Science and Engineering*, **12**, 260.
Egelstaff, P.A. and Schofield, P. (1962b). *Advances in Physics*, **11**, 203.
Einstein, A. (1905). *Annalen der Physik* (4), **17**, 549.
Ekstein, H. (1953). *Physical Review*, **89**, 490.
Emin, D. (1970). *Physical Review Letters*, **25**, 1751.
Emin, D. (1971). *Physical Review B*, **3**, 1321.
Estrade, H., Conrad, J., Langinie, P., Heitjans, P., Fujara, F., Butler, W., Kiese, G., Achermann, H., and Guerard, D. (1980). *Physica B*, **99**, 531.
Evans, R.D. (1955). *The Atomic Nucleus*. International Series in Pure and Applied Physics (L.I. Schiff ed.). McGraw Hill, London.
Eyring, H. (1935). *Chemical Review*, **17**, 65; *Journal of Chemical Physics* **3**, 107.
Eyring, H. (1936). *Journal of Chemical Physics*, **4**, 283.
Fan, J.D., Reiter, G., and Moss, S.C. (1990). *Physical Review Letters*, **64**, 188.
Farley, T.W.D., Hayes, W., Hull, S., Hutchings, M.T., Alba, M., and Vrtis, M. (1989). *Physica B*, **155 & 156**, 99.
Fassbender, J., (1967). *Einführung in die Reaktorphysik*, Thiemig, München, or any other textbook on reactor physics.
Faux, D.A. and Ross, D.K. (1987). *Journal of Physics C: Solid State Physics*, **20**, 1441.
Fernandez, J.F., Kemali, M., Johnson, M.R., and Ross, D.K. (1997). *Physica B*, **234–236**, 903.
Filabozzi, A., Deriu, A., and Andreani, C. (1996). *Physica B*, **226**, 56.
Fischer, W.E. (1995). Proceedings of ICANS-XIII and ESS-PM4, PSI-Proceedings 95-02.
Fisher, M.E. and Burford, R.J. (1967). *Physical Review*, **156**, 583.

Flottmann, T. and Vogl, G. (1987). *Nuclear Instruments and Methods in Physics Research, Section A*, **260**, 165.
Flynn, C.P. and Stoneham, A.M. (1970). *Physical Review B*, **1**, 3966.
Freilander, P., Heitjans, P., Ackermann, H., Bader, B., Kiese, G., Schirmer, A., and Stöckmann, H.J. (1986). *Zeitschrift für Naturforschung*, **41a**, 109.
Frenkel, I. (1926). *Zeitschrift für Physik*, **35**, 652.
Frick, B. and Richter, D. (1995). *Science*, **267**, 1939.
Frick, B., Magerl, A., Blanc, Y., and Rebesco, R. (1997). *Physica B*, **234–236**, 1177.
Fujii, S. (1979). *Journal of the Physical Society of Japan*, **46**, 1833.
Fukai, Y. (1993). *The Metal-Hydrogen System*. Springer Series in Materials Science 21, Springer, Berlin.
Funke, K. (1987). *Zeitschrift für Physikalische Chemie, Neue Folge*, **154**, 251.
Funke, K. (1989). Fast Ion Dynamics Studied by Neutron Scattering and High Frequency Conductivity, in A.L. Laskar and S. Chandra (eds) *Superionic Solides and Solid Electrolytes*. Academic Press, Boston, p. 269.
Funke, K. (1991). *Berichte der Bunsengesellschaft, Physikalische Chemie*, **95**, 955.
Funke, K. (1993). *Progress in Solid State Chemistry*, **22**, 111.
Funke, K., Wilmer, D., Lauxtermann, T., Holzgreve, R., and Bennington, S. M. (1996). *Solid State Ionics*, **86–88**, 141.
Gabrys, B. and Schärpf, O. (1992). *Physica B*, **180 & 181**, 495.
Gay-Lussac, R. (1846). *Annales de Chimie et Physique*, **17**, 221.
Gefen, Y., Aharony, A., and Alexander, S. (1983). *Physical Review Letters*, **50**, 77.
Gehring, P.M., Brocker, C.W., and Neumann, D.A. (1995). *Materials Research Society Symposium Proceedings*, **376**, 113.
Gillan, M.J. (1986). *Journal of Physics C: Solid State Physics*, **19**, 6169.
Gillan, M.J. and Wolf, D. (1985). *Physical Review Letters*, **55**, 1299.
Gissler, W. and Rother, H. (1970). *Physica*, **50**, 380.
Gissler, W., Alefeld, G., and Springer, T. (1970). *Journal of Physics and Chemistry of Solids*, **31**, 2361.
Gissler, W., Fay, B., Rubin, A., and Vinhas, L.A. (1973). *Physics Letters A*, **43**, 279.
Glass, L. and Rice, S.A. (1968). *Physical Review*, **165**, 186.
Gocke, E., Schöllhorn, R., Aselmann, G., and Müller-Warmuth, W. (1987). *Inorganic Chemistry*, **26**, 1805.
Goff, J.P., Hayes, W., Hull, S., Hutchings, M.T., and Ward, R.C.C. (1992). *Physica B*, **182**, 307.
Göltz, G., Heidemann, A., Mehrer, H., Seeger, A., and Wolf, D. (1980). *Philosophical Magazine A*, **41**, 723.
Groh, I. and v. Hevesy, G. (1920). *Annalen der Physik*, **65**, 216.
Grüning, U., Magerl, A., and Mildner, D.F.R. (1992). *Nuclear Instruments and Methods in Physics Research A*, **314**, 171.
Gurevich, I.I., Meleshko, E.A., Muratova, I.A., Kikolsky, B.A., Roganov, V.S., Selivanov, V.I., and Sokolov, B.B. (1972). *Physical Review Letters*, **40A**, 143.
Haken, H. (1982). *Synergetik*, Springer, Berlin.
Halban, H. and Preiswerk, P. (1936). *Comptes Rendus Hebdomadaires des Seances de l'Academie des Sciences, Paris*, **203**, 73.
Harris, J.H., Curtin, W.A., and Tenhover, M.A. (1987). *Physical Review B*, **36**, 5784.
Härtl, W. and Versmold, H. (1984). *Journal of Chemical Physics*, **80**, 1387.
Hartmann, O., Karlsson, E., Norlin, L.O., Niinikoski, T.O., Kehr, K.W., Richter, D., Welter, J.-M., Yaouanc, A., and Le Herecy, J. (1980). *Physical Review Letters*, **40**, 337.

Hartmann, O., Norlin, L.O., Yaouanc, A., Le Hericy, J., Karlsson, E., and Niinikoski, T.O. (1981). *Hyperfine Interactions*, **8**, 533.

Hartmann, O., Karlsson, E., Wäckelgård, E., Richter, D., Hempelmann, R., and Niinikoski, T.O. (1988). *Physical Review B*, **37**, 4425.

Haus, J.W. and Kehr K.W. (1987). *Diffusion in Regular and Disordered Lattices*. Physics Reports, **150**, 263.

Hayter, J.B. (1980). Matrix analysis of neutron spin echo, in F. Mezei (ed.) *Neutron Spin Echo*. Lecture Notes on Physics, Vol. 128, Springer.

Heidemann, A. and Alefeld, B. (1992). Internal Scientific ILL Report 91, HE 22G.

Heiming, A., Petry, W., Trampenau, J., Alba, M., Herzig, C., Schober, H.R., and Vogl, G. (1991). *Physical Review B*, **43**, 10 948.

Hempelmann, R. (1984). *Journal of the Less Common Metals*, **101**, 69.

Hempelmann, R. (1986). Jül-report, Jül-2096, ISSN 0366-0885.

Hempelmann, R., Richter, D., and Heidemann, A. (1983a). *Journal of the Less Common Metals*, **88**, 343.

Hempelmann, R., Richter, D., Pugliesi, R., and Vinhas, L.A. (1983b). *Journal of Physics F: Metal Physics*, **13**, 59.

Hempelmann, R., Richter, D., Faux, D.A., and Ross, D.K. (1988). *Zeitschrift für Physikalische Chemie Neue Folge*, **159**, 175.

Hempelmann, R., Richter, D., Springer, T., Dianoux A.J., and Schönfeld, C. (1992). *Physica B*, **180 & 181**, 697.

Hempelmann, R., Carlile, C.J., Bayer, D., and Kaps, Ch. (1994). *Zeitschrift für Physik B*, **95**, 49.

Hempelmann, R., Karmonik, Ch., Matzke, Th., Cappadonia, M., Stimming, U., Springer, T., and Adams, M.A. (1995). *Solid State Ionics*, **77**, 152.

Hempelmann, R., Soetratmo, M., Hartmann, O., and Wäppling, R. (1998). *Solid State Ionics*, **107**, 269.

Herzig, C. (1989). *Berichte der Bunsengesellschaft für Physikalische Chemie*, **93**, 1247.

Heumann, Th. (1992). *Diffusion in Metallen*, Werkstoff-Forschung und Technik, Vol. 10, Springer, Berlin.

Ho, P.S. (1972). *Philosophical Magazine*, **26**, 1429.

Hock, R., Vogt, T., Kulda, J., Mursic, Z., Fuess, H., and Magerl, A. (1993). *Zeitschrift für Physik B*, **90**, 143.

Holcomb, D.F. and Norberg, R.E. (1955). *Physical Review*, **98**, 1074.

Holstein, T. (1959). *Annals of Physics (NY)*, **8**, 325 and 343

Howells, W.S. (1996). *Physica B*, **226**, 78.

Hutchings, M.T., Clausen, K. Dickens, M.H., Hayes, W., Kjems, J.K., Schnabel, P.G., and Smith, C. (1984). *Journal of Physics C: Solid State Physics*, **17**, 3903.

Imry, Y. (1969). The detection of atomic tunnelling in solids, in E. Burnstein and S. Lundquist (eds), *Tunneling Phenomena in Solids*. Plenum Press, New York.

Iwahara, H. (1981). *Solid State Ionics*, **3/4**, 359.

Iwahara, H., Uchida, H., Ono, K. and Ogaki, K. (1988). *Journal Electrochemical Society*, **135**, 529.

Jäckle, J. and Kehr, K.W. (1983). *Journal of Physics F*, **13**, 753.

Jacobs, H., Haubrecht, B., Müller, P., and Bronger, W. (1982). *Zeitschrift Anorganische und Allgemeine Chemie*, **491**, 154.

Janssen, S., Mesot, J., Holitzner, L., Furrer, A., and Hempelmann, R. (1997). *Physica B*, **234–236**, 1174.

Johnson, M.W. (1974). AERE Harwell Report R 7682.

Kadono, R., Imazato, J., Nishiyma, K., Nagamine, K., Yamazaki, T., Richter, D., and Welter, J. (1984). *Hyperfine Interactions* **17–19**, 109.

Kadono, R., Imizato, J., Nishiyama, K., Nagamine, K., Amazaki, T., Richter, D., and Welter J.-M. (1985). *Physics Letters*, **109A**, 61.

Kadono, R., Imazato, J., Matsuzaki, T., Nishigama, K., Nagamine, K., Yamazaki, T., Richter, D., and Welter, J.-M. (1989). *Physical Review B*, **39**, 23.

Kadono, R., Kiefl, R.F., Ansaldo, E.J., Brewer, J.H., Celio, M., Kreitzman, S.R., and Luke, G.M. (1990). *Physical Review Letters*, **64**, 665.

Kagan, Yu. and Klinger, M.I. (1974). *Journal of Physics C*, **7**, 2791.

Kagan, Yu. and Prokof'jev, N.V. (1991). *Physics Letters A*, **159**, 289.

Kärger, J., Heitjans, P., and Haberlandt, R. (eds) (1998). *Diffusion in Condensed Matter*, Vieweg, Braunschweig.

Karlsson, E.B. (1995). *Solid State Phenomena as seen by Muons, Protons and Excited Nuclei*, Clarendon Press, Oxford.

Karlsson, E., Wäppling, R., Lidström, S.W., Hartmann, O., Kadono, R., Kiefl, R., Hempelmann, R., and Richter, D. (1995). *Physical Review B*, **52**, 6417.

Karmonik, Ch., Matzke, Th., Hempelmann, R., and Springer, T. (1995). *Zeitschrift für Naturforschung*, **50a**, 539.

Karmonik, Ch., Hempelmann, R., Cook J.C. and Güthoff, F. (1996). *Ionics*, **2**, 69.

Kehr, K. (1978). Theory of the Diffusion of Hydrogen in Metals, in G. Alefeld and J. Völkl (eds) *Hydrogen in Metals I. Topics in Applied Physics*, Springer, Berlin.

Kehr, K.W. (1984). *Hyperfine Interactions*, **17–19**, 63.

Kehr, K.W., Richter, D., and Swendsen, R.H. (1978). *Journal of Physics F: Metal Physics*, **8**, 433.

Kehr, K.W., Kutner, R., and Binder, K. (1981). *Physical Review B*, **23**, 4931.

Kehr, K.W., Richter, D., Welter, J.M., Hartmann, O., Karlsson, E., Norlin, L.O., Niinikoski, T.O., and Yaouanc, A. (1982). *Physical Review B*, **26**, 567.

Keizer, J. (1987) *Statistical Thermodynamics of Nonequilibrium Processes*, Springer, Berlin.

Khatamian, D., Stassis, C., and Beandry, B.J. (1981). *Physical Review B*, **23**, 624.

Kidson, G.V. and Kirkaldy, J.S. (1969). *Philosophical Magazine*, **20**, 1057.

Kidson, G.V. and Young, G.J. (1969). *Philosophical Magazine*, **20**, 1047.

Kiefl, R.F., Kadono, R., Brewer, J.H., Luke, G.M., Yen, H.K., Celio M., and Ansaldo, E.J. (1989). *Physical Review Letters*, **62**, 792.

Kirchheim, R. (1982). *Acta Metallugica*, **30**, 1069.

Kirchheim, R. and McLellan, R.B. (1980). *Journal of the Electrochemistry Society*, **127**, 2419.

Kjems, J., Taylor, A.D., Finney, J.L., Lengeler, H., and Steigenberger, U. (1997). *ESS: A Next Generation Neutron Source for Europe*, ESS–Council Publication.

Klafter, J., Schlesinger, M.F., and Zumofen, G. (1996). *Physics Today*, **49**(2), 33.

Kley, W. (1970). *Zeitschrift für Naturforschung*, **21a** (suppl.), 1770.

Klotsman S. M. (1983). *Physics of Metals Metallography* **55**, 82.

Knödler, D., Pendzig, P., and Dietrich, W. (1996). *Solid State Ionics*, **86–88**, 29.

Koester, L. (1965). *Zeitschrift für Physik*, **182**, 328.

Koester, L. (1977). Neutron Physics, *Springer Tracts in Modern Physics*, Vol. 80, p. 1. Springer, Berlin.

Koester, L. and Nistler, W. (1975). *Zeitschrift für Physik A—Atoms and Nuclei*, **272**, 189.

Koester, L. and Rauch, H. (1990). *Neutron Scattering Lengths* (2nd edn). IAEA Report No. 251/RB.

Koester, L., Knopf, K., Waschkowski, W., Klüver, A. (1984). *Zeitschrift für Physik A—Atoms and Nuclei*, **318**, 347.

Kondo, J. (1984). *Physica B*, **125**, 279; ibid. **126**, 377.

Kondo, J. (1991). *Berichte der Bunsengesellschaft Physikalische Chemie*, **95**, 422.

Kreuer, K.D. (1996). *Chemistry of Materials*, **8**, 610.

Kreuer, K.D., Dippel, Th., Hainovsky, N.G., and Maier, J. (1992). *Berichte Bunsengesellschaft Physikalische Chemie*, **96**, 1736.

Krist, Th., Pappes, B., Keller, Th., and Mezei, F. (1995). *Physica B*, **213 & 214**, 939.

Krivoglaz, M.A. (1969). *Theory of X-Ray and Thermal Neutron Scattering by Real Crystals*. Plenum Press, New York.

Kubo, R. (1966). *Reports on Progress in Physics*, **24**, 255.

Kudo, T. and Fueki, K. (1990). *Solid State Ionics*. VCH, Weinheim 1990.

Kuji, T. and Oates, W. A. (1984). *Journal of the Less Common Metals*, **102**, 251.

Kutner, R. (1981). *Physics Letters*, **81A**, 239.

Lamers, C., Schärpf, O., Schweika, W., Batoulis, J., Sommer, K., and Richter, D. (1992). *Physica B*, **180 & 181**, 515.

Laskar, A.L., Bocquet, J.L., Brebec, G., and Monty C. (eds) (1990). *Diffusion in Materials*, NATO ASI Series E: Applied Sciences Vol. 179, Kluwer, Dordrecht.

Le Claire, A.D. (1966). *Philosophical Magazine*, **14**, 1271.

Leadbetter, A.J. and Lechner, R.E. (1979). in J.N. Sherwood (ed.) *The Plastically Crystalline State*. Wiley, New York.

Lebsanft, E., Richter, D., and Töpfer, J. (1979). *Zeitschrift für Physikalische Chemie NF*, **116**, 175.

Lechner, R.E. (1991). KEK Report 90-25, Vol. 2, National Laboratory for High Energy Physics (Japan), p. 712.

Lechner, R.E. (1993). *Solid State Ionics*, **61**, 3.

Lechner, R. E. (1994). *Solid State Ionics*, **77**, 280.

Lechner, R. E. and Riekel, C. (1982). *Anwendungen der Neutronenstreuung in der Chemie*. Akademische Verlagsgesellschaft, Wiesbaden; *Applications of Neutron Scattering in Chemistry*, in *Neutron Scattering and Muon Spin Rotation*, Springer Tracts in Modern Physics, Vol. 101, Springer, Berlin.

Lechner, R.E., Colino, F., Dianoux, A.J., Douchin, F., Hervet, H., and Stirling, G.C. (1973). ILL Report 73L8S.

Lechner, R.E., Bleif, H.J., Dachs, H., Marx, R., and Stahn, M. and Anderson, I. (1991). *Solid State Ionics*, **46**, 25.

Lechner, R.E., Dippel, Th., Marx, R. and Lamprecht, I. (1993). *Solid State Ionics*, **61**, 47.

Lechner, R.E., v. Wallpach, R., Graf, H.A., Kasper, F.-J., and Mokrani, L. (1994). *Nuclear Instruments and Methods in Physics Research A*, **338**, 65.

Lechner, R.E., Melzer, R. and Fitter, J. (1996). *Physica* B **226**, 86.

Leibfried, G. and Breuer, N. (1978). *Point Defects in Metals I.* Springer Tracts in Modern Physics, Vol. 81, Springer, Berlin.

Li, Y. and Wahnström, G. (1992). *Physical Review B*, **46**, 14 528; *Physical Review Letters*, **68**, 3444.

Liang, K.C., Yang Du and Nowick, A.S. (1994). *Solid State Ionics*, **69**, 117.

Lottner, V., Haus, J.W., Heim, A., and Kehr, K.W. (1979a). *Journal of Physics and Chemistry of Solids*, **40**, 557.

Lottner, V., Schober, H.R., and Fitzgerald, W.J. (1979b). *Physical Review Letters*, **42**, 1162.

Lottner, V., Heim, A., and Springer, T. (1979c). *Zeitschrift für Physik B*, **32**, 157.

Lovesey, S.W. (1987). *Theory of Neutron Scattering from Condensed Matter, Vol. 1: Nuclear Scattering; Vol. 2: Polarization Effects and Magnetic Scattering*. Clarendon Press, Oxford.

Lovesey, S.W. and Springer, T. (eds) (1977). *Dynamics of Solids and Liquids by Neutron Scattering*, Topics in Current Physics Vol. 3, Springer, Berlin.

Lucazeau, G., Gavarri, J.R., and Dianoux, A.J. (1987). *Journal of Physics and Chemistry of Solids*, **48**, 57.

Lutz, H.D. (1995). *Structure and Bonding* (Berlin), **82**, 85.

Maeland, A.J. (1986). Preparation, structure and properties of glassy metal hydrides, in G. Bambakidis and R.C. Bowman Jr (eds) *Hydrogen in Disordered and Amorphous Alloys*, Plenum, New York, p. 127.

Magerl, A. (1990). Intercalate Diffusion, in *Springer Series in Materials Science. Graphite Intercalation Compounds I* (H. Zabel and S.A. Solin eds), Vol. 14, p. 221.

Magerl, A., Berry, B., and Alefeld, G. (1976). *Physica Status Solidi* (a), **36**, 161.

Magerl, A., Stump, N., Teuchert, W.D., Wagner, V., and Alefeld, G. (1977). *Journal of Physics C: Solid State Physics*, **10**, 2783.

Magerl, A., Teuchert, W.D., and Scherm, R. (1978). *Journal of Physics C: Solid State Physics*, **11**, 2175.

Magerl, A., Zabel, H. and Anderson, I.S. (1985). *Physical Review Letters*, **55**, 222.

Magerl, A., Dianoux, A.J., Wipf, H., Neumaier, K., and Anderson, I.S. (1986). *Physical Review Letters*, **56**, 159.

Maier-Leibnitz, H. (1962). *Zeitschrift für Angewandte Physik*, **14**, 738.

Maier-Leibnitz, H. (1966). *Nukleonik*, **8**, 61.

Maier-Leibnitz, H. and Springer, T. (1963). *Reactor Science and Technology (Journal of Nuclear Energy Parts A/B)*, **17**, 217.

Maier-Leibnitz, H. and Springer, T. (1966). *Annual Review of Nuclear Science*, **16**, 207.

Majkrzak, C. F. (1994). *Physica B*, **213 & 214**, 904.

Mantl, S., Petry, W., Schroeder, K., and Vogl, G. (1983). *Physical Review B*, **27**, 5313.

Marx, R., Dachs, H., and Ibberson, R. M. (1990). *Journal of Chemical Physics*, **93**, 5972.

Matzke, Th., Stimming, U., Karmonik, Ch., Soetratmo, M., Hempelmann R., and Güthoff, F. (1996). *Solid State Ionics*, **86–88**, 621.

Maurin, P.O., Dupuy-Philon, J., Jal, J.F., Asahi, N., Kaminyama, T., Kawamura, J., and Nakamura, Y. (1997). *Progress of Theoretical Physics Supplement*, **126**, 141.

Mayer-Kuckuk, T. (1988). *Kernphysik*. Teubner Studienbücher.

McKergov, M.W., Ross, D.K., Bonnet, J.E., Anderson, I.S., and Schaerpf, O. (1987). *Journal of Physics C: Solid State Physics*, **20**, 1909.

Mehrer, H. (1973). *Journal of Physics F: Metal Physics*, **3**, 543.

Mehrer, H. (1978). *Journal of Nuclear Materials*, **69** & **70**, 38.

Mehrer, H. (1991). *Diffusion in Solid Metals and Alloys*, Landolt-Börnstein, Vol. 26, Springer, Berlin.

Mesot, J., Janssen, S., Holitzner, L., and Hempelmann, R. (1996). *Journal of Neutron Research*, **3**, 293.

Messiah, A. (1961 and 1962). *Quantum Mechanics*. North-Holland, Amsterdam.

Metzger, H., Peisl, J., and Wanagel, J. (1976). *Journal of Physics F: Metal Physics*, **6**, 2195.

Mezei, F. (1972). *Zeitschrift für Physik*, **255**, 146.

Mezei, F. (1976). *Communications in Physics*, **1**, 81.

Mezei, F. (1991). Neutron Scattering and Collective Dynamics in Liquids and Glass, in J.P. Hansen, D. Levesque and J. Zinn-Justin (eds) *Liquids, Freezing and Glass Transition*, NATO-ASI in Les Houches 1989, North Holland, Amsterdam.

Mezei, F. and Dagleish, P.A. (1977). *Communications in Physics*, **2**, 41.

Mezei, F., Knaak, W., and Farago, B. (1987). *Physical Review Letters*, **58**, 57.
Mitchell, D.P. and Powers, P.N. (1936). *Physical Review*, **50**, 486.
Monkenbusch, M. (1992). *Physica B*, **180 & 181**, 935.
Morkel, C., Wipf, H., and Neumaier, K. (1978). *Physical Review Letters*, **40**, 947.
Moskvich, Yu.N., Sukhovskii, A.A., and Rozanov, O.V. (1984). *Soviet Physics Solid State*, **26**, 1984.
Müller-Warmuth, W. (1995). Stucture, Bonding, Dynamics, in W. Müller-Warmuth and R. Schöllhorn (eds) *Progress in Intercalation Research*, Kluwer, Amsterdam, pp. 339–455.
Mundy, J.N. (1971). *Physical Review B*, **3**, 2431.
Murch, G.E. (1980). *Atomic Diffusion Theory in Highly Defection Solids*, Diffusion and Defects Monograph Series, No. 6, *Trans. Tech. Publ.*, Aedermannsdorf.
Murch, G.E. and Nowick, A.S. (eds) (1984). *Diffusion in Crystalline Solids*. Academic Press, New York.
Nelin, G. and Sköld, K. (1975). *Journal of Physics and Chemistry of Solids*, **36**, 1175.
Nernst, W. (1888). *Zeitschrift für Physikalische Chemie*, **2**, 613.
Neumann, D.A., Zabel, H., Rush, J.J., Fan, Y. B., and Solin, S.A. (1987). *Journal of Physics C: Solid State Physics*, **20**, L761.
Norby, T. (1990). *Solid State Ionics*, **40 & 41**, 857.
Nowick, A. S. and Burton, J. J. (eds) (1975) *Diffusion in Solids*. Academic Press, New York.
Peisl, J. (1978). Lattice Strains Due to Hydrogen in Metals, in G. Alefeld and J. Völkl (eds) *Hydrogen in Metals I, Topics in Applied Physics*, Vol. 28, Springer, Berlin.
Peterson, N. L. (1978). *Journal of Nuclear Materials*, **69 & 70**, 3.
Petry, W. (1991). *Phase Transitions*, **31**, 119.
Petry, W., Vogl, G., Heidemann, A., and Steinmetz, K.-H. (1987). *Philosophical Magazine A*, **55**, 183.
Petry, W., Heiming, A., Alba, M., and Vogl, G. (1989). *Physica B*, **156 & 157**, 56.
Petry, W., Heiming, A., Herzig, C., and Trampenau, J. (1991a). *Defect and Diffusion Forum*, **75**, 211.
Petry, W., Heiming, A., Trampenau, J., Alba, M., Herzig, C., Schober, H. R., and Vogl, G. (1991b). *Physical Review B*, **43**, 10 933.
Philibert, J. (1985). *Diffusion et Transport de Matière dans les Solides*. Monographies de Physique, Les Éditions de Physique, Paris.
Philibert, J. (1991). *Atom Movements: Diffusion and Mass Transport in Solids*. Les Éditions de Physique, Paris.
Phillips, W.A. (1972). *Journal of Low Temperature Physics* **7**, 351.
Pionke, M., Mono, T., Schweika, W., Springer, T., and Schober, H. (1997). *Solid State Ionics*, **97**, 497.
Placzek, G. (1952). *Physical Review*, **86**, 377.
Poinsignon, C. (1989). *Solid State Ionics*, **35**, 107.
Poinsignon, C. (1997). *Solid State Ionics*, **97**, 399.
Potzel, U., Völkl, J., Wipf, H., and Margle, A. (1984). *Physica Status Solidi B*, **123**, 85.
Prager, M. (1991). *Physica B*, **174**, 218.
Press, W. (1981). *Single-Particle Rotations in Molecular Crystals*. Springer Tracts in Modern Physics Vol. 92, Springer, Berlin.
Prokof'jev, N. V. (1994). *Hyperfine Interactions*, **85**, 3.
Qian, X.W., Stump, D.R., and Solin, S.A. (1986). *Physical Review B*, **33**, 5756.

Qian, X.W., Stump, D.R., York, B.R., and Solin, S.A. (1985). *Physical Review Letters*, **54**, 1271.

Rahman, A., Singwi, K.S., and Sjölander, A. (1962). *Physical Review*, **126**, 997.

Randl, O.G., Sepiol, B., Vogl, G., and Feldwisch, R. (1994). *Physical Review B*, **49**, 8768.

Randl, O.G., Franz, H., Gerstendörfer, T., Petry, W., Vogl, G., and Magerl, A. (1997). *Physica B*, **234-236**, 1064.

Rauch, H. and Süda, M. (1974). *Physica Status Solidi (a)*, **25**, 495.

Rauch, H., Treimer, W., and Bonse, U. (1974). *Physics Letters A*, **47**, 369.

Remhof, A. and Magerl, A. (1997). *Nuclear Instruments and Methods*, **391**, 485.

Richter, D. (1983). Transport Mechanism of Light Interstitials in Metals, in *Neutron Scattering and Muon Spin Rotation*. Springer Tracts in Modern Physics, Vol. 101, Springer, Berlin.

Richter, D. (1993). Polymer Dynamics, in S.-H. Chen *et al.* (eds) *Stucture and Dynamics of Strongly Interacting Colloids and Supramolecular Aggregates in Solution*. Kluwer, Dordrecht.

Richter, D. (1998). Viscoelasticity and Microscopic Motion in Polymer Systems, in J. Kärger, P. Heitjans and R. Haberlandt (eds) *Diffusion in Condensed Matter*, Vieweg, Braunschweig/Wiesbaden.

Richter, D. and Springer, T. (1978). *Physical Review* B, **18**, 126.

Richter, D., Hempelmann, R., and Vinhas, L.A. (1982). *Journal of the Less Common Metals*, **88**, 353.

Richter, D., Driesen, G., Hempelmann, R., and Anderson, I.S. (1986). *Physical Review Letters*, **57**, 731.

Richter, D., Mahling-Ennaoui, S., and Hempelmann, R. (1989). *Zeitschrift für Physikalische Chemie*, Neue Folge, **164**, 907

Richter, D., Hempelmann, R., and Bowman, R.C. (1992). Dynamics of Hydrogen in Intermetallic Hydrides, in L. Schlapbach (ed.) *Hydrogen in Intermetallic Compounds II. Topics in Applied Physics*, Vol. 67, Springer, Berlin, p. 97.

Richter, D. and Springer, T. (1998). *A twenty years forward look at neutron scattering facilities in the OECD countries and Russia*, European Science Foundation (ed.), Organisation for Economic Co-operation and Development Megascience Forum.

Riste, T. (1970). *Nuclear Instruments and Methods*, **86**, 1.

Ritter, C., Nöldeke, C., Press, W., Stege, U. and Schöllhorn, R. (1993). *Zeitschrift für Physik B*, **92**, 437.

Roberts-Austen (1896). *Phiosophical Transactions of the Royal Society London*, A **187**, 404.

Robinson, D.S. and Salamon, M.B. (1982). *Physical Review Letters*, **48**, 156.

Röpke, G. (1987). *Statistische Mechanik für das Nichtgleichgewicht.* Physik-Verlag, Weinheim.

Ross, D.K. and Wilson, D.L.T. (1978). *Inelastic Neutron Scattering 1977*, Vol. 2, p. 383. Proceedings of the IAEA, Vienna.

Rossat-Mignod, J., Wiedemann, A., Woo, K.C., Milliken, J.W., and Fischer, J. E. (1982). *Solid State Communications*, **44**, 1339.

Rossbach, M., Schärpf, O., Kaiser, W., Graf, W., Schirmer, A., Faber, W., Duppich, J., and Zeisler, R. (1988). *Nuclear Instruments and Methods in Physics Research B*, **35**, 181.

Roth, M. and Zarzycki, J. (1947). *Journal of Applied Crystals*, **16**, 93.

Rowe, J.M., Rush, J.J., and Flotow, H.E. (1974). *Physical Review B*, **9**, 5039.

Rowe, J.M., Rush, J.J., de Graaf, L.A., and Ferguson, G. A. (1972). *Physical Review Letters*, **29**, 1250.

Rush, J.J. and Rowe, J.M. (1974). *Physical Review B*, **9**, 5039.

Rush, J.J., Udovic, T.J., Hempelmann, R., Richter, D., and Driesen, G. (1989). *Journal of Physics: Condensed Matter*, **1**, 1061.

Samuelson, E.J., Moret, R., Fuzellier, H., Klatt, M., Lelaurain, M., and Hérold, A. (1985). *Physical Review B*, **32**, 417.

Sato, H. and Kikuchi, R. (1971). *Journal of Chemical Physics*, **55**, 677 and 702.

Schärpf, O. (1989). *Physica B*, **156 & 157**, 631 and 639.

Schärpf, O. (1991). *Physica B*, **174**, 514.

Schärpf, O. (1992). *Physica B*, **182**, 376.

Schärpf, O. and Stüsser, N. (1989). *Nuclear Instruments and Methods in Physics Research A*, **284**, 208.

Schelten, J. and Alefeld, B. (1984). Backscattering spectrometer with adapted Q-resolution at the pulsed neutron source, in R. Scherm and H. Stiller (eds), Proceedings of the workshop on neutron scattering instrumentation for SNQ, Maria Laach 1984, Jül-1954 Report.

Schenk, A. (1985). *Muon Spin Rotation Spectroscopy*. Adam Hilger, Bristol.

Scher, H. and Montroll, E.W. (1975). *Physical Review B*,**12**, 2455.

Scherm, R. and Wangner, V. (1978). Some applivations of Horizontally Curved Analyzer Crystals, in *Neutron Inelastic Scattering*, IAEA, Vienna, p. 149

Scherm R., Carlile C., Dianoux A.J., Suck, J., and White, J. (1976). ILL Report 76S235.

Scherm, R., Dolling, G., Ritter, R. Schedler, E., Teuchert, W., and Wagner, V. (1977). *Nuclear Instruments and Methods*, **143**, 77.

Schirmacher, W. (1991). *Berichte Bunsengesellschaft Physikalische Chemie*, **95**, 368.

Schirmacher, W., Prem, M., Suck, J.-B., and Heidemann, A. (1990). *Europhysics Letters*, **13**, 523.

Schirmer, A. and Heitjans, P. (1995). *Zeitschrift für Naturforschung*, **50A**, 643.

Schirmer, A., Heitjans, P. Ackermann, H. Bader, B., Freiländer P., and Stöckmann, H.-J. (1988). *Solid State Ionics*, **28-30**, 717.

Schlapbach, L. (ed.) (1988, 1992). *Hydrogen in Intermetallic Compounds I, II*, Topics in Applied Physics, Vol. 63, 67; Springer, Berlin.

Schlesinger, M.F., Zaslavsky, G.M., and Frisch, U. (1995). *Lévy Flights and Related Topics in Physics*. Lecture Notes in Physics Vol. 450, Springer, Berlin.

Schmatz, W. (1973). *Treatise on Material Science and Technology*, Vol. 2, p. 105. Academic Press, New York.

Schnabel, P., Hayes, W., Hutchings, M.T., Lechner, R.E., and Renker, B. (1983). *Radiation Effects*, **75**, 73.

Schober, H.R. (1995). *Philosophical Transactions of the Royal Society London* A, **350**, 297.

Schober, H.R. (1987). Interaction of phonons and point defects and its influence on defect mobility, in T. Paszkiewicz (ed.) *Physics of Phonons*. Lecture Notes in Physics, Vol. 285, p. 188. Springer, Berlin.

Schober, H.R. and Lottner, V. (1979). *Zeitschrift für Physikalische Chemie*, **114**, 203.

Schober, H.R., Petry, W., and Trampenau, J. (1992). *Journal of Physics: Condensed Matter*, **4**, 9321.

Schöllhorn, R. (1980). *Angewandte Chemie*, **92**, 1015; *Angewandte Chemie International Edition*, **19**, 983

Schönfeld, C., Hempelmann, R., Richter, D., Springer, T., and Dianoux, A.J. (1992). *Physica B*, **180 & 181**, 697.

Schönfeld, C., Hempelmann, R., Richter, D., Springer, T., Dianoux, A.J., Rush, J.J., Udovic, T.J., and Bennington, S.M. (1994). *Physical Review B*, **50**, 853.

Schwartz, L.H. and Cohen, J.B. (1977). *Diffraction from Materials*. Academic Press, San Diego.

Schweika, W. (1993). Diffuse scattering determination of short range order in alloys, in P.E.A. Turchi and A. Gonis (eds) *Statics and dynamics of alloy phase transformations*. NATO School, Rhodes, Greece. Plenum Press.

Schweika, W. (1998). *Disordered Alloys: Diffuse Scattering and Monte Carlo Simulations*. Springer Tracts in Modern Physics, Vol. 141, Springer, Berlin.

Sears, V.F. (1982). *Physics Reports*, **82**, 1.

Sears, V.F. (1985). *Zeitschrift für Physik A—Atoms and Nuclei*, **321**, 443.

Sears, V.F. (1992). *Neutron News*, **3**, No. 3, 26.

Seeger, A. and Mehrer, H. (1970). Analysis of self-diffusion and equilibrium measurements, in A. Seeger, D. Schumacher, W. Schilling, and J. Diehl (eds) *Vacancies and Interstitials in Metals*, North-Holland, Amsterdam, p. 1.

Sepiol, B. and Vogl, G. (1993). *Physical Review Letters*, **71**, 731.

Sepiol B., Randl, O.G., Karner, C., Heiming, A., and Vogl, G. (1994). *Journal of Physics: Condensed Matter*, **6**, L43.

Shockley, W., Hollomon, J.H., Maurer, R., and Seitz, F. (1952). *Imperfections in Nearly Perfect Crystals*. Wiley, NewYork.

Shull, C.G. and Gingrich, N.S. (1964). *Journal of Applied Physics*, **35**, 678.

Silver, R.N. and Sokol, P.E. (1989). *Momentum Distributions*. Plenum Press, New York.

Singwi, K.S. (1965). *Physica*, **31**, 1257.

Singwi, K.S., and Sjölander, A. (1960). *Physical Review*, **119**, 863.

Singwi, K.S., Sköld, K., and Tosi, M.P. (1970). *Physical Review A*, **1**, 454.

Sinha, S.K. (1987). Adsorbed monolayers and intercalated compounds in D.L. Sköld and K. Price (eds) *Methods of Experimental Physics Vo. 23 B: Neutron Scattering* Academic Press, New York.

Sinha, S.K. and Ross, D.K. (1988). *Physica B*, **149**, 51.

Sinitsyn, V.V., Ponyatovskii, E.G., Baranov, A.I., Shuvalov, L.A., and Bobrova, N.I. (1988). *Soviet Physics Solid State*, **30**, 2838.

Sivia, D.S., Carlile, C.J., Howells, W.S., and König, S. (1992). *Physica B*, **182**, 341.

Sjölander, A. (1958). *Arkiv Fysik*, **14**, 315.

Sköld, K. (1967). *Physical Review Letters*, **19**, 1023.

Sköld, K. and Nelin, G. (1967). *Journal of Physics and Chemistry of Solids*, **28**, 2369.

Sköld, K., and Price, D.L. (eds) (1986). *Methods of Experimental Physics*, Vol. 23 Part A, B, C: Neutron Scattering, Academic Press, New York.

Skripov, A.V., Pionke, M., Randl, O., and Hempelmann, R. (1999). *Journal of Physics: Condensed Matter*, **11**, 1489

Skripov, A.V., Cook, J.C., Karmonik, C., and Hempelmann, R. (1997). *Journal of Alloys and Compounds*, **253–254**, 432.

Skripov, A.V., Cook, J.C., Sibirtsev, D.S., Karmonik, C., and Hempelmann, R. (1998). *Journal of Physics: Condensed Matter*, **10**, 1787.

Slade, R. C. T., Hirst, P. R., West, B. C., Ward, R. C., and Magerl, A. (1989). *Chemical Physics Letters*, **155**, 305.

Smidt, D. (1976). *Reaktortechnik*, G. Braun, Karlsruhe; or any other textbook on reactor techniques.

Somenkov, V.A., Entin, I.R., Chervyakov, A.Y., Shil'stein, S.Sh., and Chertkov, A.A. (1972). *Soviet Physics Solid State*, **13**, 2178.

Spring, W. (1880). *Bulletin de l'Acadamic Royale de Belgique*, **49**, 323.

Springer, T. (1972). *Quasielastic Neutron Scattering for the Investigation of Diffusive Motions in Solids and Liquids*, Springer Tracts in Modern Physics Vol. 64, Springer, Berlin.

Springer, T. (1978). Investigation of Vibrations in Metal Hydrides by Neutron Spectroscopy, in G. Alefeld and J. Völkl (eds) *Hydrogen in Metals I. Basic Properties*, Topics in applied physics. Vol. 28, Springer, Berlin.

Springer, T. (1979). *Mémoires Scientifiques Revue Métallurgie*, 545.

Springer, T., and Richter, D. (1987). Hydrogen in metals, in D. Price and K. Sköld (eds) *Neutron Scattering Part B, Methods in Experimental Physics Vol. 23B*, Academic Press, San Diego, p. 131.

Squires, G.L. (1978). *Introduction to the Theory of Thermal Neutron Scattering*. Cambridge University Press.

Stahn, M., Lediner, R.E., Dachs, H., and Jacobs, H.E. (1983). *Journal of Physics C: Solid State Physics*, **16**, 5073.

Stanley, H.E. (1971). *Introduction to Phase Transition and Critical Phenomena*. Oxford University Press.

Stauffer, D. and Aharony, A. (1992). *Introduction to Percolation Theory*. Taylor & Francis, London.

Stearn, A.E. and Eyring, H. (1940). *Journal of Physical Chemistry*, **44**, 955.

Steinbinder, D., Wipf, H., Magerl, A., Richter, D., Dianoux, A.-J., and Neumaier, K. (1988). *Europhysics Letters*, **6**, 535.

Steinbinder, D., Wipf, H., Dianoux, A.-J., Magerl, A., Neumaier, K., Richter, D., and Hempelmann, R. (1991). *Europhysics Letters*, **16**, 211.

Steinmetz, K.-H., Vogl, G., Petry, W. and Schroeder, K. (1986). *Physical Review B*, **34**, 107.

Stoneham, A.M. (1969). *Reviews of Modern Physics*, **41**, 82.

Stoneham, A.M. (1972). *Journal of Physics F*, **2**, 417.

Stuhr, U., Steinbinder, D., Wipf, H., and Frick, B. (1992). *Europhysics Letters*, **20**, 117.

Suda, S., Kobayashi, N., and Yoshida, K. (1980). *Journal of the Less Common Metals*, **73**, 119.

Switendick, A.C. (1979). *Zeitschrift für Physikalische Chemie NF*, **117**, 89.

Szökefalvi-Nagy, A., Filipek, S., and Kirchheim, R. (1987). *Journal of Physics and Chemistry of Solids*, **48**, 613.

Teichler, H. and Seeger A. (1981). *Physics Letters A*, **82**, 91.

Tewary, V.K. (1973). *Journal of Physics F: Metal Physics*, **3**, 704.

Töpfer, J., Lebsanft, E., and Schätzler, R. (1978). *Journal of Physics F: Metal Physics*, **8**, L25.

Touzain, P. (1979/80). *Synthetic Metals*, **1**, 3.

Trampenau, J., Heiming, A., Petry, W., Alba, M., Herzig, C., Miekeley, W., and Schober, H.R. (1991). *Physical Review B*, **43**, 10 963.

Uchida, H., Yoshikawa, H., Esaka, T., Ohtsu, S., and Iwahara, H. (1989). *Solid State Ionics*, **36**, 89

Udovic, T. J., Rush, J. J., Huang, Q., and Anderson, I. S. (1997). *Journal of Alloys and Compounds*, **253–254**, 241.

van Hove, L. (1954). *Physical Review*, **95**, 249.

van Hove, L. (1958). *Physica*, **24**, 404.

van Kampen, N.G. (1981, 1992). *Stochastic Processes in Physics and Chemistry*. North-Holland, Amsterdam 1981, new edition 1992.

Vineyard, G.H. (1957). *Journal of Physics and Chemistry of Solids*, **3**, 121.

Vineyard, G.H. (1958). *Physical Review*, **110**, 999.

Vogl, G. (1996). *Physica B*, **226**, 135.

Vogl, G. and Feldwisch, R. (1998). The elementary diffusion step in metals studied by methods from nuclear solid state physics, in J. Kärger, P. Heitjans, and R. Haberlandt (eds) *Diffusion in Condensed Matter*. Vieweg, Braunschweig.

Vogl, G., Miekeley, W., Heidemann, A., and Petry, W. (1984). *Physical Review Letters*, **53**, 934.
Vogl, G., Petry, W., Flottmann, T., and Heiming, A. (1989). *Physical Review B*, **39**, 5025.
Vogl, G., Randl, O.G., Petry, W., and Hünecke, J. (1993). *Journal of Physics: Condensed Matter*, **5**, 7215.
Vogl, G., Kaisermayr, M. and Randl, O. (1996). *Journal of Physics: Condensed Matter*, **8**, 4727.
Volino, F. and Dianoux, A.J. (1980). *Molecular Physics*, **41**, 271.
Völkl, J. and Alefeld, G. (1978). Diffusion of hydrogen in metals, in G. Alefeld and J. Völkl (eds) *Hydrogen in Metals I*. Topics in Applied Physics, Vol. 28, Springer, Berlin.
von Grotthuss, C.J.D. (1806). *Annales de Chimie*, **58**, 54.
Wagner, B., Essmann, R., Bock, J., Jacobs, H., and Fischer, P. (1992). *European Journal of Solid State Inorganic Chemistry*, **29**, 1217.
Wagner, C. (1938). *Zeitschrift für Physikalische Chemie* Abt. B, **38**, 325.
Wagner, C. and Schottky, W. (1930). *Zeitschrift für Physikalische Chemie* Abt. B, **11**, 163.
Wagner, H. (1978). Elastic interacton and phase transition in coherent metal–hydrogen alloy, in G. Alefeld and J. Völkl (eds) *Hydrogen in Metals I*, Topics in Applied Physics, Vol. 28, Spinger, Berlin.
Wagner, H. and Honer, H. (1974). *Advances in Physics*, **23**, 587.
Wagner, M. and Schirmacher, W. (1991). *Berichte der Bunsengesellschaft Physikalische Chemie*, **95**, 983.
Wagner W., Bauer, G.S., Duppich, J., Janssen S., Lehmann, E., Lühty, M., and Spitzer, H. (1998). *Journal of Neutron Research*, **6**, 249
Wakabayashi, N., Alefeld, B., Kehr, K.W., and Springer, T. (1974). *Solid State Communications*, **15**, 503.
Wang, J. and Angell, C. A. (1973). *Journal of Non-Crystalline Solids*, **12**, 402.
Watanabe, N., and Kiyanagi, Y. (1992). *Physica B*, **180 & 181**, 893.
Wegener, B., Essmann, R., Bock, J., Jacobs, H., and Fischer, P. (1992). *European Journal of Solid State Inorganic Chemistry*, **29**, 1217.
Weigelt, J. (1991). Thesis, University of Rostock.
Weinberg, A. and Wigner, E. (1958). *The Physical Theory of Neutron Chain Ractors*. University of Chicago Press.
Welter, J.-M., Richter, D., Hempelman, R., Hartmann, O., Karlsson, E., Norlin, L.O., Niinikoshi, T.O., and Lenz, D. (1983). *Zeitschrift für Physik B*, **52**, 303.
Westlake, D.G. (1983). *Journal of the Less Common Metals*, **90**, 251.
Willaime, F., and Massobrio, C. (1991). *Physical Review B*, **43**, 11 653.
Williams, W.G. (1988). *Polarized Neutrons*. Clarendon Press, Oxford.
Willis, B.T. (Ed.) (1973). *Chemical Application of Thermal Neutron Scattering*, Oxford University Press.
Willis, B.T.M. and Pryor, A.W. (1975), *Thermal Vibrations in Crystallography*. Cambridge University Press.
Windsor, C.G. (1981). *Pulsed Neutron Scattering*. Taylor & Francis, London.
Wipf, H. (1991). *Berichte der Bunsengesellschaft Physikalische Chemie*, **95**, 438.
Wipf, H. (1997). Diffusion of hydrogen in metals, in H. Wipf (ed.) *Hydrogen in metals III, Topics in Applied Physics*, Vol. 73, Springer, Berlin.
Wipf, H., and Neumaier, K. (1984). *Physical Review Letters*, **52**, 1308.
Wipf, H., Magerl, A., Shapiro, S.M., Satija, S.K., and Anderson, I. (1981). *Physical Review Letters*, **46**, 947.

Wipf, H., Steinbinder, D., Neumaier, K., Gutsmiedl, P., Magerl, A., and Dianoux, A.-J. (1987). *Europhysics Letters*, **4**, 1379.

Wipf, H., Völkl, J., and Alefeld, G. (1989). *Zeitschrift für Physik B*, **76**, 353.

Wolf, D. (1977a). *Physical Review B*, **15**, 37.

Wolf, D. (1977b). *Solid State Communications*, **23**, 853.

Wolf, D. (1983), in F. Bénière C.R.A. and Catlow (eds) *Mass Transports in Solids*, NATO Series P, Vol. 97, Plenium Press, London, p. 149.

Yamada, K. (1984). *Progress in Theoretical Physics*, **72**, 195.

Yamashita, J. and Kurosawa, T. (1958). *Physics and Chemistry of Solids*, **5**, 34.

Yang Du and Nowick, A.S. (1996). *Solid State Ionics*, **91**, 85.

Zabel, H. (1990). Lattice dynamics: neutron studies, in H. Zabel and S.A. Solin (eds) *Graphite Intercalation Compounds I: Structure and Dynamics*, Springer Series in Materials Science. Vol. 14, Springer, Berlin, p. 101.

Zabel, H. and Solin S.A. (eds.) (1990). *Graphite Intercalation Compounds I: Structure and Dynamics*. Springer Series in Materials Science Vol. 14, Springer, Berlin.

Zabel, H., Magerl, A., Dianoux, A.J., and Rush, J.J. (1983). *Physical Review Letters*, **50**, 2094.

Zabel, H., Hardcastle, S.E., Neumann, D.A., Suzuki, M., and Magerl, A. (1986). *Physical Review Letters*, **57**, 2041.

Zabel, H., Magerl, A., Rush, J.J., and Misenheimer, M.E. (1989). *Physical Review B*, **40**, 7616.

Zachariassa, W.H. (1945). *Theory of X-Ray Diffraction in Crystals*. Wiley, London.

Zh Qi, Völkl, J., Lässer, R., and Wenzl, H. (1983). *Journal of Physics F: Metal Physics*, **13**, 2053.

Zippelius, A., Halperin, B.I., and Nelson, D.R. (1980). *Physical Review B*, **22**, 2514.

Züchner, H. (1970). *Zeitschrift für Naturforschung*, **25a**, 1490.

Index

Lightning Source UK Ltd.
Milton Keynes UK
UKOW06n0410031217
313754UK00003B/387/P

9 780198 517436